CFD Modelling and Simulation of Water Turbines

CFD Modelling and Simulation of Water Turbines

Editors

Santiago Lain
Omar Dario Lopez Mejia

MDPI • Basel • Beijing • Wuhan • Barcelona • Belgrade • Manchester • Tokyo • Cluj • Tianjin

Editors
Santiago Lain
Universidad Autónoma de Occidente
Colombia

Omar Dario Lopez Mejia
Universidad de los Andes
Colombia

Editorial Office
MDPI
St. Alban-Anlage 66
4052 Basel, Switzerland

This is a reprint of articles from the Special Issue published online in the open access journal *Processes* (ISSN 2227-9717) (available at: https://www.mdpi.com/journal/processes/special_issues/CFD_water_turbine).

For citation purposes, cite each article independently as indicated on the article page online and as indicated below:

LastName, A.A.; LastName, B.B.; LastName, C.C. Article Title. *Journal Name* **Year**, *Volume Number*, Page Range.

ISBN 978-3-0365-6015-1 (Hbk)
ISBN 978-3-0365-6016-8 (PDF)

About the Editors

Santiago Lain

Santiago Lain (PhD. Dr.-Ing. Habil.) is a professor of Fluid Mechanics at the Department of Mechanical Engineering at Universidad Autónoma de Occidente Cali, Colombia. He is Physicist, Mathematician and PhD in Physical Sciences from Universidad of Zaragoza, Spain and Dr.-Ing. Habil. from Martin Luther University Halle-Wittenberg, Germany. His research interests are focused on numerical methods and computational fluid dynamics applied to turbulence modelling and simulation, multi-phase flow and renewable energy systems.

Omar Dario Lopez Mejia

Omar Dario Lopez Mejia (PhD) is an associate professor in the Department of Mechanical Engineering at Universidad de los Andes in Bogotá, Colombia. He received his PhD degree in mechanical engineering from The University of Texas in Austin in 2009. His research interests are focused on the application of computational fluid dynamics (CFD) in thermal-fluid systems. Specifically, he has focused on those problems involving turbulence modelling applied to aerodynamics and hydrodynamics.

Preface to "CFD Modelling and Simulation of Water Turbines"

Numerical simulations for the performance prediction of water turbines are becoming a standard tool in the engineering practice. This book is focused on the modelling and simulation of water turbines for both conventional and unconventional hydropower energy. The scope of this book includes state-of-the-art Topics in the field of hydropower energy production, such as: modelling of pump-turbines, simulation of horizontal and vertical axis turbines for hydrokinetic applications and modelling of hydropower plants. Finally, this book promises to be a good reference for researchers and graduate students alike who share an interest in reliably predicting the performance of water turbines based on numerical simulations.

Santiago Lain and Omar Dario Lopez Mejia
Editors

Editorial

Special Issue on "CFD Modelling and Simulation of Water Turbines"

Santiago Lain [1,*] and Omar Darío Lopez Mejia [2]

[1] PAI+ Group, Mechanical Engineering Department, Faculty of Engineering, Universidad Autónoma de Occidente, Cali 760030, Colombia
[2] Department of Mechanical Engineering, Universidad de los Andes, Cra 1 Este N 19A-40, Bogotá 111711, Colombia
* Correspondence: slain@uao.edu.co; Tel.: +57-6023188000

Citation: Lain, S.; Lopez Mejia, O.D. Special Issue on "CFD Modelling and Simulation of Water Turbines". *Processes* **2022**, *10*, 2410. https://doi.org/10.3390/pr10112410

Received: 7 November 2022
Accepted: 13 November 2022
Published: 15 November 2022

Publisher's Note: MDPI stays neutral with regard to jurisdictional claims in published maps and institutional affiliations.

Climate change and the energy crisis are two main problems that humanity is currently facing and that require immediate action. The global average surface temperature has been increasing over the last few decades, and this does not seem to be changing in the coming years [1]. In addition, the increase in CO_2 emissions will be more than 1 billion tons in 2022 despite the incremental increase in the use of renewable energy technologies around the world, including electric vehicles [2]. One of the solutions that has been proposed to mitigate these problems is related to the energy transition from conventional (nonrenewable) to renewable energy sources and the intense use of highly efficient technologies for this purpose. It is known that the most common sources for renewable energy are: solar energy, wind energy, geothermal energy, bioenergy, water (hydropower and hydrokinetics) and ocean energy. Hydropower (including hydrokinetics) and ocean energy are of interest in this Special Issue due to their high percentage of contribution to the total amount of energy generation in the world and higher growth in comparison with other renewable energy sources. Furthermore, hydropower energy generates low emissions (between solar photovoltaic and wind energies) and low costs in comparison to wind and solar energies. Some important facts that corroborate these statements include: In 2010, Tomabechi reported that out of the maximum usable hydropower energy (0.059 ZJ/year), only 0.001 ZJ/year was generated, which is in comparison to wind energy with a potential estimated to be 0.7 ZJ/year, but only 0.00038 ZJ/year was actually generated [3]. In 2019, 3.6% out of the 11.2% of renewable energy that was consumed in the world was generated with hydropower [2]. In 2020, 29% of the energy that was generated globally was from renewable sources, and out of this, 16.8% was generated from hydropower [4]; in addition, the global installed capacity of hydropower grew 1.6%, reaching 1330 GW [5]

Large, conventional hydropower generation plants (greater than 10 MW) majorly contribute to the total installed capacity. These plants require a large dam that typically impacts the physical characteristics of a river and has important biological (ecosystems and river habitat) and social (communities that live on the river shores) impacts. Small hydropower plants or SHPs (less than 10 MW) have fewer environmental impacts than large hydropower plants and are less expensive despite still requiring a dam. The installed capacity of SHPs in the world grew from 71 GW to 78 GW from 2013 to 2019 [6]. On the other hand, unconventional hydropower generation does not require a dam construction and could use the kinetic energy available in water currents such as rivers, waves and tides. Examples of these kinds of hydropower generation are based on hydrokinetic and ocean energy, which are in different stages of development but both have had very interesting growth in recent years. Independent of the type of hydropower generation, the main, common device that transforms the kinetic energy available in the water to mechanical energy is the *turbine*, which is the focus of this Special Issue.

The design and performance analysis of a water turbine have been historically carried out with low-order models and physical modelling. In recent decades, the use of simulation tools such as computational fluid dynamics (CFD) have significantly increased up to the point that, in recent years, physical modelling is no longer used. All the contributions to the present Special Issue are related to the use of different kinds of models and simulation tools applied to water turbines in different contexts. The contributions to this Special Issue can be organized into three main topics:

1. Pump-turbine modelling and simulation [7–9].
2. Modelling and simulation of horizontal- and vertical-axis turbines for hydrokinetic applications [10–14].
3. Modelling of hydropower plants [15,16].

For the first topic, three contributions are related to the use of CFD in the simulation of turbines working as pumps. One of the new trends in hydropower is energy storage (pumped-storage hydropower or PSH); for this purpose, the turbine must work as a pump (typically overnight or in low-energy-demanding time-frames). For this strategy to be considered renewable, extensive research and development must be conducted to understand the performance analysis of the turbine working as a pump. For example, Deng et al. performed a CFD simulation of a model turbine as a pump in order to quantify the energy losses and the entropy generation [7]. Despite CFD simulations assuming that the problem is isothermal, entropy generation can be computed for the viscous dissipation as a postprocess. The proposed analysis is important in order to improve the new designs of pump-turbines with the objective of minimizing the entropy generation and the energy losses for PSH to be feasible. On the other hand, Li et al. [8] investigated the effects of the change in the guide vane airfoil on the flow characteristics in the draft tube of a water pump-turbine. The CFD numerical simulations show that the proposed movable guide vane is able to improve the energy recovery of the draft tube, reducing the vibrations' intensity and enhancing the turbine's stable operation. Moreover, Peng et al. [9] developed an integrated optimization design method of multiphase pumps from both hydrodynamic and structural points of view. Their method was applied to increase the efficiency of a pump, obtaining an enhancement of around 1% with regard to the original model; this numerical conclusion was validated against the experimental tests, allowing them to conclude that the proposed integrated design–meshing–numerical methodology is effective.

A great number of contributions cover the second topic, which is highly related to the development of unconventional hydropower energy. All of these contributions use CFD in different problems, ranging from vertical-axis turbines to horizontal-axis turbines, but all have the common characteristic of the inclusion of more details (either geometrical or physical) in the simulation. For example, Niebuhr et al. proposed a simplified model for the calculation of a small horizontal-axis turbine that includes backwater effects [10]. The model uses the data generated from CFD simulations that include not only the presence of the walls (channel flow), but also the free surface using a multiphase model. The result is a simple model that can be used for the engineering prediction of backwater effects without the need for complex simulations.

The effects of considering the free surface and support structure on the performance of a horizontal-axis water turbine were considered in [11]. The CFD computations clearly show that for shallow immersion, the presence of the free surface induces a reduction in the power coefficient, as well as it being a constraint for the wake developing, which consequently, recovers at a slower rate than in free-flow conditions, a fact that must be taken into account in the design stage of farms based on hydrokinetic turbines. On the other hand, in order to make progress in the optimization of hydraulic, marine, tidal and hydrokinetic turbines, dynamic mesh models have to be used in CFD computations to deal with the complex turbulent flow developing around such turbines. Therefore, the study performed in [12] compares the performance of the standard sliding mesh and the newer overset (chimera) mesh techniques in a vertical-axis water turbine. Advantages and disadvantages of both methodologies are clearly illustrated and thoroughly discussed. Such results are useful to establish best practice guidelines for CFD simulations of this kind of system.

Xiong et al. [13] investigated the hydrodynamic performance enhancement of a hydrofoil when a flapping motion is added to its trailing edge. Such an idea was bio-inspired by examining the flapping movement of a whale tail fin. The authors made CFD simulations as well as particle image velocimetry (PIV) experiments, obtaining a good agreement between them. As a result, the hydrofoil with a clockwise flap presents a better hydrodynamic performance than the basic hydrofoil for small angles of attack, and the counter-clockwise flap increases the critical stall angle, improving the hydrofoil's navigation stability. Finally, it was concluded that adding a short flap to the original hydrofoil improves its performance; moreover, it can be applied to various operating conditions by adjusting the angle of the flap.

Li et al. performed CFD simulations of a tubular turbine (axial-flow) prototype using different clearances or gaps between the blade tip and the shroud [14]. It is important to mention that these gaps are in the order of 5 mm to 20 mm, which is very small in comparison to the runner diameter, which is 7.5 m. This is a very challenging aspect of the CFD simulations since the mesh should be fine enough to correctly capture the details of the flow in the gap. Numerical results show that the size of the gap has an important impact on the axial momentum and the leakage flow, which, in the end, have a great influence on the performance and efficiency of the turbine.

Finally, two contributions are related to the modelling of hydropower plants. At this level, high-order models such as CFD are not practical due to the high computational cost. Saeed et al. proposed a low-order model to identify the fault occurrence in a hydropower plant, which consists of three ponds, a surge tank and the turbine [15]. The dynamic model has the capability to predict the water level in the ponds and the surge tank, as well as the total output power of the turbine. The dynamic model is then coupled with a controller and a fault diagnosis model. Low-order models such as the one proposed in this contribution are useful for the next generation of smart grids based on different types of renewable energy, including conventional and unconventional hydropower. The contribution of Yoosefdoost and Lubitz [16] deals with nonstandard hydropower plants based on Archimedean screw generators. These machines possess several advantages, such as being fish friendly, operating under a wide range of flow heads and generating power from any flow, even wastewater. These authors propose a simple method for estimating the number and geometry of such devices considering installation site properties, river flow characteristics and technical issues. The introduced methodology can be applied to easily evaluate the potential of green and renewable energy generation based on the Archimedes screw.

In general, all the contributions of this Special Issue demonstrate the importance of modelling and simulation in hydropower generation. This new generation of models and simulations will play a major role in the global energy transition and energy crisis, and, of course, in the mitigation of climate change. As the design of turbines relies more on modelling and simulation, researchers and engineers must provide reliable simulation methods and tools. It is important to mention the role that CFD currently plays in the design of water turbines, as its use and implementation is very common in the hydropower industry. Despite this, CFD still has a long way to go in simulating specific details of water turbines such as, for instance, in the simulation of turbines used for unconventional hydropower generation, in the inclusion of geometrical and physical details, and in the correct computation of performance and energy losses.

Funding: This research received no external funding.

Conflicts of Interest: The authors declare no conflict of interest.

References

1. Climate Change: Global Temperature. Available online: https://www.climate.gov/news-features/understanding-climate/climate-change-global-temperature (accessed on 29 October 2022).

2. Defying Expectations, CO_2 Emissions from Global Fossil Fuel Combustion Are Set to Grow in 2022 by Only a Fraction of Last Year's Big Increase. Available online: https://www.iea.org/news/defying-expectations-co2-emissions-from-global-fossil-fuel-combustion-are-set-to-grow-in-2022-by-only-a-fraction-of-last-year-s-big-increase (accessed on 29 October 2022).

3. Tomabechi, K. Energy Resources in the Future. *Energies* **2010**, *3*, 686–695. [CrossRef]

4. Center for Climate and Energy Solutions. Available online: https://www.c2es.org/content/renewable-energy/ (accessed on 29 October 2022).

5. 2021 Hydropower Status Report. Available online: https://www.hydropower.org/publications/2021-hydropower-status-report (accessed on 29 October 2022).

6. World Small Hydropower Development Report 2019. Available online: https://www.unido.org/sites/default/files/files/2020-07/Executive%20Summary.pdf (accessed on 29 October 2022).

7. Deng, Y.; Xu, J.; Li, Y.; Zhang, Y.; Kuang, C. Research on Energy Loss Characteristics of Pump-Turbine during Abnormal Shutdown. *Processes* **2022**, *10*, 1628. [CrossRef]

8. Li, Q.; Xin, L.; Xie, G.; Liu, S.; Wang, Q. Influence of Guide Vane Profile Change on Draft Tube Flow Characteristics of Water Pump Turbine. *Processes* **2022**, *10*, 1494. [CrossRef]

9. Peng, C.; Zhang, X.; Chen, Y.; Gong, Y.; Li, H.; Huang, S. A Method for the Integrated Optimal Design of Multiphase Pump Based on the Sparse Grid Model. *Processes* **2022**, *10*, 1317. [CrossRef]

10. Niebuhr, C.M.; Hill, C.; Van Dijk, M.; Smith, L. Development of a Hydrokinetic Turbine Backwater Prediction Model for Inland Flow through Validated CFD Models. *Processes* **2022**, *10*, 1310. [CrossRef]

11. Benavides-Morán, A.; Rodríguez-Jaime, L.; Laín, S. Numerical Investigation of the Performance, Hydrodynamics, and Free-Surface Effects in Unsteady Flow of a Horizontal Axis Hydrokinetic Turbine. *Processes* **2022**, *10*, 69. [CrossRef]

12. Lopez Mejia, O.D.; Mejia, O.E.; Escorcia, K.M.; Suarez, F.; Laín, S. Comparison of Sliding and Overset Mesh Techniques in the Simulation of a Vertical Axis Turbine for Hydrokinetic Applications. *Processes* **2021**, *9*, 1933. [CrossRef]

13. Xiong, P.; Deng, J.; Chen, X. Performance Improvement of Hydrofoil with Biological Characteristics: Tail Fin of a Whale. *Processes* **2021**, *9*, 1656. [CrossRef]

14. Li, X.; Li, Z.; Zhu, B.; Wang, W. Effect of Tip Clearance Size on Tubular Turbine Leakage Characteristics. *Processes* **2021**, *9*, 1481. [CrossRef]

15. Saeed, A.; Khan, A.U.; Iqbal, M.; Albogamy, F.R.; Murawwat, S.; Shahzad, E.; Waseem, A.; Hafeez, G. Power Regulation and Fault Diagnostics of a Three-Pond Run-of-River Hydropower Plant. *Processes* **2022**, *10*, 392. [CrossRef]

16. Yoosef Doost, A.; Lubitz, W.D. Design Guideline for Hydropower Plants Using One or Multiple Archimedes Screws. *Processes* **2021**, *9*, 2128. [CrossRef]

processes

MDPI

Article

Effect of Tip Clearance Size on Tubular Turbines Leakage Characteristics

Xinrui Li [1], Zhenggui Li [1,*], Baoshan Zhu [2] and Weijun Wang [3]

[1] Key Laboratory of Fluid and Power Machinery, Ministry of Education, Xihua University, Chengdu 610039, China; lixinrui@stu.xhu.edu.cn
[2] State Key Laboratory of Hydroscience and Engineering, Tsinghua University, Beijing 100084, China; bszhu@mail.tsinghua.edu.cn
[3] AVIC Chengdu CAIC Electronics Co., Ltd., Chengdu 610091, China; liwangxu@stu.xhu.edu.cn
* Correspondence: lzhgui@mail.xhu.edu.cn; Tel.: +86-028-8772-6600

Abstract: To study the effect of tip clearance on unsteady flow in a tubular turbine, a full-channel numerical calculation was carried out based on the SST k–ω turbulence model using a power-plant prototype as the research object. Tip leakage flow characteristics of three clearance δ schemes were compared. The results show that the clearance value is directly proportional to the axial velocity, momentum, and flow sum of the leakage flow but inversely proportional to turbulent kinetic energy. At approximately 35–50% of the flow direction, velocity and turbulent kinetic energy of the leakage flow show the trough and peak variation law, respectively. The leakage vortex includes a primary tip leakage vortex (PTLV) and a secondary tip leakage vortex (STLV). Increasing clearance increases the vortex strength of both parts, as the STLV vortex core overlaps Core A of PTLV, and Core B of PTLV becomes the main part of the tip leakage vortex. A "right angle effect" causes flow separation on the pressure side of the tip, and a local low-pressure area subsequently generates a separation vortex. Increasing the gap strengthens the separation vortex, intensifying the flow instability. Tip clearance should therefore be maximally reduced in tubular turbines, barring other considerations.

Keywords: tip leakage flow; tubular turbine; clearance discipline; numerical calculation

---

Citation: Li, X.; Li, Z.; Zhu, B.; Wang, W. Effect of Tip Clearance Size on Tubular Turbines Leakage Characteristics. *Processes* **2021**, *9*, 1481. https://doi.org/10.3390/pr9091481

Academic Editor: Santiago Lain

Received: 18 July 2021
Accepted: 22 August 2021
Published: 24 August 2021

Publisher's Note: MDPI stays neutral with regard to jurisdictional claims in published maps and institutional affiliations.

1. Introduction

As energy consumption increases, pollution and its reduction are urgent challenges faced by human society. As a resource for sustainable development, green energy has attracted increasing attention from researchers and industries [1–3]. Therefore, in recent years, the trend of developing renewable and clean energy has been increasing, with hydropower playing an important role and developing at an unprecedented rate [4–6]. Among them, the tubular turbine has the advantages of a short construction period, small investment, large excess flow, high efficiency, good cavitation performance, and fewer blades. Thus, it is widely used in the development of low-head water resources [7–9].

In the turbine design process, a small clearance, as shown in Figure 1, is needed between the blade tip and the shroud to avoid interference between them. The fluid in the clearance forms a leakage flow because of the pressure difference between the blade pressure side and suction side, accompanied by the generation of vortices. Leakage flow is an irregular fluid movement with complex boundary conditions and high turbulence intensity, and seriously affects the stable operation of the turbine and reduces the effective head and efficiency [10–12]. Therefore, it is necessary to study and understand the tip leakage characteristics to predict the power generation and structural load to assess the applicability and economy of a power station before deployment. Presently, many scholars have studied the formation mechanism and geometric influence of the tip leakage vortex (TLV) through numerical calculations and experimental methods [13–16]. Taha et al. [17] studied the performance of a Wells turbine under different tip clearances using the computational

---

fluid dynamics method. The results showed that although the performance of a turbine with non-uniform tip clearance was similar to that of a turbine with uniform tip clearance in terms of torque coefficient, input power coefficient, and efficiency, a turbine with uneven tip clearance seemed to provide better overall performance. Guenette et al. [18] analyzed and studied tubular turbines with tip, hub, and no clearance and found that the influence of hub clearance on runner performance was much smaller than that of tip clearance at the outer edge of the blade. Zhang et al. [19] used a combination of numerical simulation and high-speed photography to study the TLV structure of axial flow pumps based on different tip clearances. The results show that the starting point of the TLV appears near the leading edge under low-flow conditions. Under high flow conditions, the starting point of the TLV shifts to the middle of the blade chord, and the direction is parallel to the blade wing. As the blade tip size was increased, the starting point of the TLV trajectory moved towards the center of the blade chord, and the minimum pressure in the vortex core gradually decreased. Xiao et al. [20] measured the distribution of pressure, velocity, and runner loss in an experimental study of an axial runner. Studies have found that the action of suction on the vortex increases the load near the blade tip. Total pressure loss, as well as reduction of total pressure, occurred in the clearance leakage vortex region. Lemay et al. [21] used laser Doppler velocimetry (LDV) technology to find two vortices with opposite rotation directions in the area near the blade tip of the suction surface.

The above studies are focused on a fixed gap or small gap size, and there are few studies on the tip leakage flow characteristics in large gap size. In this study, based on the SST k–ω turbulence model, the full-channel unsteady numerical calculation of a tubular turbine with one normal and two large clearances is carried out under the rated flow condition. The leakage flow characteristics in the clearance were analyzed, and the distribution and intensity of the leakage vortex were analyzed using the Q criterion vortex-dynamics method. The influence of the TLV on the energy conversion characteristics of the tubular turbine under a large clearance structure is revealed, providing a basis for the design optimization of tubular turbines.

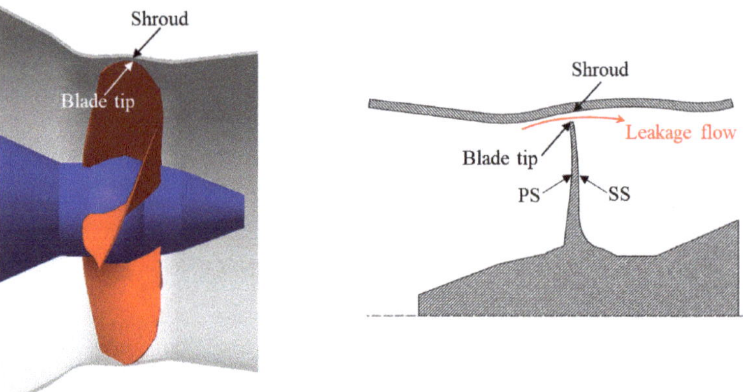

Figure 1. Clearance between blade tip and shroud.

2. Numerical Model

2.1. Computational Method

In this study, the governing equations include the continuity equation of an incompressible fluid and the Reynolds-averaged Navier-Stokes (RANS) equation.

$$\frac{\partial \rho}{\partial t} + \frac{\partial (\rho u_i)}{\partial x_i} = 0 \tag{1}$$

$$\frac{\partial(\rho u_i)}{\partial t} + \frac{\partial(\rho u_i u_j)}{\partial x_i} = -\frac{\partial p'}{\partial x_i} + \frac{\partial}{\partial x_j}\left[\mu_{eff}\left(\frac{\partial u_i}{\partial x_j} + \frac{\partial u_j}{\partial x_i}\right)\right] \tag{2}$$

The ANSYS CFX 17.0 commercial computational fluid dynamics software was used to solve the three-dimensional unsteady flow in a hydraulic turbine. The discretization of the control equation is based on the finite volume method, and a control volume composed of internal unit nodes was considered. The SST k–ω model was chosen as the turbulence model to capture the separated flow near the tip clearance more accurately [22]. Compared with the standard k–ε and RNG k–ε models, this model considers the viscosity of the inner wall of the model and has better turbulent shear stress transmission. The algorithm is more stable and has a better simulation performance for flow in a narrow space [23]. The corresponding turbulent kinetic energy and frequency equations are as follows [24]:

$$\frac{\partial(\rho k)}{\partial t} + \frac{\partial(\rho u_i k)}{\partial x_i} = \frac{\partial}{\partial x_i}\left[(\mu + \sigma_k \mu_k)\frac{\partial k}{\partial x_i}\right] + \widetilde{P}_k - 0.09\rho k\omega \tag{3}$$

$$\frac{\partial(\rho\omega)}{\partial t} + \frac{\partial(\rho u_i \omega)}{\partial x_i} = \frac{\partial}{\partial x_i}\left[(\mu + \sigma_\omega \mu_t)\frac{\partial \omega}{\partial x_i}\right] + (1 - F_1)\rho\sigma_{\omega 2}\frac{1}{\omega}\frac{\partial k}{\partial x_i}\frac{\partial \omega}{\partial x_i} \\ + \left[\frac{5}{9}F_1 + 0.44(1 - F_1)\right]\frac{1}{\nu}\widetilde{P}_{k_t} - \beta\rho\omega^2 \tag{4}$$

2.2. Computational Model

To improve the reliability of the numerical results, this study considers the tubular turbine of a power station as the research object and uses UG NX 12.0 commercial software to complete the model design. The main design parameters are listed in Table 1. The calculation domain mainly includes four components, namely, the water inlet section, water guide mechanism, impeller, and draft tube, as shown in Figure 2.

Table 1. Main design parameters of research objects.

Parameter	Unit	Value
Rated speed (N_r)	r/min	68.18
Rated flow (Q_r)	m^3/s	375.2
Maximum head (H_{max})	m	11
Minimum head (H_{min})	m	2.6
Rated head (H_r)	m	7.8
Runner diameter (D_1)	m	7.25
Blade (Z)	—	4
Guide vane (Z_0)	—	16

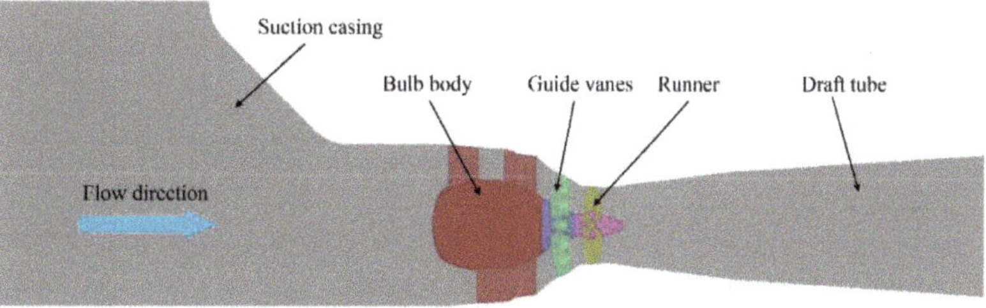

Figure 2. Schematic of full channel of tubular turbine.

Considering the platform configuration (Intel Xeon E5-2650 @ 2.3 processor with 64 GB of RAM) and the entire channel structure, this study used the commercial software, ICEM CFD 17.0, to mesh the computational domain. Parts with narrow geometric structure, such as the guide vane, impeller, and tip clearance, were locally encrypted with an "O-shaped" grid to ensure the uniform transition between the tip clearance and the grids of other parts, and to accurately capture the complex flow details around the tip clearance. The entire flow passage and local grid are shown in Figure 3. To ensure grid quality, grid independence verification analysis was carried out, and five schemes with different grid numbers were verified and analyzed with turbine efficiency as the judgment standard. The results are presented in Figure 4. It can be seen from Figure 4 that when the number of grids increases from 9.3 to 11.73 million, the efficiency increment of the hydraulic turbine is less than 0.01%, which fulfills the requirement of grid independence. Finally, the scheme with 9,379,600 grids was selected for the numerical calculation. The number of grids for the inlet section, the movable guide vane, the runner and the draft tube were 1,760,634; 3,762,800; 2,849,380 and 1,006,787, respectively. The y+ distribution near the blade and guide vane is shown in Figure 5, which meets the requirements of the SST k–ω near-wall turbulence model [25].

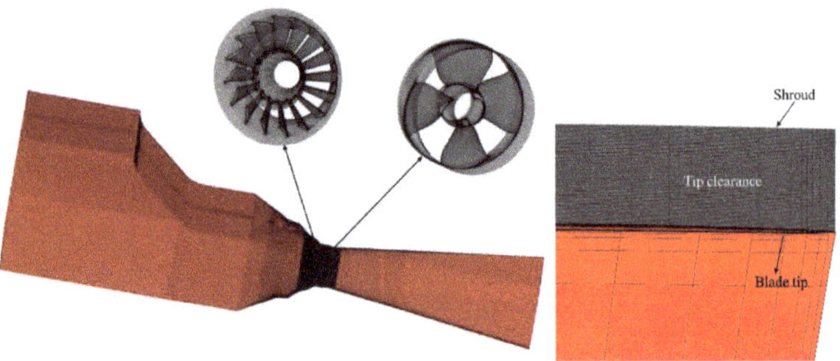

Figure 3. Computational domain grid.

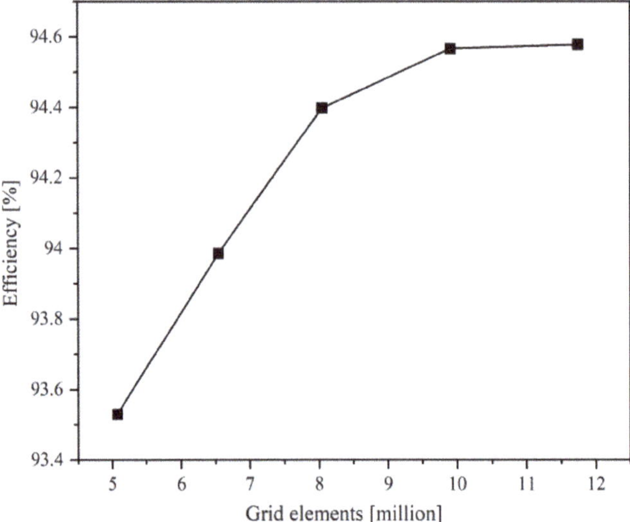

Figure 4. Grid independence verification.

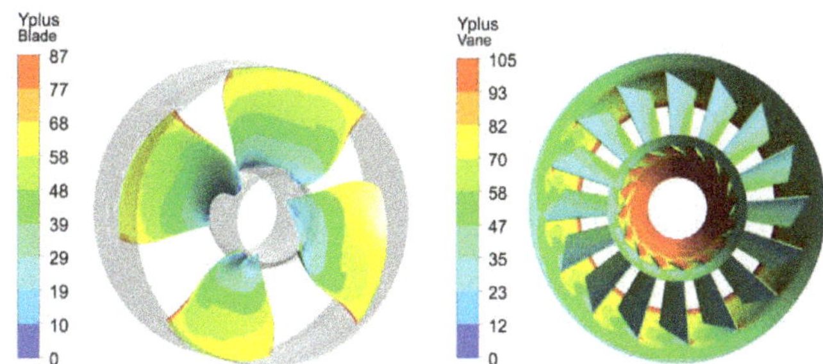

Figure 5. y+ distribution on blade and guide vane surface.

The corresponding boundary conditions were set according to operating conditions. In this study, the inlet was set as the mass flow, the outlet as the static pressure, and the wall as a non-slip wall. The turbulent flow on the wall of each flow component was determined using the standard wall function. The static reference coordinate was selected as the internal flow coordinate of the fixed part, and the rotating reference coordinate as the internal flow coordinate of the rotating part. Because the SST k–ω turbulence model was used to solve the flow state at the gap, a high-resolution mode was selected. The time step of the steady numerical calculation was set to $1/\omega$, that is, the reciprocal of the rotational angular velocity and the dynamic and static interface was set to the frozen rotor interface. For unsteady numerical calculation, to further improve the solution accuracy, high resolution and second-order backward Euler were selected. The measurement of the time step was set as each rotation by the runner by $2°$ (0.004889 s), and the total calculated time was 15.84 s, which accounted for 18 rotation cycles. The static-dynamic interface was set as the transient rotor/stator interface. The interfaces between the various flow components are shown in Figure 6, where 1 is the interface between the water inlet section and movable guide vane, 2 is the interface between the movable guide vane and runner, 3 is the interface between the runner and draft tube, and 2 and 3 are rotary interfaces.

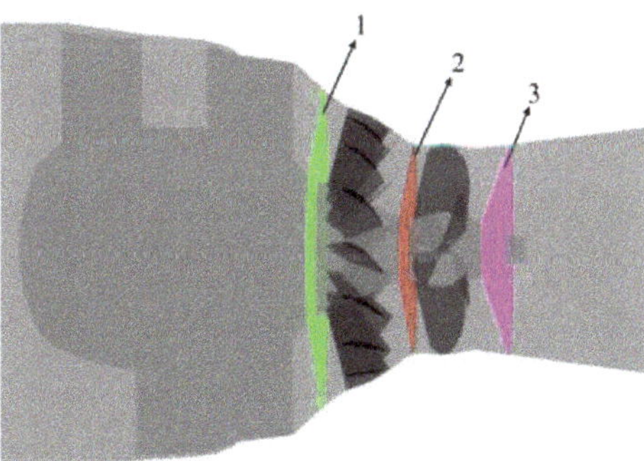

Figure 6. Interface between overcurrent components.

3. Numerical Results and Discussion

To study the leakage flow and leakage vortex characteristics of the hydraulic turbine under different tip clearances, the unsteady numerical calculation results of three structures with tip clearance $\delta_1 = 5$ mm, $\delta_2 = 10$ mm, and $\delta_3 = 20$ mm were analyzed under the rated flow condition Q_r (guide vane opening 55.6° and slurry vane opening 20°).

3.1. Tip Leakage Flow Characteristics

3.1.1. Axial Momentum and Flow

The tip leakage flow was formed under the pressure difference between the blade pressure side and suction side, as shown in Figure 1, and the axial momentum of the leakage flow can be used to describe the intensity of the tip leakage flow. An axial momentum balance existed between the main flow and tip leakage flow [26]. Under a certain clearance, the cumulative axial momentum of the tip leakage flow is defined as follows [27]:

$$M_{\text{axial}} = \int_0^z \int_0^\delta \rho(\mathbf{v} \cdot \mathbf{n}) w \, dr \, dz, \tag{5}$$

where z is the axial distance along the blade chord, δ is the tip clearance height, $\mathbf{v}$ is the velocity vector, $\mathbf{n}$ is the unit vector normal to the blade, and w is the axial velocity.

To study the axial momentum and flow rate of the leakage flow under different tip clearances, a section from the leading edge to the trailing edge of the blade was set in the clearance channel between the tip and the runner chamber, which was divided into 19 parts along the flow direction. The axial momentum and leakage flow were calculated for the 19 small sections, and the results are shown in Figures 7 and 8. It can be observed from Figure 7 that the axial momentum is maximum at the leading edge of the blade, decreases gradually along the flow direction, and reaches a minimum at the trailing edge; this is due to the maximum pressure difference between the pressure side and the suction side near the leading edge of the blade, which results in the maximum leakage flow velocity. With an increase in clearance, the axial momentum increases as a whole. The axial momentum difference at the leading-edge position of the blade was the largest, and that at the trailing edge position was the smallest under different clearances. It can be predicted that the influence of the tip clearance on the leakage flow is the strongest near the leading edge and weakest near the trailing edge of the blade. The axial momentum suddenly drops by approximately 60% in the flow direction when the tip clearance is $\delta_3 = 20$ mm, and drops by approximately 65% and 70% when the clearance is $\delta_2 = 10$ mm and $\delta_1 = 5$ mm, respectively. Especially in a small gap ($\delta_1 = 5$ mm), the axial momentum is almost linearly changing. It can be seen that the axial momentum of the tip leakage flow decreases significantly at a certain point along the flow direction, and the turning point will be closer to the middle position of the tip when the gap is larger. Figure 8 shows the leakage flow across the gap surface under different gaps, and the change rule is almost the same as the axial momentum. The leakage vortex is formed by the interaction between the leakage flow and the main flow. It is inferred from the above analysis that the TLV is mainly formed from the leading edge to the middle of the blade, and the leakage vortex generated near the trailing edge is weak.

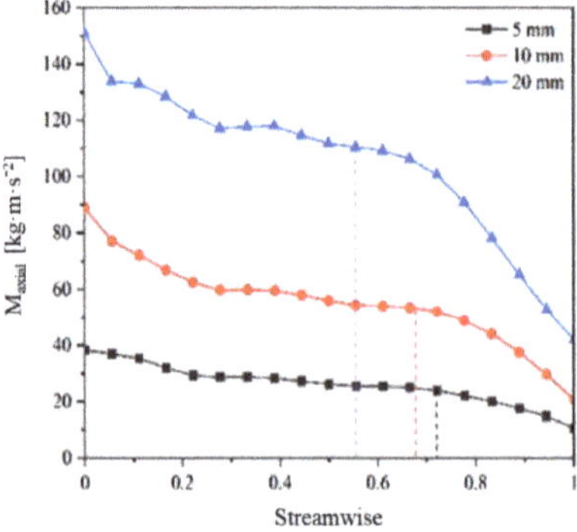

Figure 7. Axial momentum of leakage flow.

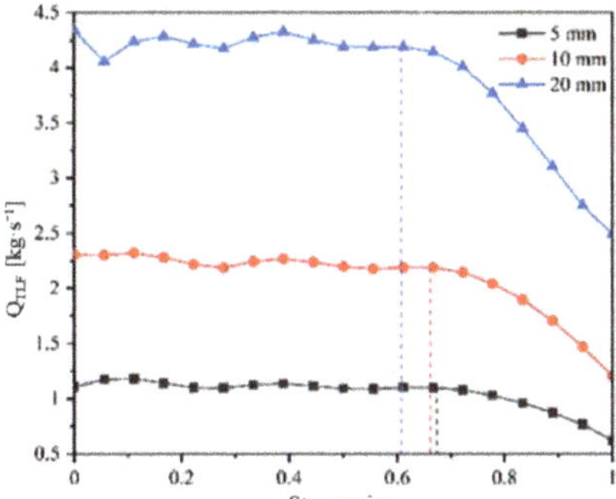

Figure 8. Leakage flow.

3.1.2. Leakage Velocity and Turbulent Kinetic Energy

The tip leakage flowed out from the suction side of the blade and then met with the main flow in the flow channel, interacting with each other to form a leakage vortex. The velocity of the leakage flow differs from that of the leakage vortex produced by the mainstream impact; therefore, the velocity of the leakage flow is also a vital index for measuring the TLV intensity. Figure 9 shows the velocity distribution of the leakage flow for different gaps. It can be seen that the leakage velocity decreases along the flow direction and increases with the increase in the gap, but there are differences in the velocity distribution at the leading edge, trailing edge, and middle position. Specifically, the maximum leakage velocity near the leading edge of the blade (the reason mentioned in Section 3.1.1) is due to the maximum pressure difference between the pressure side and suction side of the blade at the inlet position. Moreover, near the leading edge, the leakage flow velocity of small gap size (δ_1 = 5 mm) is the largest, while that of large gap size (δ_3 = 20 mm) is the smallest because, under certain pressure, the "jet" velocity increases with the gap size. The leakage velocity near the trailing edge of the blade is the smallest, and the gap size has little effect on the leakage velocity. This is because the flow in the flow channel tends to be stable along the flow direction, and the pressure difference between the two sides of the blade reaches the minimum, that is, the power source that produces the tip leakage flow is the weakest. Within the range 35–50% along the flow direction (within the green dotted line box in Figure 9), it can be clearly seen that the change in leakage velocity presents a trough shape and reaches a minimum value, which is inversely proportional to the gap size; that is, the larger the gap size, the smaller the leakage velocity. This phenomenon may be caused by the twisted blade structure. The curvature of the blade changes significantly near the range 35–50% along the flow direction, whereas the flow always follows the tangential direction of the blade. This leads to a greater tendency for the main flow to break away from the blade on the pressure side, thus reducing the leakage velocity accordingly. Overall, the distribution law of leakage velocity and the variation law of leakage volume are similar in that they always decrease along the flow direction. Figure 10 depicts the variation law of the turbulent kinetic energy of the leakage flow in the gap. Compared with the leakage velocity curve, it can be seen that the turbulent kinetic energy of the leakage flow is inversely proportional to the gap size, which is contrary to the change law of leakage velocity that the larger the gap size, the weaker the turbulent kinetic energy. Similarly, in the range 35–50% along the flow direction (within the green dotted line in Figure 10), the turbulent kinetic energy changes locally, and the maximum turbulent kinetic energy occurs in the 10 mm gap size. In addition, in the 20 mm gap size, the turbulent kinetic energy of the gap-flow changes almost linearly along the flow direction.

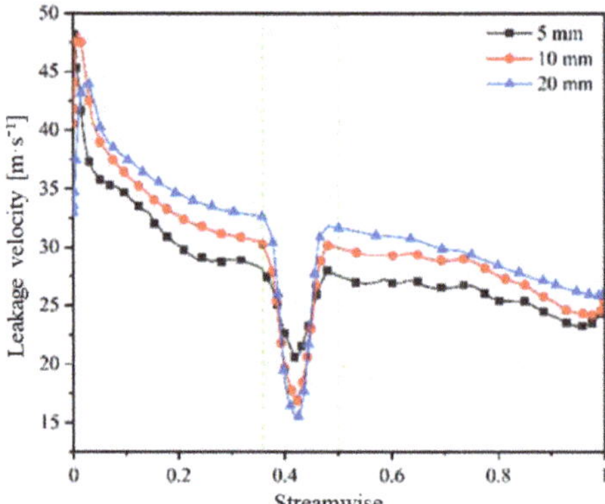

Figure 9. Leakage speed.

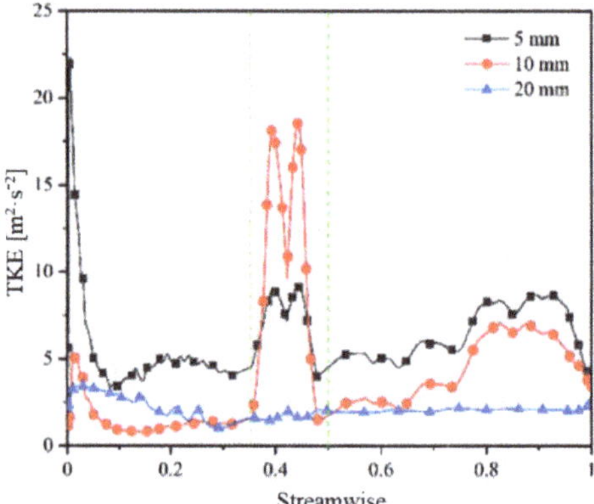

Figure 10. Turbulent kinetic energy of leakage flow.

3.2. Tip Leakage Vortex Characteristics

The leakage flow in the tip clearance enters the impeller passage, interacts with the main flow, and finally evolves into a TLV, which complicates the flow pattern near the tip clearance. The Q criterion, proposed by Hunt et al. [28], defines a vortex as the region where the magnitude of the vortex is greater than the shear strain rate, which is also called the region where the second invariant of the velocity gradient tensor is positive. Compared with the vorticity criterion, it can identify the local low-pressure region more effectively. To show the vortex at the gap more accurately, this study uses the Q criterion to analyze the LTV structure and evolution law, and its expression is as follows:

$$Q = \|\mathbf{\Omega}\|^2 - \|\mathbf{S}\|^2,\tag{6}$$

where $\mathbf{\Omega}$ is vorticity and $\mathbf{S}$ is the deformation rate tensor.

3.2.1. TLV Structure

Figure 11 shows a schematic of the TLV structure generated at the blade tip. It can be seen that TLV is generated at the leading edge (LE) of the blade tip, extends to the trailing edge (TE), and is mainly distributed on the suction side. The vortex on the suction side is divided into two parts, which are called the primary tip leakage vortex (PTLV) and secondary tip leakage vortex (STLV) by Tan et al. [5], corresponding to the green and blue parts in Figure 10. A continuous sheet STLV was attached to the suction side of the tip and extended from the leading edge to the trailing edge, while a columnar PTLV extended along the flow direction in the flow channel. In Figure 11, x represents the distance from the separation point of the PTLV on the suction side to the leading edge of the blade, and α represents the separation angle. Figure 12 shows the TLV distribution under the three tip clearances. It can be seen that when the clearance is $\delta_1 = 5$ mm, TLV is generated on both the pressure side) and the suction side and separated at the leading edge, and the distribution is consistent with that described in Figure 11. When the clearance increased to $\delta_2 = 10$ mm, the cylindrical vortex band on the suction side grew, and the separation point was still at the leading-edge position, but the separation angle α decreased. In addition, the STLV separates from the blade surface along the blade tip from the leading edge to the trailing edge under the clearance. When the gap was further increased to $\delta_3 = 20$ mm, the vortex band on the pressure side disappeared, while the columnar vortex band on the suction side continued to grow. The separation point shifts to x away from the leading edge of the blade, and the separation angle α further decreases. Under this gap, the STLV separates from the blade surface and exits in the flow passage. With an increase in the gap size, the leakage vortex intensity increases, and the vortex intensity decreases along the flow direction in a specific gap size. This can be explained by the axial momentum and leakage velocity mentioned earlier, that is, the axial momentum and velocity of the leakage flow decreases continuously along the flow direction, and the disturbance effect caused by the action of the main stream gradually weakens; thus, the strength of the leakage vortex decreases continuously.

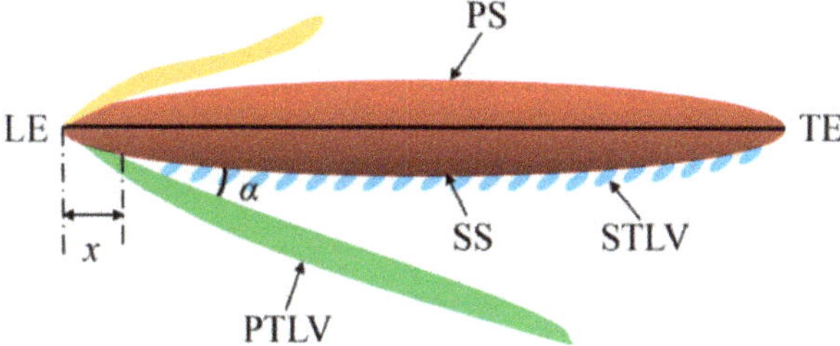

Figure 11. TLV structure diagram.

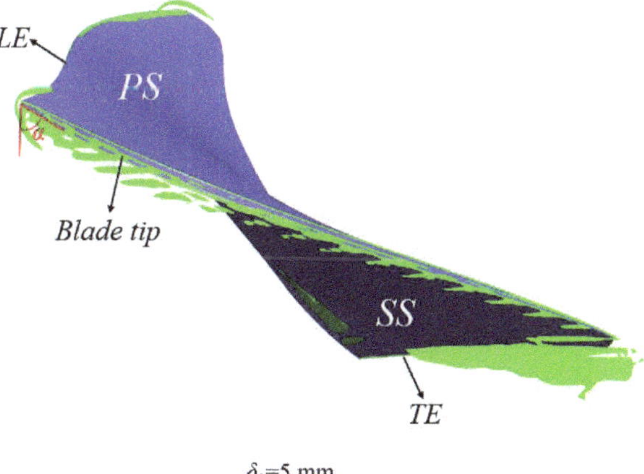

$\delta_1 = 5$ mm

Figure 12. *Cont.*

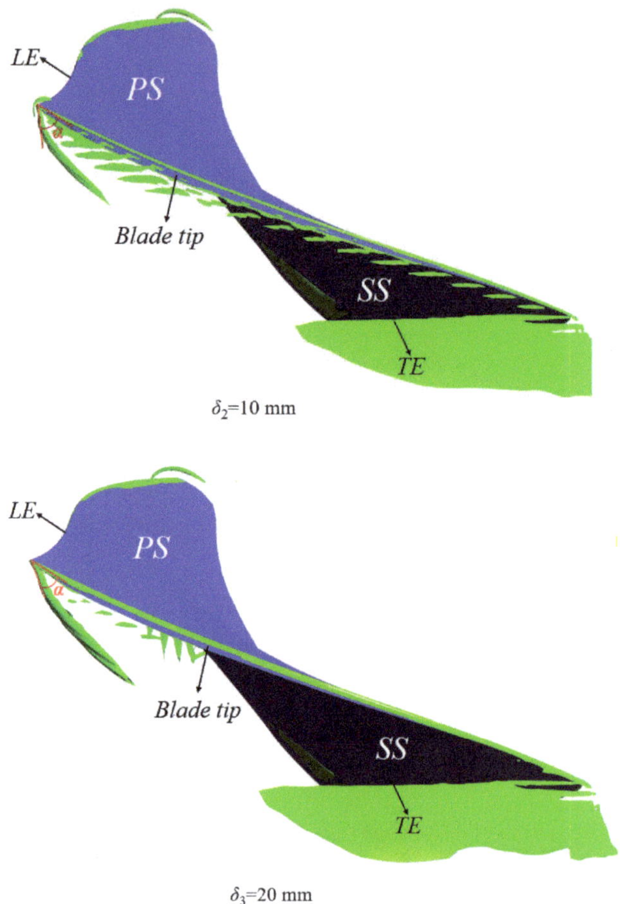

Figure 12. TLV structure under different gaps.

3.2.2. TLV Strength under Different Gaps

It can be inferred from the above results that the tip clearance has a significant influence on the TLV under the rated flow conditions. To further study the evolution law of the TLV under different gap sizes, five sections perpendicular to the streamline were uniformly set along the flow direction from the leading to the trailing edge of the blade, as shown in Figure 13.

Figure 14 shows the distribution law of vortex intensity on five characteristic sections under three blade tip clearances. When the tip leakage flow enters the impeller runner, the flow velocity slows down and interacts with the main flow in the runner to form a TLV. The TLV was mainly distributed near the tip of the suction side and inside the runner, as shown by the dashed box in Figure 14, which corresponds to STLV and PTLV, shown in Figure 11. Furthermore, PTLV has two core areas in the flow channel, corresponding to the areas indicated by the two black dots, A and B in Figure 14. Overall, the leakage vortex intensity decreases along the flow direction, and the gap size is proportional to the leakage vortex intensity in the same section; that is, the leakage intensity is significant in the impeller structure with a large gap. From Section 1 to Section 5, the core area of the PTLV vortex decreases continuously, and the distance between Cores A and B and the blade tip increases, which evolves into the flow channel. The STLV extends along the inner wall of the runner chamber to the inside of the runner. The vortex core reaches its maximum in Section 3, decreases in Section 4, and effectively disappears in Section 5. In addition, it can be clearly seen that with the increase in the gap, Core A of PTLV gradually approaches the STLV and almost overlaps along the flow direction. In Section 1, that is, the inlet position, with the increase in clearance, PTLV approaches the suction side of the tip, while STLV gradually disappears. The reason for this phenomenon may be that at the inlet end, part of the high-pressure fluid directly enters the tip clearance without passing through the runner, and the local high-pressure zone formed by the high-pressure fluid flows with the pressure and suction side, respectively, which hinders the leakage flow from the pressure side to the suction side. Overall, with the increase in clearance, more leakage flows pass through the tip clearance, the leakage vortex becomes stronger, and the leakage vortex disturbs the inner flow of the runner more strongly, which makes the inner streamline of the runner chaotic, thereby increasing the pressure fluctuation in the runner.

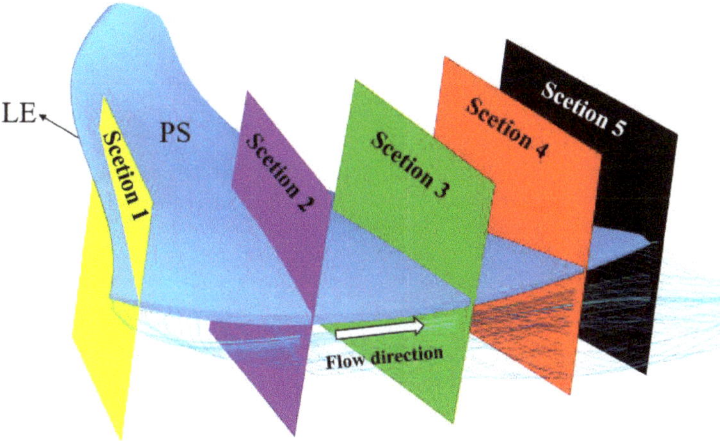

Figure 13. Schematic cross-sectional view along the flow direction.

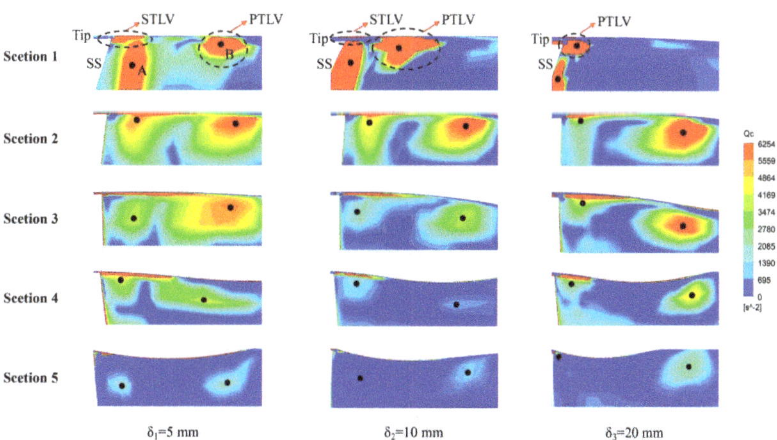

Figure 14. Vortex intensity distribution at each section under different gaps.

3.2.3. Flow Characteristics in Gaps

Figure 15 shows the pressure and velocity vector distribution nephograms of the flow field at different tip gaps under rated-flow conditions. Because there is a pressure difference between the pressure and suction side of the blade, the flow on the pressure side of the blade is pulled into the narrow channel and leaked. When the leakage flow enters the tip clearance, owing to the right-angle tip geometry, the flow separates at a right angle near the pressure side of the blade, resulting in a local low-pressure area near the pressure side in the clearance. In this area, the reverse pressure gradient reduces the velocity of the fluid particles near the separation point, and finally, the reverse flow of the fluid particles forms the tip separation vortex, which increases the possibility of cavitation in the clearance. To eliminate the corner separation vortex generated there, the tip of the pressure side can be rounded [29]. The leakage flow flows out of the gap channel on the suction side and interacts with the main flow to generate a leakage vortex. The influence of the gap on the TLV intensity has been analyzed in Section 3.2.2, so it is not repeated here. Along the flow direction, from Section 1 to Section 5, the proportion of the low-pressure area in the tip clearance near the blade pressure side gradually expands, and the separation vortex moves toward the blade suction side in the clearance. This expansion further enlarges the disturbance range to the surrounding basin, which explains the flow disorder in the tip clearance near the blade working face. In particular, the low-pressure zone in the gap in Section 5 fills half of the gap flow channel, and a large-scale flow separation occurs. With an increase in the tip clearance, the range of the low-pressure area near the pressure side in the same cross-section clearance increases, therefore, the large clearance structure is more likely to promote the generation of a separation vortex and complicate the flow in the flow and leakage channel. Considering the previous analysis of the TLV, the separation vortex in the gap and the leakage vortex in the flow passage can be effectively reduced by minimizing the tip clearance when the design conditions permit.

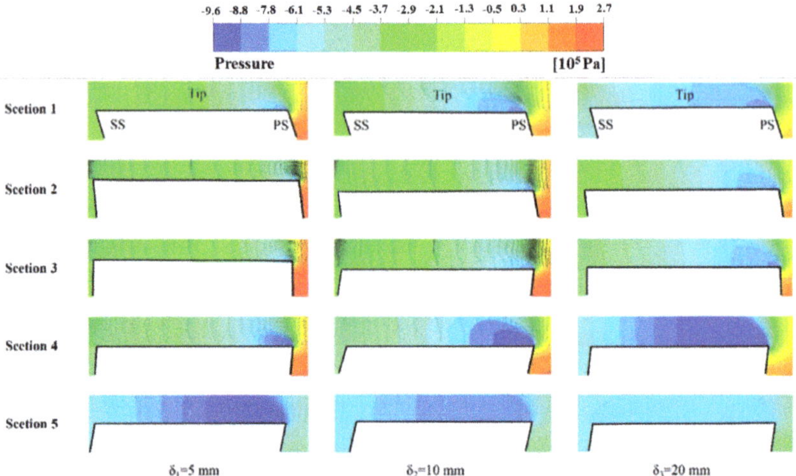

Figure 15. Vector distribution of pressure and velocity in gap.

4. Conclusions

In this study, the SST k–ω turbulence model was used to study the leakage flow and leakage vortex characteristics of a low-head tubular turbine under the blade tip clearance. Based on the numerical results, the main conclusions are as follows:

(1) The gap size is proportional to the axial momentum and flow rate of the leakage flow. With an increase in the gap size, both the axial momentum and leakage amount

increase. The gap size is proportional to the leakage flow velocity and inversely proportional to the turbulent kinetic energy. The variation of velocity and turbulent kinetic energy at 35–50% of the flow direction shows the law of trough and peak, respectively.

(2)   The leakage vortex is divided into two parts: the PTLV and STLV. The vortex intensities of these two parts increase with an increase in clearance. However, for a large clearance structure ($\delta_3 = 20$ mm), the vortex core of the STLV overlaps with Core A of the PTLV, and Core B of the PTLV becomes the main part of the TLV.

(3)   The existence of the "right angle effect" of the blade tip causes the flow separation phenomenon near the pressure side of the blade tip, resulting in a local low-pressure zone, thus generating a local separation vortex at this position. With an increase in the clearance, the strength of the separation vortex also increased, which aggravated the flow instability.

Author Contributions: Conceptualization, B.Z. and Z.L.; writing—original draft preparation and methodology, X.L.; writing—review and editing, W.W.; supervision, Z.L. All authors have read and agreed to the published version of the manuscript.

Funding: This research was funded by the National Natural Science Foundation of China (Grant No. 52079118), National Key R&D Program of China (Grant No. 2018YFE0128500) and International Cooperation Project of Sichuan Science and Technology Department (Grant No. 2020YFH0135).

Institutional Review Board Statement: Not applicable.

Informed Consent Statement: Not applicable.

Data Availability Statement: Data from this study can be made available upon request.

Acknowledgments: The authors would like to thank all staff of the treatment plants for making the data available. Useful suggestions given by Rameshwar Adhikari from Applied Science and Technology Research Center, Tribhuvan University, Nepal.

Conflicts of Interest: The authors declare no conflict of interest.

References

1.   Weitemeyer, S.; Kleinhans, D.; Vogt, T.; Agert, C. Integration of renewable energy sources in future power systems: The role of storage. *Renw. Energy* **2015**, *75*, 14–20. [CrossRef]

2.   Guney, M.S. Solar power and application methods. *Renew. Sustain. Energy Rev.* **2016**, *57*, 776–785. [CrossRef]

3.   Giannuzzi, A.; Diolaiti, E.; Lombini, M.; Rosa, A.; Marano, B. Enhancing the efficiency of solar concentrators by controlled optical aberrations: Method and photovoltaic application. *Appl. Energy* **2015**, *145*, 211–222. [CrossRef]

4.   Yang, C.X.; Lu, M.T.; Zheng, Y.; Tian, Q.X.; Zhang, Y.Q. Inlet passage's development and optimization of new tidal unit-shaft tubular turbine. *Appl. Mech. Mater.* **2014**, *607*, 312–316. [CrossRef]

5.   Liu, Y.B.; Han, Y.D.; Tan, L.; Wang, Y.M. Blade rotation angle on energy performance and tip leakage vortex in a mixed flow pump as turbine at pump mode. *Energy* **2020**, *206*, 118084. [CrossRef]

6.   Chang, X.L.; Liu, X.H.; Zhou, W. Hydropower in China at present and its further development. *Energy* **2010**, *35*, 4400–4406. [CrossRef]

7.   Elbatran, A.H.; Yaakob, O.B.; Ahmed, Y.M.; Shabara, H.M. Operation, performance and economic analysis of low head micro-hydropower turbines for rural and remote areas: A review. *Renew. Sustain. Energy Rev.* **2015**, *43*, 40–50. [CrossRef]

8.   Behrouzi, F.; Nakisa, M.; Maimun, A.; Ahmed, Y.M. Global renewable energy and its potential in Malaysia: A review of Hydrokinetic turbine technology. *Renew. Sustain. Energy Rev.* **2016**, *62*, 1270–1281. [CrossRef]

9.   Kramer, M.; Wieprecht, S.; Terheiden, K. Minimizing the air demand of micro-hydro impulse turbines in counter pressure operation. *Energy* **2017**, *133*, 1027–1034. [CrossRef]

10.   Denton, J.D. The 1993 IGTI scholar lecture: Loss mechanisms in turbomachines. *Turbomach* **1993**, *115*, 621–656. [CrossRef]

11.   Thapa, B.S.; Dahlhaug, O.G.; Thapa, B. Sediment erosion induced leakage flow from guide vane clearance gap in a low specific speed Francis turbine. *Renew. Energy* **2017**, *107*, 253–261. [CrossRef]

12.   You, D.H.; Wang, M.; Moin, P.; Mittal, R. Vortex dynamics and low-pressure fluctuations in the tip-clearance flow. *Fluids Eng.* **2007**, *129*, 1002–1014. [CrossRef]

13.   Cheng, H.Y.; Bai, X.B.; Long, X.P.; Ji, B.; Peng, X.X.; Farhat, M. Large Eddy Simulation of the Tip-leakage Cavitating flow with an insight on how cavitation influences vorticity and turbulence. *Appl. Math. Model.* **2019**, *77*, 788–809. [CrossRef]

14. Shi, L.; Zhang, D.S.; Zhao, R.J.; Shi, W.D.; Jin, Y.X. Effect of blade tip geometry on tip leakage vortex dynamics and cavitation pattern in axial-flow pump. *Sci. China Technol. Sci.* **2017**, *60*, 1480–1493. [CrossRef]
15. Guo, Q.; Zhou, L.; Wang, Z. Numerical evaluation of the clearance geometries effect on the flow field and performance of a hydrofoil. *Renew. Energy* **2016**, *99*, 390–397. [CrossRef]
16. Liu, Y.B.; Tan, L. Influence of C groove on suppressing vortex and cavitation for a NACA0009 hydrofoil with tip clearance in tidal energy. *Renew. Energy* **2020**, *148*, 907–922. [CrossRef]
17. Taha, Z.; Sugiyono; Tuan Ya, T.M.Y.S.; Sawada, T. Numerical investigation on the performance of Wells turbine with non-uniform tip clearance for wave energy conversion. *Appl. Ocean Res.* **2011**, *33*, 321–331. [CrossRef]
18. Guénette, V.; Houde, S.; Ciocan, D.; Dumas, J.; Deschênes, C. Numerical prediction of a bulb turbine performance hill chart through RANS simulations. *IOP Conf. Ser. Earth Environ. Sci.* **2012**, *15*, 032007. [CrossRef]
19. Zhang, D.S.; Shi, W.D.; Wu, S.Q.; Pan, D.Z. Numerical and experimental investigation of tip leakage vortex trajectory and dynamics in an axial flow pump. *Comput. Fluids* **2013**, *112*, 61–71. [CrossRef]
20. Xiao, X.W.; McCarter, A.A.; Lakshminarayana, B. Tip Clearance Effects in a Turbine Rotor: Part I—Pressure Field and Loss. *J. Turbomach.* **2001**, *123*, 296–304. [CrossRef]
21. Lemay, S.; Aeschlimann, V.; Fraser, R.; Ciocan, G.D.; Deschenes, C. Velocity field investigation inside a bulb turbine runner using endoscopic PIV measurements. *Exp. Fluids* **2015**, *56*, 120. [CrossRef]
22. Shi, G.T.; Liu, Z.K.; Xiao, Y.X.; Yang, H.; Li, H.L.; Liu, X.B. Effect of the inlet gas void fraction on the tip leakage vortex in a multiphase pump. *Renew. Energy* **2020**, *150*, 46–57. [CrossRef]
23. Menter, F.R.; Rumsey, C.L. Assessment of Two-Equation Turbulence Models for Transonic Flows. *AIAA* **1994**, *94*, 2343.
24. Ding, A.; Ren, X.; Li, X.; Gu, C. Numerical investigation of turbulence models for a superlaminar journal bearing. *Adv. Tribol.* **2018**, *2018*, 2841303. [CrossRef]
25. Erler, E.; Vo, H.D.; Yu, H. Desensitization of axial compressor performance and stability to tip clearance zize. *J. Turbomach.* **2015**, *138*, 031006. [CrossRef]
26. Menter, F.R. Review of the shear-stress transport turbulence model experience from an industrial perspective. *Int. J. Comput. Fluid Dyn.* **2009**, *23*, 305–316. [CrossRef]
27. Han, Y.D.; Tan, L. Influence of rotating speed on tip leakage vortex in a mixed flow pump as turbine at pump mode. *Renew. Energy* **2020**, *162*, 144–150. [CrossRef]
28. Hunt, J.C.R.; Wray, A.A.; Moin, P. Eddies, streams, and convergence zones in turbulent flows. *Stud. Turbul. Using Numer. Simul. Databases* **1988**, *1*, 193–208.
29. Laborde, R.; Chantrel, P.; Mory, M. Tip clearance and tip vortex cavitation in an axial flow pump. *J. Fluids Eng.* **1997**, *119*, 680–685. [CrossRef]

Article

Performance Improvement of Hydrofoil with Biological Characteristics: Tail Fin of a Whale

Pan Xiong [1,2], **Jianghong Deng** [1,2,*] and **Xinyuan Chen** [1,2,*]

1   Key Laboratory of Metallurgical Equipment and Control Technology, Wuhan University of Science and Technology, Ministry of Education, Wuhan 430081, China; whxpmechanical@163.com
2   Hubei Key Laboratory of Mechanical Transmission and Manufacturing Engineering, Wuhan University of Science and Technology, Wuhan 430081, China
*   Correspondence: djhmechanical@163.com (J.D.); cxywust@163.com (X.C.)

Abstract: In order to improve the hydrodynamic performance of hydrofoils, this paper shows excellent hydrodynamic performance according to the flapping motion of fish through the tail fin. The Naca66 hydrofoil is used as the original hydrofoil and the trailing edge flap configuration is added. Ansys-fluent is used to analyze the relationship between the structural parameters (length and angle) of the flap and the hydrodynamic performance of the hydrofoil, the reliability of CFD numerical simulation is verified by PIV experiment. It is found that the hydrofoil, with clockwise rotating short flap, can significantly improve the hydrodynamic performance of a hydrofoil at a small angle of attack; at a high angle of attack, the hydrofoil with counterclockwise flap can increase the critical stall angle and slightly improve the hydrodynamic performance of the hydrofoil. The hydrodynamic performance of hydrofoil with rotatable short flaps reported in this paper can provide valuable information for the design and optimization of this kind of hydrofoil.

Keywords: biological; flap; hydrodynamic performance; stall; CFD

Citation: Xiong, P.; Deng, J.; Chen, X. Performance Improvement of Hydrofoil with Biological Characteristics: Tail Fin of a Whale. *Processes* **2021**, *9*, 1656. https://doi.org/10.3390/pr9091656

Academic Editors: Santiago Lain and Omar Dario Lopez Mejia

Received: 26 August 2021
Accepted: 10 September 2021
Published: 14 September 2021

Publisher's Note: MDPI stays neutral with regard to jurisdictional claims in published maps and institutional affiliations.

1. Introduction

Hydrofoil, as an auxiliary component of high-speed ships, is widely used in hydrofoil ships and planing boats. By installing a hydrofoil on the bottom of hydrofoil ships, it can generate huge lifting force to lift the ship off the water at high speed, thus greatly reducing the drag of water to the ship and wave-making drag, and also reducing the interference of waves to the stability of the ship [1,2]. Therefore, in order to improve the performance of hydrofoil ships, many scholars have carried out research on hydrofoil lift and drag reduction.

Compared with airfoils, the flow medium of a hydrofoil is water, and the density and viscosity of water are much greater than that of air. Therefore, the lift generated by a hydrofoil with the same shape and moving state is much greater than that generated by a wing. At the same time, the friction resistance and differential pressure resistance of hydrofoils are also increased accordingly. However, scholars' research on hydrofoils and airfoils has the same goal of improving performance (increasing lift and reducing drag). Scholars have undertaken much research on increasing lift and reducing drag of airfoils, mainly including bionics [3,4], vortex generators [5,6], synthetic jet [7,8], and multi-element airfoil [9,10]. However, the working environment of the hydrofoil is quite different from that of the wing, and not necessarily applicable. In the research of hydrofoils, inspired by bird and fishes, the applications of flapping foils as energy harvesting devices have gradually attracted attention in recent decades [11,12]. For example, dolphins and sharks exhibit excellent hydrodynamic performance, including high cruising speed, high efficiency, and low noise through the flapping motion of their caudal fins [13]. Johari [14] found that a full-span wing with leading-edge tubercles increases the lift coefficient by as much as 50% in the post-stall regime at a Reynolds number of Re = 1.83×10^5; the idea was originally motivated by the physiological structures of humpback whales. But Rostamzadeh [15] and Skillen [16] also shed light on the implementation of leading-edge tubercles when they were used on aerofoils and wings immersed in free-streams. Both studies found that the tubercles led to the formation of secondary flows due to the presence of strong spanwise pressure gradients, and Chang Cai [17] also found that modified foils performed worse than the baseline foil at pre-stall angles, while the lift coefficients at high angles of attack of the modified foils were increased. In addition to applying bionics to the design of hydrofoil structure to improve hydrodynamic performance, there are also methods to optimize the structure according to the flow field. Eun Jung Chae [18] conducted numerical simulation analysis on the flexible hydrofoil and found that the compliant hydrofoil can well adapt to the fluid flow conditions, but many hydrofoils in the hydrofoil boat are rigid, so they can not meet the design requirements of the hydrofoil boat. Manhar Dhanak [19] analyzed the performance characteristics of a shallowly submerged hydrofoil with an internal slot that allowed flow ventilation from the pressure side of the hydrofoil to the suction side and found that significant improvements in hydrofoil performance in terms of lift, drag, and lift-to-drag ratio at high angles of attack. Belamadi [20] studied the effect of a straight internal slot on an S809 airfoil using numerical simulations. CFD analysis was performed for different configurations by varying slot location, width, and slope at Reynolds number of 10^6. Results showed that aerodynamic improvement was found only in the specific range of 10–20°. Although these studies have improved the hydrodynamic performance of hydrofoils to some extent, many are only applicable to fixed working conditions or high angle of attack. In a small angle of attack, the hydrodynamic performance of some hydrofoils will even decrease. To solve this problem, researchers have proposed biomimetic methodologies by observing the motion of fish in the sea. During the long periods of evolution, fish have developed mature flow control mechanisms that can be applied to a wide variety of engineering designs. CD Wilga [21] found that sharks change the direction of vortex shedding by swinging their fins, which may increase the shark's vertical maneuverability; George V. Lauder [22] found that the median fins of fishes consist of the dorsal, anal, and caudal fins and have long been thought to play an important role in generating locomotor force during both steady swimming and maneuvering; Michael Sfakiotakis [23] found that fish swim either by the body and/or caudal fin movements or using median and/or paired fin propulsion. Some fish swim upward or downward in the sea by swinging their tail fins. For example, when the whale swims upward, the tail fin is downward. When the whale swims downward, the tail fin is upward, as shown in Figure 1.

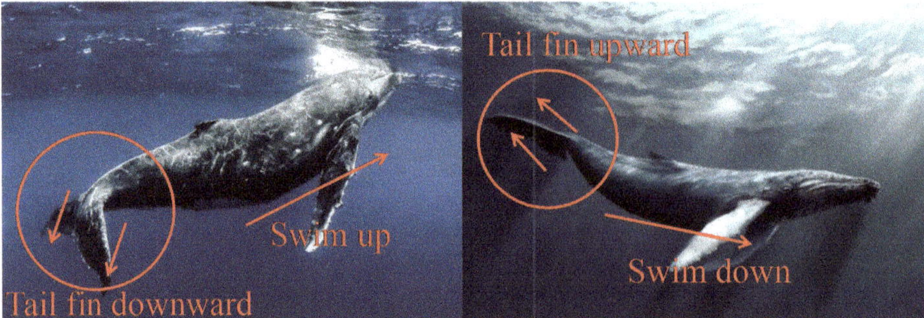

Figure 1. Whale swimming and tail fin flapping.

In this paper, the excellent hydrodynamic performance is reflected by learning from the flapping movement of the whale through the tail fin. Taking the naca66 hydrofoil as the original model, the tail edge flap is added to improve the hydrodynamic performance of the hydrofoil. Ansys-fluent is used to analyze the influence of the angle and length of trailing edge flap and different Reynolds numbers on the hydrodynamic performance of hydrofoil, and PIV experiment is used to verify the reliability of CFD numerical simulation.

2. Model

2.1. Calculation Model

A NACA66 hydrofoil was used in the present research. The hydrofoil has a relative maximum thickness of 12% at 45% chord length from the leading edge and a relative maximum camber of 2% at 50% from the leading edge. Flap modification of naca66 hydrofoil trailing edge. The effects of flap length (F_L) and flap angle (F_A) on hydrofoil performance under different Reynolds numbers are analyzed by CFD. Whenever the flap angle or flap length is changed, it needs to be re-modeled and meshed. The total chord length of the hydrofoil is C = 75 mm. Select flaps with different lengths to rotate (clockwise rotation: $F_A > 0$, counterclockwise rotation: $F_A < 0$) to obtain hydrofoils with different angle flaps. The reliability of CFD simulation is verified by PIV experiment and ICEM-CFD is used to mesh the structure of the hydrofoil, regardless of the flap shape, the number of grid nodes per unit length remains almost unchanged, so the total number of grids will not change greatly, in order to ensure the calculation accuracy, y + of the hydrofoil is less than 1, the fluid domain and fluid mesh are shown in Figure 2.

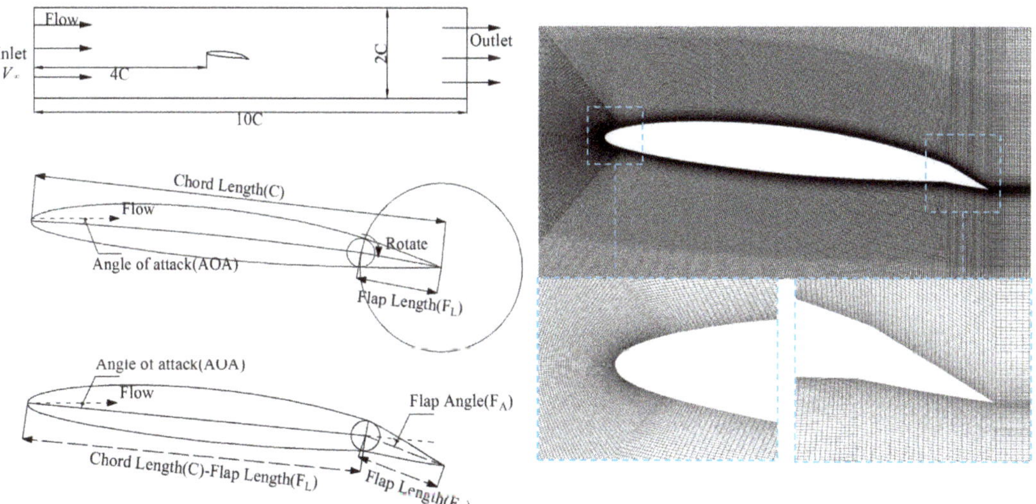

Figure 2. Fluid domain and fluid mesh.

Using ANSYS-Fluent as the numerical simulation tool, k-ω SST Turbulence model can capture the flow field around the dynamic hydrofoil for various ranges of Reynolds numbers [24–26]. Pressure and velocity coupling algorithms adopt SIMPLEC; spatial discretization scheme is QUICK; the convergence accuracy is less than 10^{-4}; time step is 10^{-4}; specific CFD setting conditions are shown in Table 1. The lift coefficient C_l and drag coefficient C_d are monitored to compare the effects of flap on hydrofoil hydrodynamic performance.

$$C_l = \frac{l}{0.5\rho V_\infty^2 C} \quad C_d = \frac{d}{0.5\rho V_\infty^2 C} \tag{1}$$

where l is lift force; d is drag force; p_{inlet} is Inlet static pressure; $0.5\rho V_\infty^2$ is dynamic pressure.

Table 1. CFD model.

Type	State	Type	State
Fluid density	1000 kg·m^{-3}	1st layer thickness (mm)	0.01
Turbulence model	SST k-ω	Growth ratio	1.1
Turbulence intensity	2%	Chord (C/mm)	75
Inlet	Velocity inlet	Flap Length (F$_L$: ×C)	0, 0.1, 0.15, 0.2, 0.25, 0.3
Outlet	Pressure outlet	Angle of attack (AOA)	0, 3, 6, 9, 12, 15, 18, 21
Reynolds number (×10^5)	0.7, 2.1, 3.5, 7, 15	Flap Angle (F$_A$)	−10, −5, 0, 5, 10
Number of grids (million)	0.9	y+	≤1

2.2. Independence of the Number of Grids and Reliability Verification

As shown in Figure 3, when the number of meshes reached a certain value, further increases in the number of meshes had no significant effect on the calculation results but increased the calculation time. Considering the balance between solution accuracy and calculation time, the number of grids selected for this study was approximately 0.9 million.

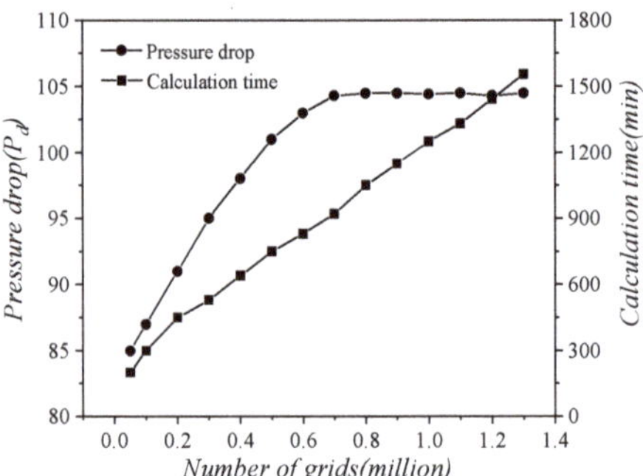

Figure 3. Independence of the number of grids.

This paper uses the PIV experiment to verify the stator blade shape to ensure the accuracy of CFD simulation. The experimental equipment is shown in Figure 4. In the PIV experiment, in order to reduce the error caused by the double refraction caused by the laser, the entire cascade runner is made of plexiglass. Its refractive index is close to water, The interface between the flow channel and the pipe is made by 3D printing. In order to avoid the reflection of the laser on the surface of the non-flow channel and affect the image capture, black processing is performed around the shooting area. The main equipment used for PIV measurement experiment includes: Dual cavity UV laser; PIV Camera; Timing circuit; DynamicStudio.

In order to verify the accuracy of the CFD simulation flow state, the vector diagram measured by PIV is shown in Figure 5a, which is a CFD simulation trace diagram with the same boundary conditions. It can be seen from the figure that the vortex position simulated by CFD is very close to the vortex position measured by PIV. Then extract the velocity distribution data at the trailing edge of the hydrofoil (F$_L$ = 0.30 C; F$_A$ = 10°; AOA = 6°); as shown in Figure 5b, the velocity distribution is also close. This shows that the simulation has high reliability.

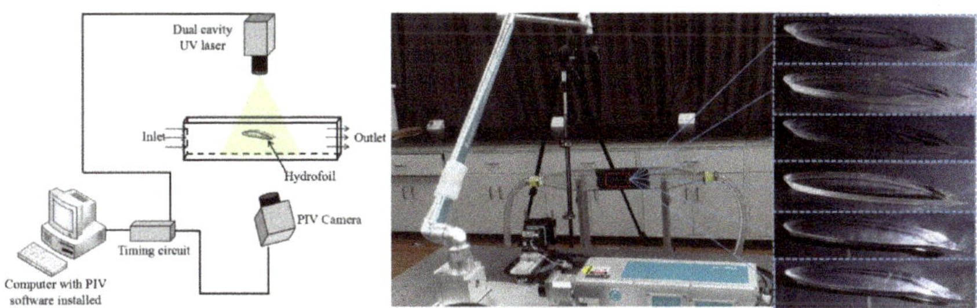

Figure 4. PIV experimental equipment.

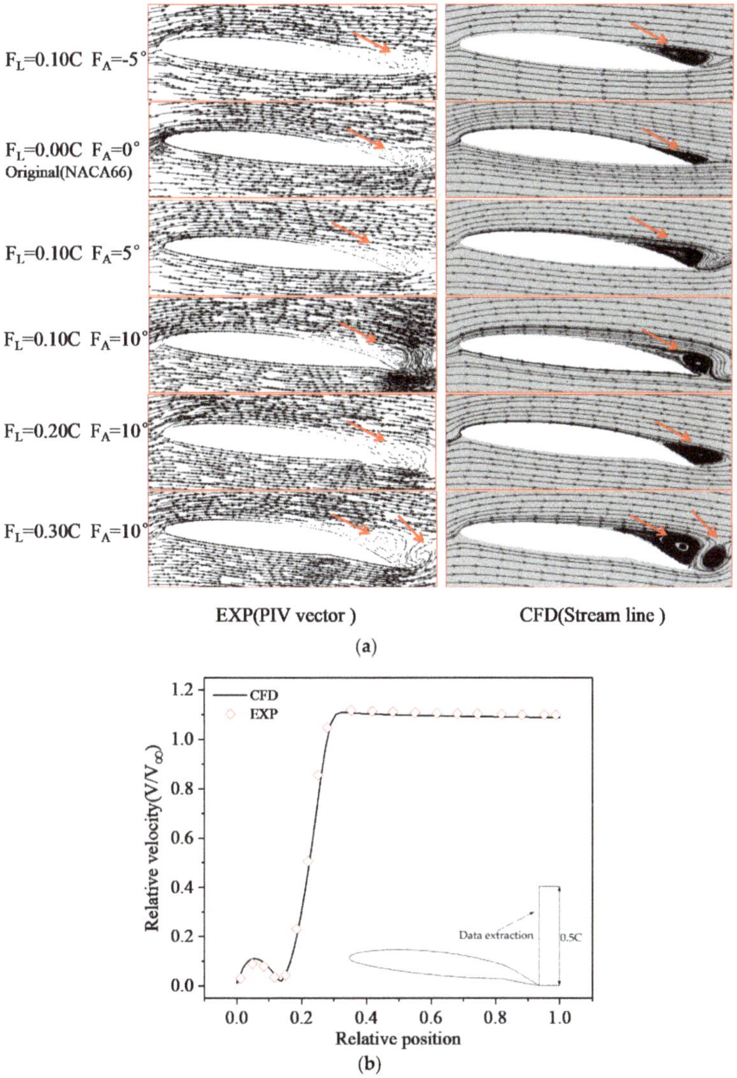

(a)

(b)

Figure 5. Reliability verification (AOA = 6; Re = 3.75×10^5): (**a**) velocity vector diagram; (**b**) relative velocity.

3. Results and Analysis

3.1. Influence of Flap Angle (F_A) on Hydrodynamic Performance of Hydrofoil

Figure 6 shows the hydrodynamic performance and streamline of hydrofoils with different flap angles. This paper mainly analyzes the influence of flap on hydrofoil according to hydrodynamic performance, so the flow field is not analyzed in detail, However, as can be seen in Figure 6a, the lift coefficient of the hydrofoil increases with the increase of the angle of attack, then decreases and then increases again, this shows that stall and deep stall effects appear in this range of attack angle. In order to explain this phenomenon, the flow field is analyzed for AOA = 9°, AOA = 12°, AOA = 15°, as shown in Figure 6d. and there will be no more analysis later.

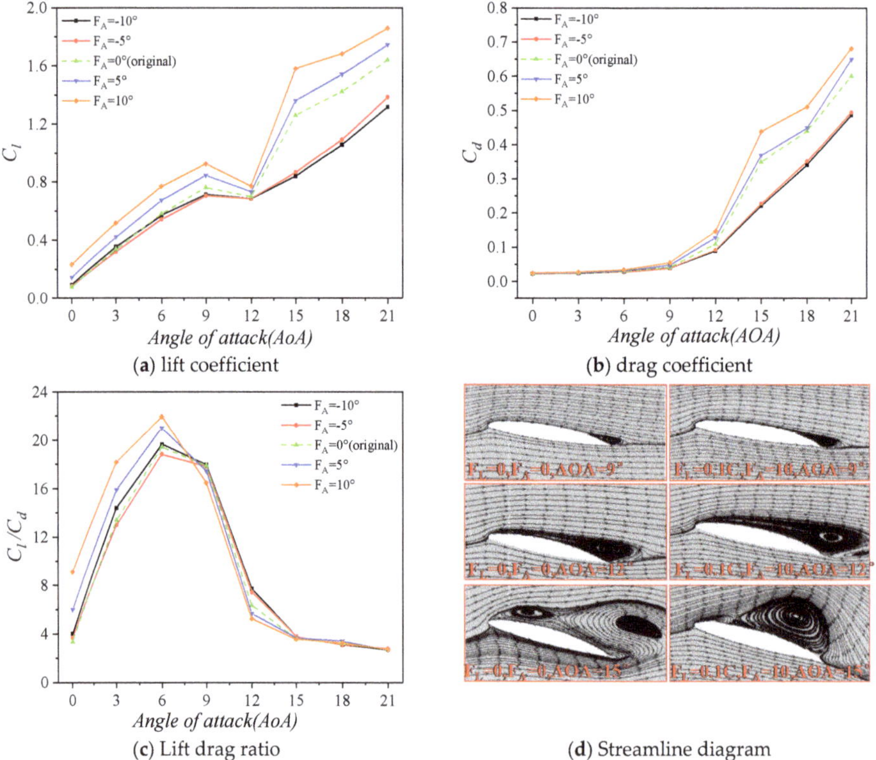

Figure 6. Influence of F_A on hydrodynamic performance of hydrofoil (Re = 3.5×10^5; F_L = 0.1C).

Combined with Figure 6a,d, it can be found that when the angle of attack is small, Due to the Coanda effect, the flow of the hydrofoil is an attached flow. With the increase of the angle of attack, the suction surface of the hydrofoil produces a pressure gradient, resulting in flow separation at the trailing edge of the hydrofoil and small-scale trailing edge vortex (AOA = 9°); when the angle of attack increases, the pressure gradient increases, the hydrofoil has complete flow separation, resulting in large-scale trailing edge vortex (AOA = 12°) and the hydrofoil stalls; With the increase of the angle of attack again, the leading edge vortex and double separation vortex appear in the hydrofoil (AOA = 15°), the hydrofoil has a deep stall and the lift coefficient rises again. It can be seen that the size of vortex affects the hydrodynamic performance of the hydrofoil. The larger the vortex, the worse the hydrodynamic performance.

It can be seen from Figure 6a–c that the hydrodynamic performance of the hydrofoil with clockwise flap rotation($F_A = 5°$, $F_A = 10°$) is obviously better than that of the hydrofoil with counterclockwise flap rotation($F_A = 0°$, $F_A = -5°$, $F_A = -10°$) at a small angle of attack (AOA $\leq 6°$), this is because when the flap rotates clockwise, it increases the differential pressure up and down the trailing edge and increases the lift coefficient, but it also increases the differential pressure resistance at the leading edge and trailing edge, so the resistance coefficient will also increase. Because the angle of attack is small (AOA $\leq 6°$), the influence of the flap on lift coefficient is greater than that on the drag coefficient, so the hydrodynamic performance of hydrofoil with a clockwise flap is better. With the increase of the angle of attack, the influence of the flap on the lift coefficient and the drag coefficient is almost the same, so the lift drag ratio of the hydrofoil is very close with or without flap. When the flap rotates counterclockwise, part of the water flow at the trailing edge of the hydrofoil will change the direction and change the pressure distribution due to the Coanda effect, the existence of flap not only reduces the pressure difference between the upper and lower trailing edges, resulting in the decrease of lift coefficient, but also reduces the pressure difference between the leading and trailing edges of hydrofoil, resulting in the decrease of the drag coefficient. Although the counterclockwise flap will reduce the hydrodynamic performance of hydrofoil, the lift decreases little under the condition of critical stall angle, indicating that we can try to change the flap angle to increase the critical stall angle.

3.2. Influence of Flap Length (F_L) on Hydrodynamic Performance of Hydrofoil

It is concluded in the previous section that the clockwise rotation of the flap can improve the hydrodynamic performance of the hydrofoil. In this section, two models ($F_A = 5°$, $F_A = 10°$) with clockwise flap rotation are used to analyze the influence of flap length (F_L) on the hydrofoil's hydrodynamic performance, as shown in Figure 7. It can be seen from the figure that changing the flap length has little effect on the critical stall angle (the slope is almost constant). Under the same angle of attack, the drag coefficient increases with the increase of flap length, and the lift coefficient increases with the increase of flap length before stall; When AOA $\geq 12°$, the lift coefficient decreases with the increase of flap length, but it is always greater than the original hydrofoil. The influence of flap length on the hydrofoil with a larger flap angle is more obvious. This is because the larger the flap angle is, the increase of flap length will increase the pressure difference between the upper and lower ends of the hydrofoil trailing edge (lift coefficient increases), and also increase the pressure difference resistance at both ends of the hydrofoil's leading edge and trailing edge (drag coefficient increases). However, with the change of angle of attack, the influence of flap length on lift coefficient and the drag coefficient is different (slope is different). As can be seen from Figure 7e,f, compared with the original hydrofoil, when AOA $= 0°$ and AOA $= 3°$, the influence of flap on lift coefficient is greater than that on drag coefficient, and the hydrodynamic performance with flap is always better than that of the original hydrofoil; With the increase of angle of attack (AOA $= 6°$), the influence of long flap ($F_L = 0.25C$, $F_L = 0.3C$) on drag coefficient is greater than that on lift coefficient. The hydrodynamic performance of a hydrofoil with a long flap is worse than that of the original hydrofoil; Until AOA $= 9°$, the hydrodynamic performance of the original hydrofoil is the best, and the flap will reduce the hydrodynamic performance of the hydrofoil. It shows that the flap can improve the hydrodynamic performance of the hydrofoil in the small angle of attack range, a hydrofoil with a short flap has a better angle of attack characteristics and can maintain a higher hydrodynamic performance in a wider angle of attack range.

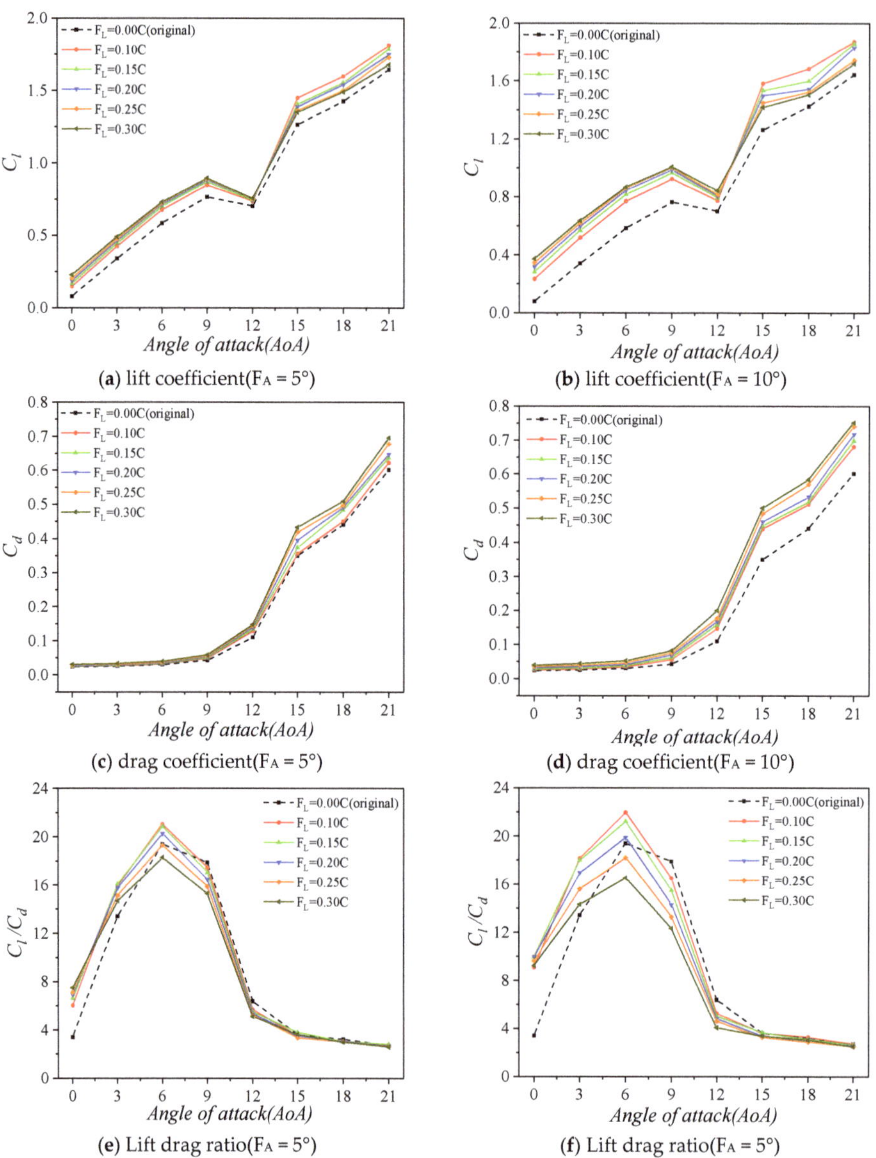

(**a**) lift coefficient(F_A = 5°)

(**b**) lift coefficient(F_A = 10°)

(**c**) drag coefficient(F_A = 5°)

(**d**) drag coefficient(F_A = 10°)

(**e**) Lift drag ratio(F_A = 5°)

(**f**) Lift drag ratio(F_A = 5°)

Figure 7. Influence of F_L on hydrodynamic performance of hydrofoil (Re = 3.5×10^5).

3.3. Influence of Flap on Hydrodynamic Performance of Hydrofoil at Different REYNOLDS Numbers

Through the analysis of two sections, it was found that when AOA = 6°, the lift-drag ratio of the hydrofoil reached the maximum and the hydrodynamic performance of the hydrofoil was the best. Therefore, in this section, we took the angle of attack as quantitative (AOA = 6°) and Reynolds number and flap length (F_L) as variables to analyze the influence of different Reynolds numbers on the hydrodynamic performance of hydrofoils with flaps of different lengths, as shown in Figure 8. It can be seen from the figure that the lift coefficient increases with the increase of the Reynolds number and the drag coefficient decreases with the increase of Reynolds number, but the slopes of both decrease with the increase of the Reynolds number, which causes the slope of the lift–drag ratio to decrease with the increase of Reynolds number. By analyzing the lift-drag ratio diagram (Figure 6e,f), it can be found that when the Reynolds number is small (Re $\leq 2.1 \times 10^5$), the hydrodynamic performance of the hydrofoil with a flap was better than that of the original hydrofoil. With the increase of Reynolds number, the hydrodynamic performance of the hydrofoil with a long flap ($F_L = 0.2C$; $F_L = 0.25C$; $F_L = 0.3C$) is worse than that of the original hydrofoil. It shows that the hydrofoil with a short flap has a better Reynolds number characteristics and can maintain a higher hydrodynamic performance in a wider Reynolds number range.

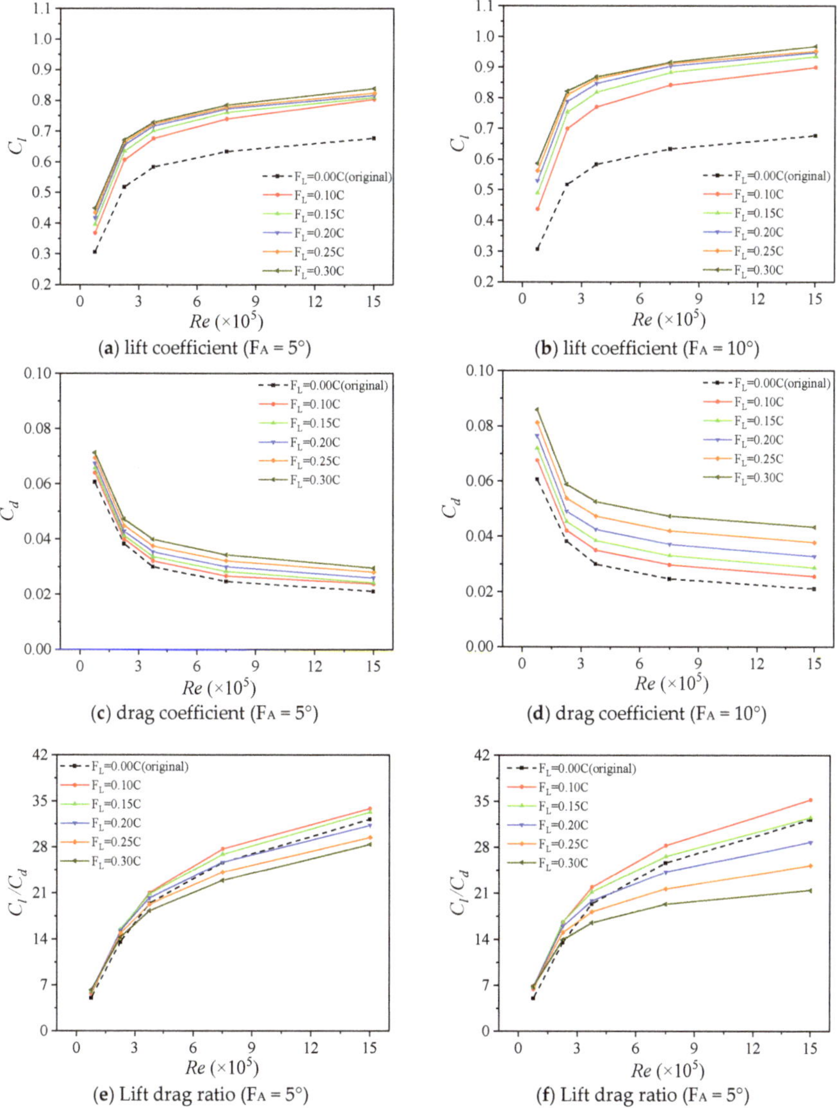

(**a**) lift coefficient ($F_A = 5°$)

(**b**) lift coefficient ($F_A = 10°$)

(**c**) drag coefficient ($F_A = 5°$)

(**d**) drag coefficient ($F_A = 10°$)

(**e**) Lift drag ratio ($F_A = 5°$)

(**f**) Lift drag ratio ($F_A = 5°$)

Figure 8. Influence of Re on hydrodynamic performance of hydrofoil (AOA = 6°).

4. Conclusions

In this paper, ansys-fluent numerical calculation software is used to analyze the influence of flap on the hydrodynamic performance of naca66 hydrofoil. The accuracy of CFD simulation was verified by the PIV experiment, and the influence of flap length (F_L) and flap angle (F_A) on the hydrodynamic performance of the hydrofoil was studied through the lift coefficient and drag coefficient of the hydrofoil. The main conclusions include:

1. By comparing the vector diagram of the PIV experiment with the trace diagram of CFD, it is found that the size and position of the vortex measured in the experiment are close to that obtained by CFD, which shows that the simulation has high reliability.
2. By comparing the streamline diagrams of AOA = 9°, AOA = 12°, and AOA = 15°, it is found that when AOA = 9°, partial flow separation occurs at the tail of the hydrofoil; With the increase of the angle of attack (AOA = 12°), the hydrofoil appears complete flow separation, the hydrofoil stalls and the lift coefficient decreases; As the angle of attack continues to increase (AOA = 15°), the hydrofoil appears double separation vortex. At this time, the hydrofoil has a deep stall effect and the lift increases again.
3. By analyzing the influence of flap angle (F_A) on the hydrodynamic performance of hydrofoil, it is found that the hydrofoil with clockwise flap can have better hydrodynamic characteristics at a small angle of attack (AOA $\leq$ 6°) under the same Reynolds number and flap length (F_L). Although the counterclockwise flap will reduce the hydrodynamic characteristics of the hydrofoil, it will increase the critical stall angle to improve the navigation stability of the hydrofoil.
4. By analyzing the influence of flap length (F_L) on the hydrofoil hydrodynamics, it is found that hydrofoil with flap has better hydrodynamic characteristics at a small angle of attack(AOA $\leq$ 6°), and hydrofoil with a short flap has a better angle of attack characteristics, which can maintain a higher hydrodynamic performance in a wider range of angle of attack.
5. By analyzing the influence of different Reynolds numbers on the hydrodynamic performance of hydrofoil, it is found that under the same small angle of attack, the hydrofoil with a short flap has better Reynolds number characteristics and can maintain a higher hydrodynamic performance in a wider range of Reynolds numbers.
6. Compared with the original hydrofoil, the short flap improves the hydrodynamic performance of the hydrofoil at a small angle of attack. The hydrofoil can also be applied to various working conditions by adjusting the angle of the flap. Therefore, when designing the hydrofoil, a rotatable short flap can be added at the tail of the hydrofoil to enable the hydrofoil to navigate in a more complex flow environment.

Author Contributions: Conceptualization, P.X. and J.D.; methodology, P.X. and X.C.; software, P.X.; validation, P.X. and X.C.; formal analysis, P.X.; resources, X.C.; data curation, P.X. and J.D.; writing—original draft preparation, P.X.; writing—review and editing, J.D.; supervision, X.C. All authors have read and agreed to the published version of the manuscript.

Funding: This research received no external funding.

Institutional Review Board Statement: Not applicable.

Informed Consent Statement: Not applicable.

Data Availability Statement: Data is contained within the article.

Conflicts of Interest: The authors declare no conflict of interest.

References

1. Nachtane, M.; Tarfaoui, M.; Goda, I.; Rouway, M. A review on the technologies, design considerations and numerical models of tidal current turbines. *Renew. Energy* **2020**, *157*, 1274–1288. [CrossRef]
2. Xu, W.; Xu, G.; Duan, W.; Song, Z.; Lei, J. Experimental and numerical study of a hydrokinetic turbine based on tandem flapping hydrofoils. *Energy* **2019**, *174*, 375–385. [CrossRef]

3. Hao, W.; Li, C. Performance improvement of adaptive flap on flow separation control and its effect on VAWT. *Energy* **2020**, *213*, 118809. [CrossRef]

4. Yan, H.; Su, X.; Zhang, H.; Hang, J.; Zhou, L.; Liu, Z.; Wang, Z. Design approach and hydrodynamic characteristics of a novel bionic airfoil. *Ocean Eng.* **2020**, *216*, 108076. [CrossRef]

5. Velte, C.M.; Hansen, M. Investigation of flow behind vortex generators by stereo particle image velocimetry on a thick airfoil near stall. *Wind Energy* **2013**, *16*, 775–785. [CrossRef]

6. Hussain, S.; Liu, J.; Wang, L.; Sundén, B. Suppression of endwall heat transfer in the junction region with a symmetric airfoil by a vortex generator pair. *Int. J. Therm. Sci.* **2019**, *136*, 135–147. [CrossRef]

7. Itsariyapinyo, P.; Sharma, R.N. Large Eddy simulation of a NACA0015 circulation control airfoil using synthetic jets. *Aerosp. Sci. Technol.* **2018**, *82–83*, 545–556. [CrossRef]

8. Tousi, N.M.; Coma, M.; Bergadà, J.M.; Pons-Prats, J.; Mellibovsky, F.; Bugeda, G. Active Flow Control Optimisation on SD7003 Airfoil at Pre and Post-stall Angles of Attack using Synthetic Jets. *Appl. Math. Model.* **2021**, *98*, 435–464. [CrossRef]

9. Somers, D.M. *An Exploratory Investigation of a Slotted, Natural-Laminar-Flow Airfoil*; Langley Research Center: Hampton, VA, USA, 2012.

10. Coder, J.G.; Dan, M.S. Design of a slotted, natural-laminar-flow airfoil for commercial transport applications. *Aerosp. Sci. Technol.* **2020**, *106*, 106217. [CrossRef]

11. Xiao, Q.; Zhu, Q. A review on flow energy harvesters based on flapping foils. *J. Fluid Struct.* **2014**, *46*, 174–191. [CrossRef]

12. Young, J.; Lai, J.C.S.; Platzer, M.F. A review of progress and challenges in flapping foil power generation. *Prog. Aerosp. Sci.* **2014**, *67*, 2–28. [CrossRef]

13. Triantafyllou, M.S.; Techet, A.H.; Hover, F.S. Review of Experimental Work in Biomimetic Foils. *IEEE J. Ocean. Eng.* **2004**, *29*, 585–594. [CrossRef]

14. Johari, H.; Henoch, C.; Custodio, D.; Levshin, A. Effects of Leading-Edge Protuberances on Airfoil Performance. *AIAA J.* **2007**, *45*, 2634–2642. [CrossRef]

15. Rostamzadeh, N.; Hansen, K.L.; Kelso, R.M.; Dally, B.B. The formation mechanism and impact of streamwise vortices on NACA 0021 airfoil's performance with undulating leading edge modification. *Phys. Fluids* **2014**, *26*, 51–60. [CrossRef]

16. Skillen, A.; Revell, A.; Pinelli, A.; Piomelli, U.; Favier, J. Flow over a Wing with Leading-Edge Undulations. *AIAA J.* **2014**, *53*, 464–472. [CrossRef]

17. Cai, C.; Zuo, Z.; Liu, S. Numerical investigations of hydrodynamic performance of hydrofoils with leading-edge protuberances. *Adv. Mech. Eng.* **2015**, *7*. [CrossRef]

18. Chae, E.J.; Akcabay, D.T.; Lelong, A.; Astolfi, J.A.; Young, Y.L. Numerical and experimental investigation of natural flow-induced vibrations of flexible hydrofoils. *Phys. Fluids* **2016**, *28*, 075102. [CrossRef]

19. Ni, Z.; Dhanak, M.; Su, T.C. Performance of a slotted hydrofoil operating close to a free surface over a range of angles of attack. *Ocean Eng.* **2019**, *188*, 106296. [CrossRef]

20. Belamadi, R.; Djemili, A.; Ilinca, A.; Mdouki, R. Aerodynamic performance analysis of slotted airfoils for application to wind turbine blades. *J. Wind Eng. Ind. Aerodyn.* **2016**, *151*, 79–99. [CrossRef]

21. Wilga, C.D.; Lauder, G.V. Biomechanics—Hydrodynamic function of the shark's tail. *Nature* **2004**, *430*, 850. [CrossRef]

22. Lauder, G.V.; Nauen, J.C.; Drucker, E.G. Experimental Hydrodynamics and Evolution: Function of Median Fins in Ray-finned Fishes. *Integr. Comp. Biol.* **2002**, *42*, 1009–1017. [CrossRef] [PubMed]

23. Sfakiotakis, M.; Lane, D.M.; Davies, J. Review of fish swimming modes for aquatic locomotion. *IEEE J. Ocean. Eng.* **1999**, *24*, 237–252. [CrossRef]

24. Karbasian, H.R.; Moshizi, S.A.; Maghrebi, M.J. Dynamic Stall Analysis of S809 Pitching Airfoil in Unsteady Free Stream Velocity. *J. Mech.* **2016**, *32*, 227–235. [CrossRef]

25. Karbasian, H.R.; Esfahani, J.A.; Barati, E. Effect of acceleration on dynamic stall of airfoil in unsteady operating conditions. *Wind Energy* **2014**, *19*, 17–33. [CrossRef]

26. Wang, S.; Ingham, D.B.; Ma, L.; Pourkashanian, M.; Tao, Z. Numerical investigations on dynamic stall of low Reynolds number flow around oscillating airfoils. *Comput. Fluids* **2010**, *39*, 1529–1541. [CrossRef]

 *processes*

Article

Comparison of Sliding and Overset Mesh Techniques in the Simulation of a Vertical Axis Turbines for Hydrokinetic Applications

Omar D. Lopez Mejia [1,*], Oscar E. Mejia [1], Karol M. Escorcia [1], Fabian Suarez [2] and Santiago Laín [3]

[1] Department of Mechanical Engineering, Universidad de los Andes, Cra 1 Este N 19A-40, Bogotá 111711, Colombia; oe.mejia10@uniandes.edu.co (O.E.M.); km.escorcia10@uniandes.edu.co (K.M.E.)

[2] e.Ray Europa GmBh, Hilpertstraße 31, 64295 Darmstadt, Germany; fabian.suarez@e-ray.org

[3] PAI+ Group, Energetics & Mechanics Department, Faculty of Engineering, Universidad Autónoma de Occidente, Cali 760030, Colombia; slain@uao.edu.co

* Correspondence: od.lopez20@uniandes.edu.co; Tel.: +57-1-339-4949

Abstract: The application of Computational Fluid Dynamics (CFD) to energy-related problems has increased in the last decades in both renewable and conventional energy conversion processes. In recent years, the application of CFD in the study of hydraulic, marine, tidal, and hydrokinetic turbines has focused on the understanding of the details of the complex turbulent flow and also in improving the prediction of the performance of these devices. There are several complexities involved in the simulation of Vertical Axis Turbine (VAT) for hydrokinetic applications. One of them is the necessity of a dynamic mesh model. Typically, the model used in the simulation of these devices is the sliding mesh technique, but in recent years the fast development of the overset (also known as chimera) mesh technique has caught the attention of the academic community. In the present paper, a comparison between these two techniques is done in order to establish their advantages and disadvantages in the two-dimensional simulation of vertical axis turbines. The comparison was done not only for the prediction of performance parameters of the turbine but also for the capabilities of the models to capture complex flow phenomena in these devices and computational costs.

Keywords: Computational Fluid Dynamics; vertical axis water turbine; overset mesh; sliding mesh

Citation: Lopez Mejia, O.D.; Mejia, O.E.; Escorcia, K.M.; Suarez, F.; Laín, S. Comparison of Sliding and Overset Mesh Techniques in the Simulation of a Vertical Axis Turbine sfor Hydrokinetic Applications. *Processes* **2021**, *9*, 1933. https://doi.org/10.3390/pr9111933

Academic Editor: Krzysztof Rogowski

Received: 16 September 2021
Accepted: 21 October 2021
Published: 28 October 2021

Publisher's Note: MDPI stays neutral with regard to jurisdictional claims in published maps and institutional affiliations.

1. Introduction

Due to the global concern related to climate change, some worldwide challenges and international warnings revolve around the usage of fossil fuels. The reduction in the consumption of fossil fuels and the increment in the use of renewable energies is a primary objective to mitigate this global concern. The present work focused on the use of small-scale Vertical Axis Turbine (VAT) to convert the kinetic energy available in river streams (hydrokinetic energy) into mechanical work. In recent years, extensive research work has been done to understand the dynamics of the flow around VATs and also in the improvement of the efficiency of these devices [1–4]. Typically, the study of these devices can be done with three different techniques: experimental, analytical, and computational. Regarding experimental techniques, the most common are performed in scale prototypes either in water tunnels/channels [5–7] or towing tanks [8–10]. These experimental techniques require large-scale facilities and precise measurement systems, which, in general, are expensive. Analytic and semi-empirical techniques such as the Streamtube model [11,12], free wake vortex models [13–15], and cylinder and line actuator models [16–18], among others, are considered simple to implement but highly dependent on experimental data. The most common computational technique used in the simulation of Vertical Axis Turbines (VAT) is Computational Fluid Dynamics (CFD) in which the governing equations of the dynamics of flow are solved in a computational domain that needs to be discretized in elements (Control volumes).

In the literature, a great deal of work can be found related to the simulation of vertical axis turbines for wind, marine, and hydrokinetic applications; but, in few of them the details of the correct configuration of the numerical methods and models that are typically used in this kind of simulations are addressed. Some of the details and parameters that influence the numerical results and that are typically studied are related to the correct use of two- or three-dimensional models, the space and time discretization, and turbulence modelling. Balduzzi et al. studied the influence of several different parameters in two-dimensional models of VAT such as the Turbulence model, computational domain dimensions, and number of revolutions required for convergence of the torque coefficient. It was found that the most suitable Unsteady Reynolds Navier–Stokes (URANS) turbulence model for this application is the k-ω SST. For the computational domain size, it is recommended that it should be significantly extended in order to reduce the influence of the boundary conditions, and for the rotating domain (when using sliding mesh technique) it is always less than twice the turbine diameter in order to reduce computational costs [19]. Rezaiha et al. found that the size of the rotating domain is negligible and a difference of less than 1% was observed for 2D simulations of Vertical Axis Wind Turbines (VAWT) using diameter of the rotating domain between 1.25D and 2D where D is the diameter of the turbine [20]. A convergence criterion for the torque coefficient was proposed so that in two consecutive revolutions the difference in this parameter should be lower than 0.1%, which is typically achieved after eight revolutions when using the sliding mesh techniques [19]. Maitre et al. studied the influence of the near wall refinement as the precision of the prediction of the global performance of the turbine. It was found that too-coarse grids tend to overestimate the stall condition of the blades and, in general, to underestimate the power coefficient. Maitre et al. also studied the accuracy of two-dimensional CFD simulations in the prediction of the power coefficient of VAT and it was observed that the computational results were always higher than the experimental measurements except for small tip speed ratios (λ). This overprediction of two-dimensional models is important for small-scale VAT in which the losses due to the blade tips and arm-blade junctions are not included [21].

One of the most challenging aspects of using CFD in the simulation of a vertical axis turbine is related with the dynamic mesh required due to the motion of the blades of the turbine. Remeshing methods are practically prohibited for this application due to the large motion of the blades, which represents a very high computational cost. Dynamic mesh techniques that do not involve remeshing, such as sliding and overset (also known as chimera) mesh, are suitable for this application. In both techniques, two or more computational domains that are meshed independently can be merged in order to generate the relative motion between them. In the sliding mesh technique, the meshes do not overlap but they are connected through a boundary that is called interface. At every time step, the meshes slide relative to each other along the interface, typically with a prescribed motion (for VATs rotation is involved). The governing equations are solved at every time step in both domains but information between meshes must be interchanged along the interface, which requires the estimation of the fluxes across interfaces, which, in general, are non-conformal [22]. The overset mesh technique consists of creating independent meshes and allows overlapping between them. Typically, one or more near body meshes and at least one background mesh are required to apply this technique. The governing equations are solved on both meshes and on the overlapping region (also known as overset boundary). The solution is interpolated and shared between the meshes in order to impose the correct boundary conditions [23]. The main advantages of this method are that the meshes can be generated independently, reducing the complexity of mesh generation and the total number of elements. One restriction is that the size of the elements of each mesh on the overlapping boundaries must be similar. To summarize, even though in both techniques the meshes are generated independently, they must be merged before running the simulation. The main differences between OM and SM are: (1) The meshes overlap in OM but do not overlap in SM; (2) a clear and fixed interface is generated between the two meshes in SM and interchange of information is performed every time step; and (3) in

OM, an overlapping region is generated every time step and the solution of both meshes is interpolated in this region.

Most of the CFD studies of VAT (both for wind and hydrokinetic applications) reported in the literature were performed using sliding mesh (e.g., [24,25]) and in recent years the application of the overset method for VAT has increased. Kozak studied the unsteady effects of vertical axis wind turbines with overset meshes, proposing some recommendations to correctly configure the overset overlapping [26]. Mclean studied the performance improvement of vertical axis wind turbine with active blade pitch control using CFD and polygonal overset meshes. It was concluded that a fully coupled solver has a better performance when using the overset mesh technique [27]. Lei et al. used the overset mesh method to study an offshore vertical axis wind turbine in pitch and surge motion. Only one rotating overset domain and one background mesh were used. A convergence analysis was done, in which six different meshes were generated and the number of elements in both domains were changed simultaneously. The overset mesh method showed a good ability to capture the complex flow dynamic when both pitch and surge motions were included [28]. Regarding the simulation of VAT for hydrokinetic applications, Kinsey and Dumas studied the effect of the wake blockage in the performance of crossflow turbines (three blades with high solidity). Three meshes were used for each blade and one mesh was used as background; all the meshes used were generated with polyhedral elements. A convergence study using three different numbers of total elements was performed, but no detail of the number of elements in each mesh was given. Numerical results show very good agreement with experimental data and the application of overset mesh was significant since only the background mesh was changed in order to study different blockage ratios [29]. Gorle et al. studied flow control based in circulation control of a vertical axis hydrokinetic turbine in order to improve its performance. A two-dimensional CFD simulation was performed, in which the overset technique was used in order to model not only the rotation of the blades but also the pitching motion. Numerical results showed very interesting findings about the controlled flow and the versatility of overset mesh in the study of flow control for VAT based on circulation control; nevertheless, no details of a convergence analysis and the number of meshes were provided [30].

In the present paper, a detailed comparison between the sliding mesh (SM) technique and overset mesh (OM) technique is shown and discussed. It is important to clarify that the objective of the present work was not focused on finding which is the best strategy, SM or OM, to reproduce a particular set of experimental data, but better to point out the differences, advantages, and disadvantages of each of them in the simulation of VATs. To achieve this objective, a vertical axis turbine design was simulated with both techniques and the results of the performance parameters were compared between them. Moreover, a comparison of pressure, velocity, and vorticity fields obtained by both techniques was performed; this allowed obtaining a better idea of the capabilities of the models to capture complex flow phenomena. Finally, a mesh convergence analysis was done for both techniques so that the number of elements was similar when using either technique.

2. Configuration of Study

The study case was based on a vertical axis hydrokinetic turbine that was designed by e.Ray Europa Gmbh [31]. This turbine was designed with the purpose of being used in rivers of low to medium stream current velocities. The objective was that the turbine generates electricity (approximately 1 kW at 2.2 m/s) for a floating station that measures several river parameters (level, velocity, temperature, etc.) for management and monitoring. The turbine design expects a peak performance at a Tip Speed Ratio (TSR) of 2. Figure 1 shows the floating station and Table 1 lists the geometric characteristics of the turbine.

Table 1. Aerodynamic profile and turbine's dimensions.

Description	Dimension
Span	750 mm
Diameter (D)	650 mm
Chord	150 mm
Number of blades	3
Blade profile	GOE222

Figure 1. Floating station.

The performance of a vertical axis turbine is characterized by three nondimensional parameters: Tip Speed Ratio (*TSR*), moment coefficient (C_M), and power coefficient (C_P). *TSR* is the ratio between the tangential velocity of the turbine to the free stream velocity of the flow, and it is given by Equation (1), where ω is the angular speed of the turbine, R is the radius of the turbine, and U_∞ is the velocity of the free stream.

$$TSR = \frac{\omega R}{U_\infty} \tag{1}$$

The moment coefficient is the nondimensional form of the torque generated by the turbine when it operates at a given *TSR*. The torque is nondimensionalized with respect to the maximum torque that the turbine could generate with the available kinetic energy (see Equation (2)),

$$C_M = \frac{T}{\frac{1}{2}\rho U_\infty^2 AR} \tag{2}$$

where T is the torque generated by the turbine, ρ is the density of the fluid, and A is the frontal area of the turbine (i.e., Span $\times$ diameter). Finally, the power coefficient is the nondimensional form of the mechanical power generated by the turbine at a given *TSR* (see Equation (3)).

$$C_P = \frac{T\omega}{\frac{1}{2}\rho U_\infty^3 A} \tag{3}$$

For Vertical Axis Wind Turbines, C_M and C_P are only function of *TSR* since the Reynolds number (based on the free stream velocity) of this application is high (O(10^6) or more); but for hydrokinetic applications, the Reynolds number is not higher than O(10^5), so that an effect of this parameter is observed on the performance curves [8,32].

3. Computational Setup

3.1. Computational Domain Size

Figure 2 shows a schematic of the two-dimensional computational domain used in the present study. The height (H) is 15 times the diameter of the turbine (D), while the length of the computational domain (L) is 20D. These dimensions satisfy the criteria mentioned in the literature. The letter C indicates the position of the axis of the turbine, which is located closer to the left boundary of the computational domain, so that the distance l is 5D. while h is 7.5D. The size of the computational domain was the same for both meshing techniques, for comparison purposes.

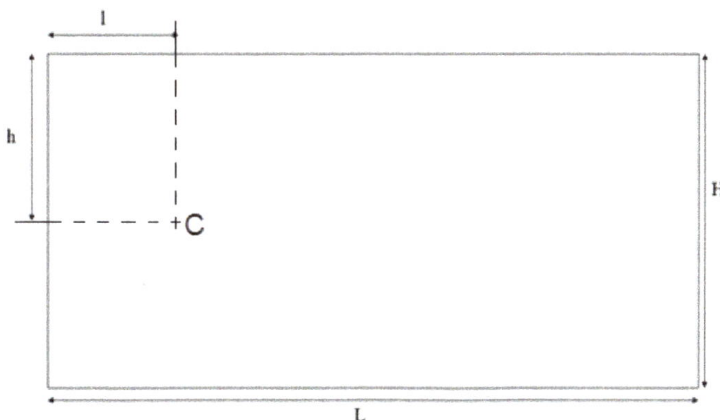

Figure 2. Computational domain.

3.2. Mesh Generation

In order to correctly resolve the turbulent boundary layer over the surface of the blades, the size of the first element of the mesh in the normal direction of the blades surface should satisfy the restriction of $y^+ < 1$ [19,21]. In the context of VAWT simulations, this criterion needs to be satisfied independently of the meshing technique that is used. The y^+ is a dimensionless parameter for the distance to the wall, which is the characteristic length scale of turbulent wall-bounded flows (see Equation (4)).

$$y^+ = \frac{u_\tau y}{\nu} \tag{4}$$

Here ν is the kinematic viscosity, y is the distance to the wall, and u_τ is the friction velocity that is defined as:

$$u_\tau = \sqrt{\frac{\tau_w}{\rho}} \tag{5}$$

where τ_w is the shear stress at the wall and ρ is the density of the fluid. A priori, the shear stress is unknown so the process to determine the correct size of the first element close to the blade surface is by trial and error. As a first approximation, typically a flat plate model of the blade can be used to generate the coarser grid used in the convergence analysis process. Once the meshes are generated and the simulations are run, then the averaged and distribution of y^+ values for the blade in one revolution should be checked.

Other parameters that are important in the generation of the mesh close to the surface of the blades (Boundary layer mesh) are the total height, the growth rate, and the total number of layers used. The total height should be enough in order to include the developed boundary layer at every time step; of course, at high angles of attack this is difficult to achieve. The growth rate follows the rules typically used in every Reynolds Averaged Navier–Stokes (RANS) simulation, in which values between 1.1 and 1.2 are acceptable. Based on the size of the first element, the growth rate, the total height, and the method (linear, geometric, exponential, hyperbolic, etc.) used in the boundary layer mesh generation, then the total number of layers can be specified. The first step to generate the boundary layer mesh is to create a database with the XY points that define the blade profile. Using these points, the upper and lower surfaces of the profile are generated and discretized using at least 200 segments for each surface. Finally, the structured mesh is generated by extrusion. For both techniques (SM and OM), the commercial software Pointwise V18 was used in the present study.

3.2.1. Sliding Mesh (SM)

When using the SM technique in VAT simulations, the computational domain must be divided in two: a stationary and a rotational domain. These two domains are connected through a numerical interface. The rotational domain is represented by a cylinder that includes a structured mesh close to the surface of the blade and an unstructured one in the rest of the domain. The diameter of the rotating domain is 1.5D. In the present study, for the structured mesh (boundary layer) a growth rate of 1.2 was used, with a total height of 14.2 mm and 22 layers. A hyperbolic extrusion algorithm with a Kinsey–Barth smoothing parameter of 1.6 was used in the generation of the structured mesh. Figure 3a shows the details of the rotating mesh in which the details of the structured mesh close to the blades of the turbine can be observed. Figure 3b shows the detail of the mesh close to the blade surface. It is clear that the resolution close to the blade was very high in order to achieve the target value of $y^+ < 1$.

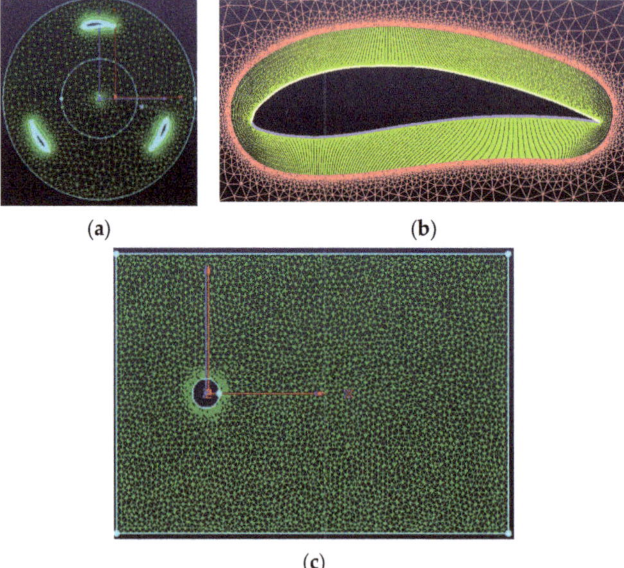

(a)   (b)

(c)

Figure 3. Dimensions and border conditions. (**a**) Rotational domain, (**b**) detail of the mesh close to the blade, (**c**) stationary domain.

Figure 3c shows the mesh used for the stationary domain; in this case the complete domain was discretized with triangles with refinements in the region close to the rotational domain.

For the convergence analysis, it was necessary to have the generation of meshes that have the number of elements between 73,125 and 256,899, as shown in Table 2. The size and estimated y+ for the first element of the boundary layer mesh is presented in Table 3. As the size of the first element decreased, the total number of elements increased in order to keep an acceptable ratio of increment in the size of the elements between all the regions of the mesh.

Table 2. Meshes used in the convergence analysis (SM).

	Mesh 1	Mesh 2	Mesh 3	Mesh 4	Mesh 5
Blade (rotating)	15,609	20,691	30,889	37,043	55,737
Cylinder (rotating)	17,757	25,337	61,067	75,429	59,783
Farfield (stationary)	8541	19,319	9897	15,623	29,905
Total	73,125	106,729	163,631	202,181	256,899

Table 3. Parameterization of the mesh of the blades (SM).

Mesh	Y+	Size of First Element [m]
Mesh 1	10	0.00014094
Mesh 2	5	0.00005705
Mesh 3	1	0.00001580
Mesh 4	0.75	0.00001057
Mesh 5	0.5	0.00000352

3.2.2. Overset Mesh (OM)

For the OM technique, it was necessary to generate the meshes (background and near body mesh) independently. Figure 4a shows the near body mesh, which was fully structured, and it was obtained from extrusion of the discretized blade surface. In this case, a growth rate of 1.2 and a total of 46 layers were used. These parameters allowed having a total height for the boundary layer mesh of 240.7 mm. This total height was higher than the value used in the SM technique, which is a requirement in the OM technique in order to avoid that the receptor cells of the background mesh are too close to the blade surface. Another aspect that is important when generating the near body mesh is that the size of the elements (approx. 3 mm) close to the overset boundary should be of the same size as those in the background mesh. Figure 4b shows the background mesh, which was also structured. A refinement zone was clearly observed in the region where the turbine was located. Since the mesh was structured, the refinement propagated to the boundaries, which had some advantages such as the improvement in the resolution in the region where the wake was expected. However, at the same time, it increased the number of elements in regions such as upstream where such a high resolution was not needed. In general, the OM technique allows using structured meshes in both near body and background, which brings important benefits such as the reduction of the total number of elements and the computational cost.

In this case, the convergence analysis consisted of the variation of the number of elements in the near body mesh based on the expected y+ while adjusting the background mesh number of elements and size in order to avoid donor cells close to the blade surface. Based on the convergence study performed for the SM method, the total number of meshes used in the OM method was reduced to three. Table 4 shows the parameters used to generate the meshes and the total number of elements. It was clear that the total number of elements of the three meshes was close to the total number of elements of the meshes 4 and 5 used in the convergence study with SM technique.

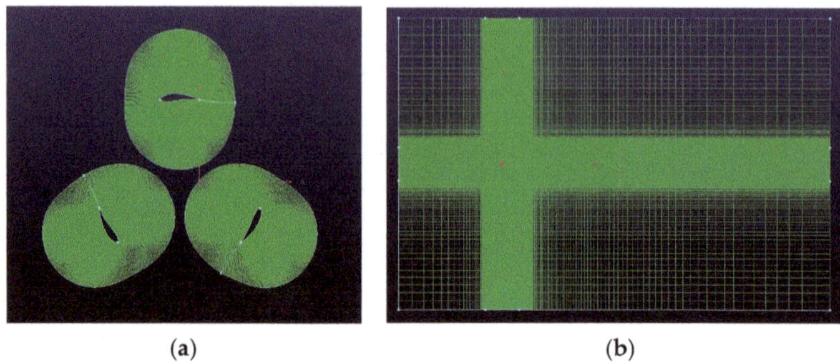

(**a**)   (**b**)

Figure 4. Meshes used in OM (**a**) Near body, (**b**) Background mesh.

Table 4. Meshes used in the convergence analysis (OM).

	Y+	Size of First Element [m]	N° of Elements Near Body	N° of Elements Background	Total
Mesh 1	20	0.000254	16,560	174,283	190,843
Mesh 2	10	0.000152	31,680	174,283	205,963
Mesh 3	1	0.000015	59,740	193,159	252,899

3.3. Boundary Conditions and Solver Configuration

The simulations were performed with the commercial software Ansys Fluent V 19. In both cases (SM and OM), the flow was considered transient, incompressible, turbulent, and Newtonian. The fluid used in the simulation was liquid water at 20 °C. The Reynolds number based on the chord length ranged between 1×10^5 to 3×10^5 and the inlet velocity was fixed to 1.5 m/s. Table 5 shows the boundary conditions used in the simulations. The boundaries named inlet, outlet, top, and bottom correspond to the left, right, upper, and lower surfaces of the computational domain, as shown in Figure 2. The inlet boundary had a uniform velocity, which was applied in the normal direction. The inflow turbulence was estimated using a turbulence intensity of 5% and a length scale of one-tenth of the chord length (15 mm). Moving wall boundary conditions were used in the upper and lower surfaces of the computational domain, while, at the outlet, a fixed gauge pressure of 0 Pa was used. Besides these boundary conditions, both dynamic mesh techniques required a definition of an interface between the stationary and rotational domains, in the case of the SM method and between the near-body and background meshes in the OM method.

Table 5. Parameters of the boundary conditions.

Boundary	Condition	Value
Inlet	Velocity Inlet	$V = 1.5 \text{ m/s}$ $k = 0.0084 \text{ m}^2/\text{s}^2$ $\omega = 6.11 \text{ m/s}$
Outlet	Pressure outlet	$P = 0 \text{ Pa}$
Top	Moving wall	$V = 1.5 \text{ m/s}$
Bottom	Moving wall	$V = 1.5 \text{ m/s}$
Blades	Wall	No slip

Table 6 shows the different cases that were simulated for different Tip Speed Ratios (TSR). The *TSR* of the turbine was changed by increasing the angular speed of the turbine from 3.46 to 11.54 rad/s while keeping the inflow velocity fixed at 1.5 m/s. To obtain stable results, each simulation needed 10 revolutions of the turbine with a time step of 0.001 s, and 30 iterations were performed per time step. According to Rezaiha, a maximum azimuthal increment ($\Delta\theta$) of 0.5° was required in the simulation of VAWT in order achieve a correct time discretization that did not impact the prediction in Cp. In the present study, a maximum $\Delta\theta$ of 0.66° was used and the average value of $\Delta\theta$ used in all the simulations was 0.43°

Table 6. *TSR* and time step used in the simulations.

TSR	Angular Speed [rad/s]	Time Steps per Revolution	Time Step [s]	$\Delta\theta$ (°)
0.75	3.46	1816	0.001	0.198
1	4.62	1361	0.001	0.264
1.25	5.77	1089	0.001	0.330
1.5	6.92	908	0.001	0.396
1.75	8.08	778	0.001	0.462
2	9.23	681	0.001	0.529
2.25	10.38	605	0.001	0.594
2.5	11.54	545	0.001	0.660

For every time step, the convergence criteria were set to 1×10^{-6} for all the residuals. The pressure-velocity coupling used in the simulation was fully COUPLED.

For comparison purposes, in both techniques (SM and OM) the turbulence model that was used was the $k - \omega\ SST$ because of its excellent prediction of the flow for this specific application, already reported in the literature. Regarding spatial and time discretization, second order schemes were used for all the equations.

4. Numerical Results

The variable used for the convergence analysis was the total torque generated by the turbine, which was computed as an average in the last (10th) rotation of the turbine. As it is shown in Figure 5, the total torque (at a $TSR = 2$) had a periodic evolution in time with three peaks per revolution. The same behavior in time was observed for the OM technique (not shown) but with a lower average value.

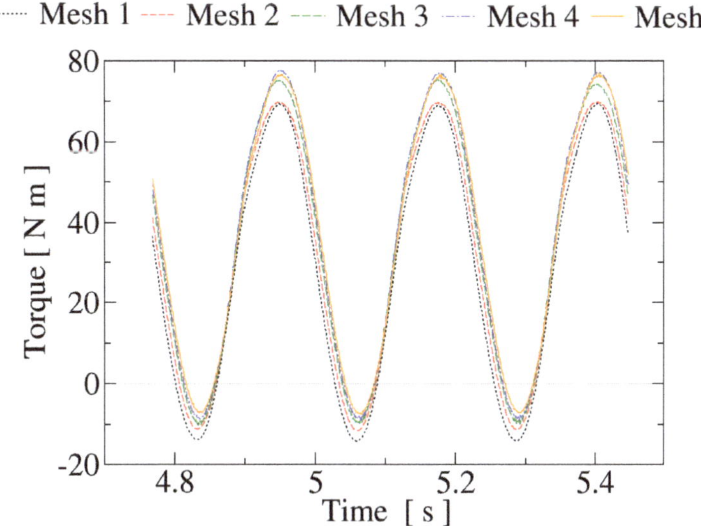

Figure 5. Evolution of the torque in time for the different meshes (SM, $TSR = 2$).

Figure 6 shows the results of the convergence analysis for both (SM and OM) techniques, which were performed for a *TSR* of 2. It was clear that both techniques showed an asymptotic tendency in their results with meshes higher that 200 k elements, but with a difference of approximately 15% in the predicted torque between the two techniques. It was also observed that the OM method had a faster convergence than SM. It was clear that both techniques arrived at the asymptotic region in the last two meshes, so that this faster convergence was not related to the number of meshes used in the convergence analysis, but in the capability of reaching the asymptotic region with few changes in the total number of elements. The influence of the refinement of the background mesh was also studied (not shown), but it was found that increasing the number of elements of the background mesh did not have a great impact on the predicted torque.

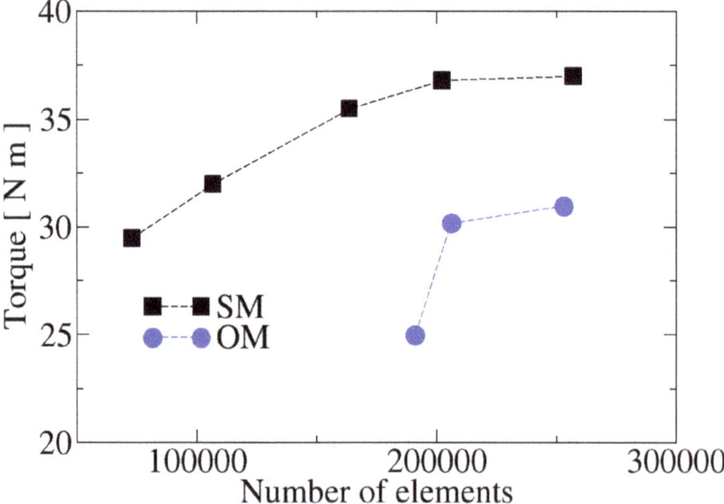

Figure 6. Convergence results for both techniques at *TSR* = 2.

Table 7 shows the difference in the predicted torque between consecutive meshes used in the convergence analysis in both techniques. The criterion for selecting the appropriated mesh was based on a difference between consecutive meshes of less than 2%. Based on this criterion, Meshes M4 and M2 were selected to perform the rest of the simulations using SM and OM techniques, respectively.

Mesh 5 (SM) and Mesh 3 (OM) were run using the same computational resource in order to establish an approximate relative computational cost. It was found that for a *TSR* = 2, using the same numerical setup, the CPU time of OM was approximately 10% higher than SM. No comparison was done with respect to computational resources used such as RAM and processor performance during the simulation.

Table 7. Differences in the average torque between the meshes.

SM	Diff	OM	Diff
M1–M2	7.8%	M1–M2	17.2%
M2–M3	9.8%	M2–M3	1.6%
M3–M4	3.5%		
M4–M5	0.5%		

Regarding the prediction of the performance of the turbine at different TSRs, both techniques predicted comparable results, as shown in Figures 7 and 8. However, the SM technique always predicted higher values of the torque and the power coefficient in comparison to OM in the range of *TSR* between 0.75 and 2. For TSRs higher than 2, the OM method predicted higher values for both torque and power coefficient (C_P) in comparison to SM. It was also clear that the SM technique predicted a peak in the torque at a *TSR* of 1.5, while for the OM method the peak occurred at a higher *TSR* (1.75). The highest torque predicted by SM was 39 Nm, while for OM it was approximately 32 Nm. For TSRs lower than 0.75, both methods predicted that the turbine did not generate a torque so that a starting mechanism was required.

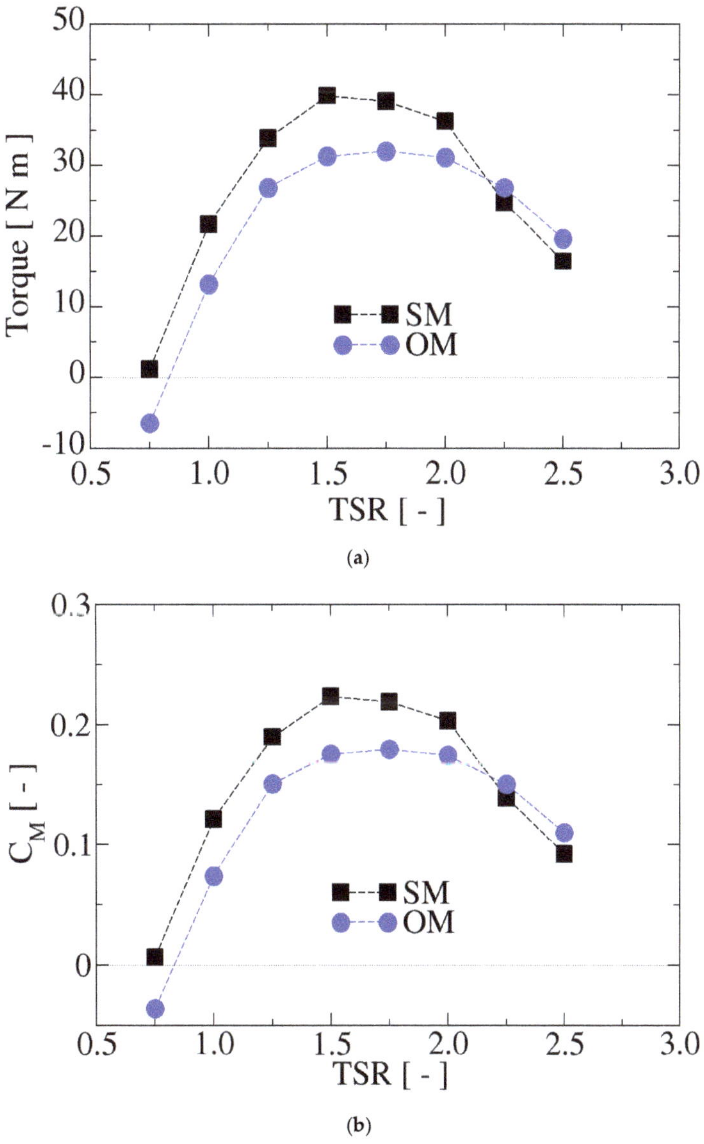

Figure 7. Turbine performance based on the generated torque (**a**) Nm, (**b**) C_M.

A very interesting result was that both methods (SM and OM) agreed in the prediction of the peak performance of the turbine in which the peak C_P was achieved close to a $TSR = 2$ (see Figure 8). It was also observed that the curves predicted by both methods had very similar tendencies and shape. The maximum power coefficient predicted by both methods was close to 0.35, which is a value typically expected for small-size VATs. By performing a simple calculation, it was possible to estimate the highest power that the design turbine could provide if it was operated at a free stream velocity of 2.2 m/s (max speed according to the design calculations). The maximum power that this turbine could generate in such condition is 1040 W according to the SM method and 910 W according to the OM method; both predictions were very close to the maximum power used in the design calculations.

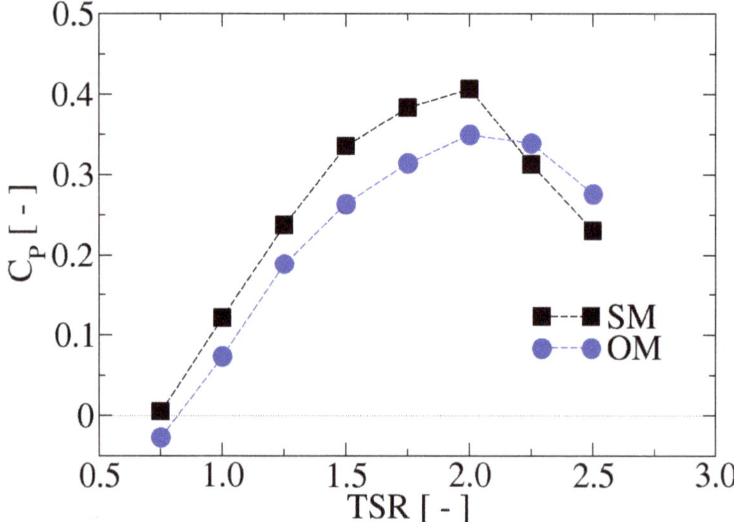

Figure 8. Power coefficient as a function of TSR.

Unfortunately, for the present case, there were not available experimental results for the single rotor, as the turbine was installed on site with the diffuser without being tested first in an equipped laboratory. Nevertheless, the complete floating station was tested in a small dam as it was towed by a boat. It is well known that the experimental data in these kinds of on-site tests have large uncertainties, but the maximum power measured at a boat speed of 1.05 m/s was approximately 99 Watts, which corresponded to a Cp value of 0.35. This value agreed very well with the peak performance predicted by the OM technique.

On the other hand, a qualitative comparison of the fluid flow visualization between both methods will be shown using the vorticity, velocity, and pressure fields near the turbine. Figures 9 and 10 show the instantaneous vorticity field close to the turbine in the last revolution for several azimuthal angles (based on the blade marked as A) for both OM and SM techniques. At a first glance, the figures marked as (a), (c), and (e) look alike in the region close to the turbine but some differences were observed in the far wake as it was convected downstream. It is clearly shown that the resolution of the vortical structures that were generated due to the rotation of the turbine was higher in the OM method than in the SM method, even though both methods used the same number of elements. This observation is related to the advantage that the OM technique has due to the structured meshes used in the background and near-body. Both methods predicted the vortex shedding at the same azimuthal angle (between 135° and 195°) with a vortex with a very similar intensity when it separates from the blade. This shed vortex was convected downstream as it diffused to form the wake; in the SM technique, the vortex diffusion happens more rapidly than in the OM technique due to the resolution of the mesh (may be also the interface connection). A very interesting observation is that both methods predicted that the shed vortex would not strongly interact with the other blades. This phenomenon is connected to the geometry of the airfoil used in the blade. In other studies in which NACA profiles were used (see references [5,21,33,34]), this interaction was very strong, which directly affected the generated power of the turbine. Although both techniques predicted the same behavior of the vorticity, the overset mesh presented a higher resolution, details of the vortices, and unsteadiness of the flow in the near and far wake.

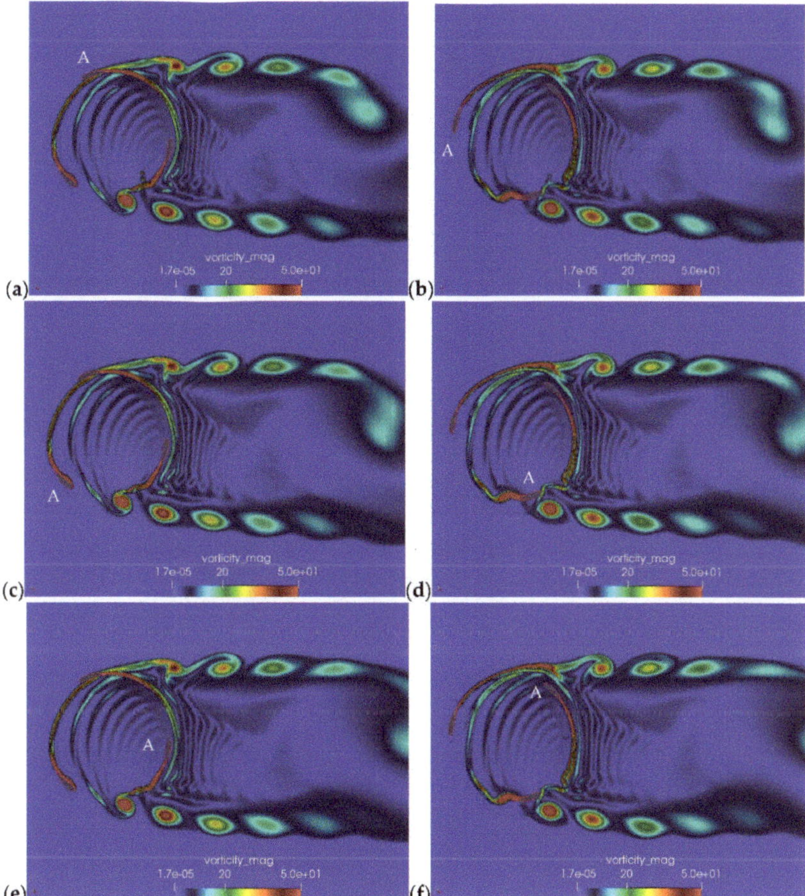

Figure 9. Instantaneous vorticity field for OM at *TSR* = 2 and different azimuthal angles: (a) 15°, (b) 75°, (c) 135°, (d) 195°, (e) 255°, (f) 315°.

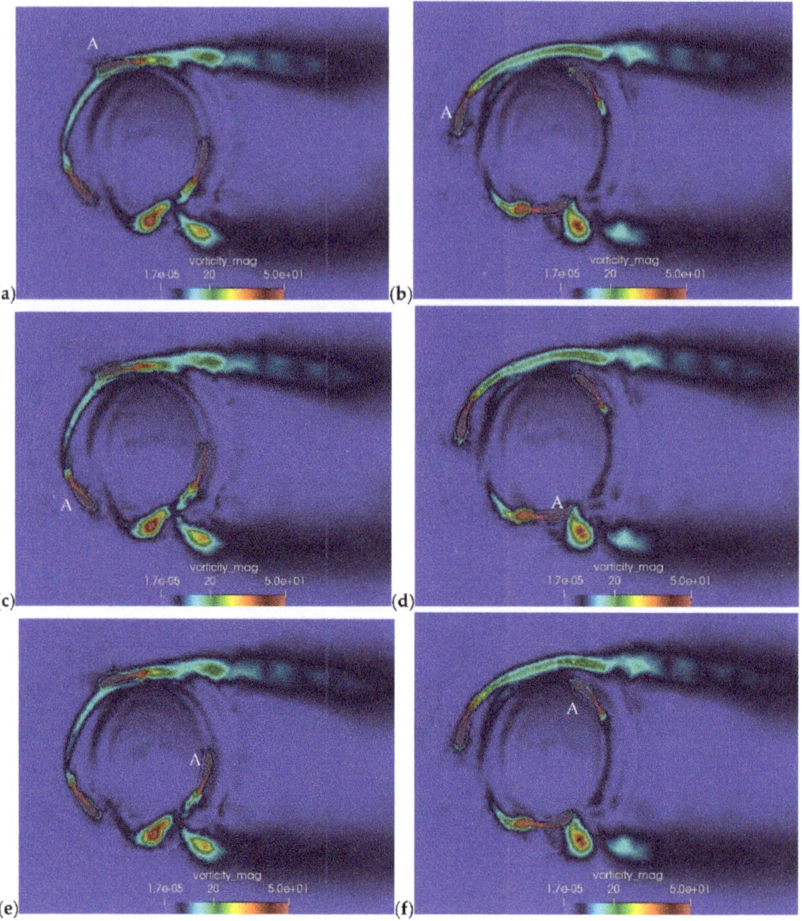

Figure 10. Instantaneous vorticity field for SM at $TSR = 2$ and different azimuthal angles: (**a**) 15°, (**b**) 75°, (**c**) 135°, (**d**) 195°, (**e**) 255°, (**f**) 315°.

Figure 11 shows a comparison of the instantaneous velocity magnitude contour of the flow near to the turbine at an azimuthal angle of 15° for both SM and OM at $TSR = 2$. It was observed that close to the surface of the blades the gradients of the velocities were correctly predicted by both methods. The main differences were related to the resolution of the wake of the blades and the wake of the turbine. OM showed a better resolution of the wake of the turbine including the unsteadiness of the shear layers in the edge of the wake. This observation was completely smeared out in the case of SM. In general, the width of the wake is always larger in SM than OM; this is in agreement to the higher C_M results in the SM technique. Similar observations can be done for the instantaneous pressure fields, which are shown in Figure 12. The distribution of pressure along the blades surface looked very similar between both models. For SM, the position of the interface between the stationary and rotating mesh seemed to have an impact on the definition of the pressure gradients. It was very clear that the definition of the low-pressure regions related to the shed vortices was better captured by the OM technique.

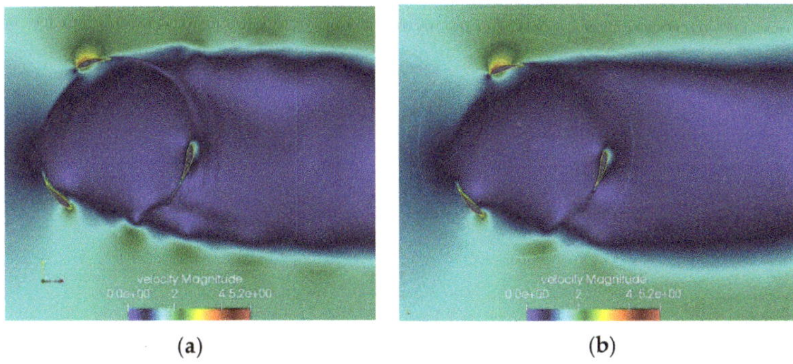

Figure 11. Instantaneous velocity magnitude field at $TSR = 2$ and azimuthal angle of 15°for (**a**) OM and (**b**) SM.

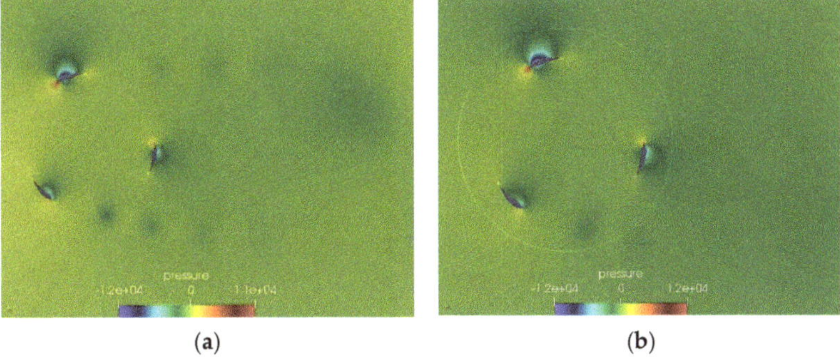

Figure 12. Instantaneous pressure field at $TSR = 2$ and azimuthal angle of 15°for (**a**) OM and (**b**) SM.

5. Conclusions

A comparison of two of the most common dynamic mesh techniques used in CFD applied to the modelling of hydrokinetic turbines was presented and discussed. For this comparison, two CFD models were implemented, one using the sliding mesh technique and the other one using the overset mesh technique. For a fair comparison, both models were two-dimensional and several parameters were kept similar in both models such as computational domain size, boundary conditions, turbulence model, turbine size, blade geometries, total number of elements, and grid resolution (y^+) close to the blades' surface. The implemented model was based on a 1-kW hydrokinetic turbine design by e.Ray Europa GmBh with the purpose to be used in rivers of low to medium stream current velocities. Since little information was found in the literature regarding best practices for mesh generation and convergence analysis when using dynamic meshes, the present paper proposed a first approximation in this direction for both SM and OM techniques.

Numerical results of the convergence analysis for both (SM and OM) techniques showed an asymptotic tendency in their results with meshes higher that 200 k elements, but with a difference of approximately 15% in the predicted torque. An interesting observation was that the OM method had a faster convergence than SM. Regarding the prediction of the performance of the turbine at different TSRs, both techniques predicted comparable results. It was observed that the SM technique tended to predict higher values of the torque and the power coefficient in comparison to OM. It was also clear that the SM technique predicted a peak in the torque at a TSR of 1.5, while, for the OM method, the peak occurred at a higher TSR (1.75). Both methods also agreed in the tendency and shape of the C_P curve, and the prediction of the peak performance of the turbine was very similar and close to a $TSR = 2$. The maximum power coefficient predicted by both methods was close to 0.35, which is a value typically expected for small-size VATs.

Regarding the resolution of flow dynamics' properties such as vorticity, velocity, and pressure fields, the OM technique showed better performance than SM. The dynamics of the wake including the width of the wake and the unsteadiness were better captured with the OM technique using the same number of elements. Regarding the computational cost, it was found that, for a $TSR = 2$, using the same numerical setup and approximately the same number of elements, the CPU time of OM was approximately 10% higher than that of SM.

Author Contributions: Conceptualization, O.D.L.M., F.S. and S.L.; validation, F.S.; investigation, O.E.M. and K.M.E.; writing—original draft preparation, O.D.L.M.; writing—review and editing, S.L.; visualization, O.E.M.; supervision, O.D.L.M. and S.L. All authors have read and agreed to the published version of the manuscript.

Funding: This research was founded by the mechanical engineering department and the Vice-Presidency for Research & Creation at Universidad de los Andes. Experimental data was provided by e.ray Europa Gmbh.

Institutional Review Board Statement: Not applicable.

Informed Consent Statement: Not applicable.

Acknowledgments: The authors acknowledge the support of the department of mechanical engineering at Universidad de los Andes for providing the computational resources (software and hardware) for the present study. All authors acknowledge financial support provided by the Vice Presidency for Research & Creation publication fund at the Universidad de los Andes to cover for APC.

Conflicts of Interest: The authors declare no conflict of interest.

References

1. Khana, M.J.; Bhuyana, G.; Iqbalb, M.T.; Quaicoe, J.E. Hydrokinetic energy conversion systems and assessment of horizontal and vertical axis turbines for river and tidal applications: A technology status review. *Appl. Energy* **2009**, *86*, 1823–1835. [CrossRef]
2. Laws, N.D.; Epps, B.P. Hydrokinetic energy conversion: Technology, research, and outlook. *Renew. Sustain. Energy Rev.* **2016**, *57*, 1245–1259. [CrossRef]
3. Ma, Y.; Hu, C.; Li, Y.; Deng, R. Research on the Hydrodynamic Performance of a Vertical Axis Current Turbine with Forced Oscillation. *Energies* **2018**, *11*, 3349. [CrossRef]
4. Patel, V.; Eldho, T.I.; Prabhu, S.V. Performance enhancement of a Darrieus hydrokinetic turbine with the blocking of a specific flow region for optimum use of hydropower. *Renew. Energy* **2019**, *135*, 1144–1156. [CrossRef]
5. Amet, E. Simulation Numerique d'une Hydrolienne à Axe Vertical de Type Darrieus. Ph.D. Thesis, Institut Polytechnique Grenoble, Grenoble, France, 2009.
6. Hoerner, S.; Abbaszadeh, S.; Maître, T.; Cleynen, O.; Thévenin, D. Characteristics of the fluid–structure interaction within Darrieus water turbines with highly flexible blades. *J. Fluids Struct.* **2019**, *88*, 13–30. [CrossRef]
7. Birjandi, A.; Bibeau, E. Frequency analysis of the power output for a vertical axis marine turbine operating in the wake. *Ocean Eng.* **2016**, *127*, 325–344. [CrossRef]
8. Bachant, P.; Wosnik, M. Effects of Reynolds Number on the Energy Conversion and Near-Wake Dynamics of a High Solidity Vertical-Axis Cross-Flow Turbine. *Energies* **2016**, *9*, 73. [CrossRef]
9. Rawlings, G. Parametric Characterization of An Experimental Vertical Axis Hydroturbine. Master's Thesis, University of British Columbia, Vancouver, DC, Canada, 2008.
10. Ouro, P.; Runge, S.; Stoesser, Q.T. Three-dimensionality of the wake recovery behind a vertical axis turbine. *Renew. Energy* **2019**, *133*, 1066–1077. [CrossRef]
11. Akbarn, M.; Mustafa, V. A new approach for optimization of Vertical Axis Wind Turbines. *J. Wind Eng. Ind. Aerodyn.* **2016**, *153*, 34–45. [CrossRef]
12. Ma, Y.; Lam, W.-H.; Cui, Y.; Zhang, T.; Jiang, J.; Sun, C.; Guo, J.; Wang, S.; Lam, S.S.; Hamill, G. Theoretical vertical-axis tidal-current-turbine wake model using axial momentum theory with CFD corrections. *Appl. Ocean Res.* **2018**, *79*, 113–122. [CrossRef]
13. Goude, A.; Ågren, O. Simulations of a vertical axis turbine in a channel. *Renew. Energy* **2014**, *63*, 477–485. [CrossRef]
14. Epps, B.; Roesler, B.; Medvitz, R.; Choo, Y.; McEntee, J. A viscous vortex lattice method for analysis of cross-flow propellers and turbines. *Renew. Energy* **2019**, in press. [CrossRef]
15. Chatelain, P.; Duponcheel, M.; Caprace, D.; Marichal, Y.; Winckelmans, G. Vortex particle-mesh simulations of vertical axis wind turbine flows: From the airfoil performance to the very far wake. *Wind Energy Sci.* **2017**, *2*, 317–328. [CrossRef]

16. de Tavernier, D.; Ferreira, C. An extended actuator cylinder model: Actuator-in-actuator cylinder (AC-squared) model. *Wind Energy* **2019**, *22*, 1058–1070. [CrossRef]
17. Mendoza, V. Aerodynamic Studies of Vertical Axis Wind Turbines using the Actuator Line Model. Ph.D. Thesis, Uppsala University, Upssala, Sweden, 2018.
18. Madsen, H.; Paulsen, U.; Vitae, L. Analysis of VAWT aerodynamics and design using the Actuator Cylinder flow model. *J. Phys. Conf. Ser.* **2014**, *555*, 012065. [CrossRef]
19. Balduzzi, F.; Bianchini, A.; Maleci, R.; Ferrara, G.; Ferrari, L. Critical issues in the CFD simulation of Darrieus wind turbines. *Renew. Energy* **2016**, *85*, 419–435. [CrossRef]
20. Rezaeiha, A.; Kalkman, I.; Blocken, B. CFD simulation of a vertical axis wind turbine operating at a moderate tip speed ratio: Guidelines for minimum domain size and azimuthal increment. *Renew. Energy* **2017**, *107*, 373–385. [CrossRef]
21. Maître, T.; Amet, E.; Pellone, C. Modeling of the flow in a Darrieus water turbine: Wall grid refinement analysis and comparison with experiments. *Renew. Energy* **2013**, *51*, 497–512. [CrossRef]
22. McNaughton, J.; Afgan, I.; Apsley, D.D.; Rolfo, S.; Stallard, T.; Stansby, P.K. A simple sliding-mesh interface procedure and its application to the CFD simulation of a tidal-stream turbine. *Int. J. Numer. Methods Fluids* **2014**, *74*, 250–269. [CrossRef]
23. Steger, L.J.; Dougherty, F.; Benek, J.A. A Chimera Grid Scheme. In *Advances in Grid Generation*; Ghia, K., Ghia, U., Eds.; American Society of Mechanical Engineers: Houston, TX, USA, 1983.
24. Laín, S.; Taborda, M.A.; López, O.D. Numerical study of the effect of winglets on the performance of a straight blade Darrieus water turbine. *Energies* **2018**, *11*, 297. [CrossRef]
25. López, O.D.; Quiñones, J.J.; Laín, S. RANS and Hybrid RANS-LES Simulations of an H-Type Darrieus Vertical Axis Water Turbine. *Energies* **2018**, *11*, 2348.
26. Kozak, P. Effects of Unsteady Aerodynamics on Vertical-Axis Wind Turbine Performance. Master's Thesis, Illinois Institute of Technology, Chicago, IL, USA, 2014.
27. McLean, D. Development of the Dual-Vertical-Axis Wind Turbine with Active Blade Pitch Control. Master's Thesis, Concordia University, Montreal, QC, Canada, 2017.
28. Lei, H.; Su, J.; Bao, Y.; Chen, Y.; Han, Z.; Zhou, D. Investigation of wake characteristics for the offshore floating vertical axis wind turbines in pitch and surge motions of platforms. *Energy* **2019**, *166*, 471–489. [CrossRef]
29. Kinsey, T.; Dumas, G. Impact of channel blockage on the performance of axial and cross-flow hydrokinetic turbines. *Renew. Energy* **2017**, *103*, 239–254. [CrossRef]
30. Gorle, J.; Chatellier, L.; Pons, F.; Ba, M. Modulated circulation control around the blades of a vertical axis hydrokinetic turbine for flow control and improved performance. *Renew. Sustain. Energy Rev.* **2019**, *105*, 363–377. [CrossRef]
31. Suarez, F. Parameter Study of a Ducted H-Darrieus Rotor in a Hydrokinetic Power Plant: Numerical Simulation. Master's Thesis, TU Darmstadt, Darmstadt, Germany, 2015.
32. Miller, M.A.; Duvvuri, S.; Kelly, W.D.; Hultmark, M. Rotor solidity effects on the performance of vertical-axis wind turbines at high Reynolds numbers. *J. Phys. Conf. Ser.* **2018**, *1037*, 052015. [CrossRef]
33. López, O.; Meneses, D.; Quintero, B.; Laín, S. Computational study of transient flow around Darrieus type cross flow water turbines. *J. Renew. Sustain. Energy* **2016**, *8*, 014501. [CrossRef]
34. Laín, S.; Cortés, P.; López, O.D. Numerical Simulation of the Flow around a Straight Blade Darrieus Water Turbine. *Energies* **2020**, *13*, 1137. [CrossRef]

processes

MDPI

Article

Design Guideline for Hydropower Plants Using One or Multiple Archimedes Screws

Arash YoosefDoost * and William David Lubitz

School of Engineering, University of Guelph, 50 Stone Rd E, Guelph, ON N1G 2W1, Canada;
wlubitz@uoguelph.ca
* Correspondence: YoosefDoost@Gmail.com

Abstract: The Archimedes/Archimedean screw generator (ASG) is a fish-friendly hydropower technology that could operate under a wide range of flow heads and flow rates and generate power from almost any flow, even wastewater. The simplicity and low maintenance requirements and costs make ASGs suitable even for remote or developing areas. However, there are no general and easy-to-use guidelines for designing Archimedes screw power plants. Therefore, this study addresses this important concern by offering a simple method for quick rough estimations of the number and geometry of Archimedes screws in considering the installation site properties, river flow characteristics, and technical considerations. Moreover, it updates the newest analytical method of designing ASGs by introducing an easier graphical approach that not only covers standard designs but also simplifies custom designs. Besides, a list of currently installed and operating industrial multi-Archimedes screw hydropower plants are provided to review and explore the common design properties between different manufacturers. On top of that, this study helps to improve one of the biggest burdens of small projects, the unscalable initial investigation costs, by enabling everyone to evaluate the possibilities of a green and renewable Archimedes screw hydropower generation where a flow is available.

Citation: YoosefDoost, A.; Lubitz, W.D. Design Guideline for Hydropower Plants Using One or Multiple Archimedes Screws. *Processes* **2021**, *9*, 2128. https://doi.org/10.3390/pr9122128

Academic Editors: Santiago Lain and Omar Dario Lopez Mejia

Received: 2 November 2021
Accepted: 24 November 2021
Published: 25 November 2021

Publisher's Note: MDPI stays neutral with regard to jurisdictional claims in published maps and institutional affiliations.

Keywords: design Archimedes screw hydropower plant; quick estimation method; Archimedean screw; fish safe/friendly; multi-ASG; hydropower plant; hydro power plant; small/micro/pico/low head hydro power plant

1. Introduction

Hydropower is among the most efficient and reliable renewable energy resources [1] and offers significant value for a sustainable future [2]. Hydropower plants can be classified as in Figure 1 [3] based on the electrical generating capacity. It is estimated that small hydropower plants generate about 10% of global hydropower [3]. Hydropower plants that utilize the natural flow of water in a run-of-river (ROR) configuration [4] usually have small or no reservoirs. The lack of a large, actively-controlled reservoir minimizes flooding land and soil destruction, greenhouse gas emissions, as well as environmental [5] and social impacts [6]. However, the resulting reduction in control of river flow can result in more variable or poorly timed power generation [6].

The Archimedes/Archimedean screw was one of the earliest hydraulic machines [7]. It is made of one or more helical arrays of blades wrapped around a central cylinder [8] and supported within a fixed trough with a small gap that allows the screw to rotate freely [6]. Using Archimedes screws as water pumps dates back many centuries. As a modern application, Archimedes screw pumps (ASP) have been regularly used in wastewater treatment plants. The earliest patent involving using a screw in hydro power plants dates back to 1922 [9], but examples of application of Archimedes screw generator (ASG) technology date back only to the 1990s [10]. The Archimedes screw is theoretically a reversible hydraulic, and there are examples of single installations where screws can be used alternately as pumps and generators [11]. The flexibility, mechanical simplicity, ease of use, low operation demands, and maintenance costs make Archimedes screws a practical choice for sites where other types of turbines may not be feasible [6]. Some of these conditions are discussed below.

Figure 1. Classification of hydropower plants based on the installation capacity.

ASGs are safer for aquatic life (especially fish) than other hydropower turbines [12–19]. Since large living organisms such as fish can pass through ASGs, there is no surprise that sediment and small debris can pass through ASGs as well. This advantage enables ASGs to generate power from almost any flow, even unconventional sources such as wastewater that are often not appropriate for conventional hydro turbines [6,20–22]. Since Archimedes screw turbines run in open channel flow, theoretically, small hydropower could be considered as an alternative to the energy dissipator hydraulic structures [23] to make use of the flow energy and offer the added value of power generation.

The Archimedes screw properties can be scaled considering the available volumetric flow rate and installation site specifications. ASGs can be designed to operate in a wide range of flow rates (currently from 0.01–10 m^3/s) and flow heads (currently from 0.1–10 m). Currently, PicoPica-10 is the smallest known commercial ASG [6,24]. PicoPica also is claimed as the world's first easily portable ASG due to its small dimensions and 17.5 kg weight [24–26]. The largest screw diameter known to the author at the time of publication is 5 m, as in the "Widdington Plant" [6]. In terms of length, the Hasselt hybrid (pump and turbine) screws are claimed as the largest hybrid Archimedean screws in the world [27]. In terms of power, manufacturers have announced single screws that can pass flow rates as high as 15 m^3/s and generate up to 800 kW of power [28]. In terms of the largest number of installed parallel screw generators, currently, the largest known number is six parallel Archimedes screws in hydropower plants such as Marengo (in Goito, Italy) [6], Rosko (in Rosko, Poland) and Steinsau (in Alsace, France) which are all listed in Table 2.

In addition, it is possible to run an ASG with flow rates less or more than the optimal design volume flow rate. Studies show that ASGs can handle flow rates even of up to 20% more than optimal filling without a significant loss in efficiency [29]. When there are large fluctuations in flow rate, or when the conditions are not perfect for a single fixed speed screw, using variable-speed ASGs or installing more than one screw are ways to potentially better utilize available flow at a wider range of sites.

The simplicity of this technology and low maintenance requirements and costs makes ASGs suitable even for remote or developing areas with no, expensive or limited access to the electricity grid [6]. Such remarkable flexibities make ASGs good candidates to upgrade or retrofit existing developments to make current developments sustainable or more sustainable by offering added values such as power generation [6]. Therefore, it dramatically increases the number of sites suitable for green and renewable hydroelectric power generation and makes ASGs among practical options in sustainable developments [6,22]. For example, there are about 280 sites with a head less than 5 m and less than 200 kW power generating capacity within Ontario [30] and using ASGs to utilize this potential could result in about 16 MW of additional total power generating capacity of [22].

Operating ASGs at their most mechanically efficient operating condition is not necessarily the most financially efficient since it may not essentially lead to generating the highest amount of overall energy [31]. Using more than one ASG may lead to easier maintenance, more flexible operation plans and utilize the available volume of flow more efficiently. However, there are still no general and easy-to-use guidelines for designing Archimedes screw power plants using more than one ASG. In fact, still there are no general and easy-to-use guidelines for designing Archimedes screw power plants and most studies focused on designing the Archimedes screws.

In terms of the design of screws, some analytical methods have been developed for Archimedes screw pumps (ASP) [32]. However, several design aspects of ASPs remain on experience [33]. In designing Archimedes screw generators (ASG), the lack of analytical guidelines was so serious that usually listed among the important ASGs disadvantage [6,33]. Currently, the designs are highly dependent on the experience of the designer [34]. In non-English design literature of ASGs, the well-known studies are Brada (1996) [29], Aigner (2008) [35], Schmalz (2010) [36], Lashofer et al. (2011) [37], and Nuernbergk's (2020) book [38]. The well-known English studies are made by Rorres [39] and Nuernbergk and Rorres [34]. However, none of these methods is easy to understand and implement [6], particularly for the initial estimations and the early stages of designing the ASG hydropower plants.

Dragomirescu (2021) proposed a method to design ASGs using a regression method [32]. This method is easier to understand and implement, but the empirical nature of the used regression method, especially for such limited case studies, may affect the generality of the model. YoosefDoost and Lubitz (2021) proposed an easy-to-understand and implement analytical model for designing Archimedes screw generators [33]. The comparison of this method with Dragomirescu (2021) for the same cases indicates a slightly better performance. Moreover, comparisons indicate a reasonable agreement of YoosefDoost and Lubitz (2021) method results with a wider range of industrial ASGs designed by different manufactures [33].

Therefore, this study focuses on addressing this important concern by offering a simple method for quick, rough estimations of the number and geometry of Archimedes screws in considering the installation site properties, river flow characteristics, and technical considerations. It provides an update to the YoosefDoost and Lubitz (2021) analytical method for designing ASGs by introducing slightly different equations based on the same concept to offer a simpler graphical approach that accelerates and eases the design of single and multi-ASG hydropower plants' initial design estimations. The new graph offers three times more ASG design combinations that not only cover standard designs but also simplifies custom designs. Then a simple method for quick rough estimations of the number and geometry of Archimedes screws is proposed considering the installation site properties, river flow characteristics, and technical considerations. This study also provides an update to the new ASG records to provide a better understanding of the advancements and the current possibilities in this technology. A list of currently operating industrial multi-Archimedes screw hydropower plants is compiled to support the exploration of the common design properties that are used by different manufacturers. This study helps us to improve one of the biggest burdens of small projects, the unscalable initial investigation costs, by enabling everyone to evaluate the possibilities of green and renewable Archimedes screw hydropower generation where a flow is available.

2. Methods and Materials

2.1. Design Parameters of Archimedes Screws

The most important dimensions and parameters required to define the Archimedes screws are represented in Figure 2 and described in Table 1. Archimedes screw design parameters can be categorized as external (D_O, L, and β) and internal (D_i, N, and S) parameters. Generally, the screw installation site location properties and the passing volumetric flow rate determine the external parameters and the screw performance can then be optimized by adjusting the internal parameters [39].

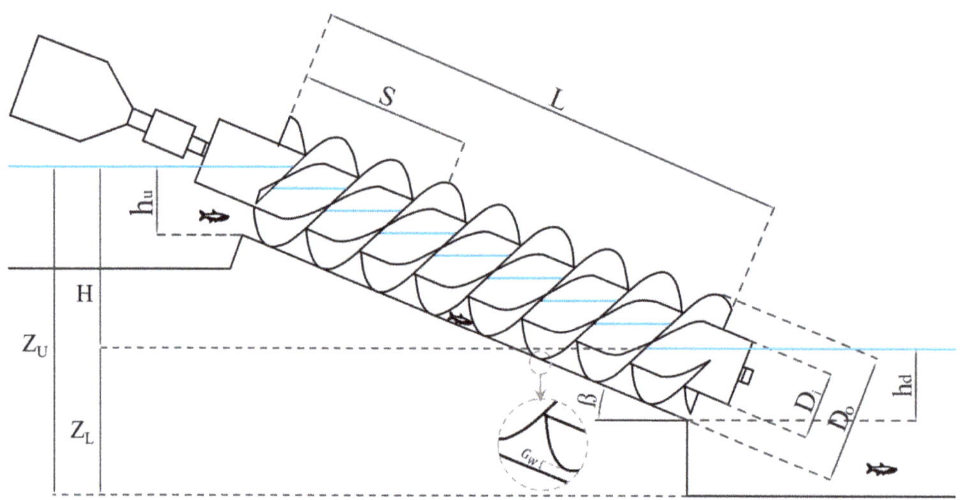

Figure 2. Required parameters to define Archimedes screw geometry [6,40].

Table 1. Archimedes screws' geometry and operating variables.

Parameter	Description	Unit	Variable	Description	Unit
L	Length of the screw	(m)	ω	Rotation speed of screw *	(rad/s)
D_O	Outer diameter of the screw	(m)	h_u	Upper (inlet) water level	(m)
D_i	Inner diameter of the screw	(m)	h_L	Lower (outlet) water level	(m)
S	Screw's pitch or period [39] (The distance along the screw axis for one complete helical plane turn)	(m)	Q	Volumetric flow rate passing through the screw	(m^3/s)

			Ratio	Description	Unit
N	Number of helical planed surfaces (also called blades, flights or starts [39])	(1)	δ	Inner to outer diameter ratio $\delta = D_i/D_O$	(1)
β	Inclination Angle of the Screw	(rad)	σ	Pitch to outer diameter ratio $\sigma = S/D_O$	(1)
G_w	The gap between the trough and screw.	(m)	Ξ	Dimensionless inlet depth $\Xi = h_u/D_O \cos \beta$	(1)

* Note: In the fixed speed Archimedes screws rotation speed is a constant.

The installation level (position) of the screw(s) relative to the dam crest or expected level of poundage (reservoir) should be defined considering the determined geometry and optimum inlet level of the Archimedes screw(s). For Archimedes screw generators, it can be shown that the head (H) is the difference of free surface elevations upstream (Z_U) and downstream (Z_L) of the screw relative to the same datum ($H = Z_U - Z_L$) [33]. Therefore, the screw length (L) will be:

$$L = H/ \sin \beta \tag{1}$$

The inclination angle of the Archimedes screw (β) may be limited by the geometry or slope of the installation site. However, Lashofer et al. [10] confirmed that many current industrial ASGs are installed at $\beta = 22°$ [10]. Inclination angles less than about 20° increase the length of screw and more than 30° lead to or considerably decrease the capacity of the screw [6].

Increasing the inclination angle also leads to a faster occurrence of the overflow leakage and increasing gap leakage. These issues could be managed by increasing the number of blades [41]. Based on CFD simulations in conjunction with laboratory-scale experiments, Dellinger et al. proposed that for their particular setup, the optimal inclination angle for N = 3 is about 15.5°. The maximum efficiency of the four- and five-bladed screws occur in inclination angles between 20° and 24.5°, and the highest power was achieved for N = 5, although it was only marginally higher than N = 4 [41]. However, due to the thickness of the blades, more blades come at the cost of reducing the bucket sizes as well as increasing manufacturing costs and challenges. Modern ASGs usually utilize three or four helical blades (N = 3 or 4). A survey of operating ASG power plants in the UK found that N = 4 was the most common configuration [42]. Lashofer (2012) observations indicate that for most of Archimedes screw installations N = 3 [10]. Based on Lyons (2014) experimental observations, there is a considerable reduction in ASGs' performance for N = 2 [43]. Rosly et al. (2016) based on CFD simulations of non-rotating screws reported that the number of helix turns (lower is better) are more important than the number of blades so that, the screws with three helix-turns and N = 3 showed the highest efficiency [44]. Lyons (2014) and Songin (2017) reported that experimental observations showed no significant increase in the Archimedes screw generators' efficiency for N > 3 [43,45]. Based on these observations, Dragomirescu (2021) concluded N = 3 as the optimal number of blades [32].

2.2. Archimedes Screws Configurations in Hydropower Plants

It is most common to have only one Archemidian screw installed in a power plant. However, two or more Archimedes screws can be installed in series or parallel to deal with technical limitations, increase the overall energy generation, easier maintenance, or more flexible operation plans. Multiple screws can be configured in series or parallel.

2.2.1. Archimedes Screws in Series Configuration

The idea of installing Archimedes screw pumps (ASPs) in series to pump the fluids into higher levels has been used in practice for a long time (e.g., [46] p. 18). However, installing ASGs in series to take advantage of low flow rate but high heads is not very usual. For higher heads, it is most common to utilize other hydropower technologies, often in combination with a long penstock. However, most hydropower turbines cannot safely pass sediment or larger objects such as living organisms, especially fish. Therefore, one can see Archimedes screws as long as 30 m [27] that are currently installed and in operation (Table 2). However, technical limitations such as bending of long shafts, weight limits of bearings, etc., can limit the maximum length of Archimedes screws. For example, for the two 19 m-long Archimedes screws in the Low Wood project, there is a significant engineering challenge supporting the considerable length and weight of the screw (40 tons) just with bearings at two contact points, one at the top and one at the bottom of the screw [47]. To deal with such limitations, theoretically, several ASGs could be installed in series instead of a very long screw, potentially as a chain of hydropower plants alongside a river or channel [6]. Such a configuration allows utilization of Archimedes screws at locations with an available high head when the available volume of flow rate is low.

2.2.2. Archimedes Screws in Parallel Configuration

The parallel installation of ASGs is the most common configuration in multi-ASG hydropower plants, especially where the volume of flow in a river is so high that a large ASG is required to make use of that amount of water or when the fluctuations in river volumetric of flow rate are considerable. Using several screws in parallel could lead to preventing the technical limitations of a large screw and offers easier maintenance, as well as more flexible operation plans besides taking advantage of a wider range of flow rates that could lead to an increase in the overall energy generation.

Table 2. Hydropower plants using several parallel Archimedes screw turbines. All screws at a plant are identical unless otherwise noted. Power and flow values are for a single screw at the plant: multiply by the number of screws for total plant flow and power.

Name		No. of ASGs	D_O (m)	H (m)	Q (m³/s)	P (kW)	Location (River)	Refs.
Totnes		2	3.7	3.45	6.5	160	Dart, UK	[48]
Hannoversch-Münden		2	2.8	2.6	2	35.455	Werra, DE	[49]
Low Wood		2	3	7.2	4	200	Leven, UK	[47]
Radyr		2	3.5	3.5	11	200	Taff, UK	[50]
Künzelsau		2	4.1	1.72	8.95	132	Kocher, DE	[51]
Ahornweg		2	2.3	1.45	2	21	Mühlbach, DE	[51]
Linton Falls		2	2.4	2.7	2.6	50	Trent, UK	[47]
Niklasdorf/Birgl and Bergmeister		2		3.9	3.6	106	Mur, AT	[51]
Hausen III Neumatt		2	3.4	5.8	5.5	235	Wiese, DE	[51]
Höllthal		2	4.3	2.22	10.5	220	Alz, DE	[52]
Gunthorpe Weir		2	4.3	2.03	14.15	165	Trent, UK	[53]
Solvay		2	2.3	2	2.5	35	Saja, ES	[54]
Linton Lock	Linton Plant	2	3	3.2	4.5	110	Ouse, UK	[6,55]
	Widdington Plant		5	3	14.5	335		
Plana		3	4.1	3.5	8.73	220	Vltava, CZ	[51]
Crescenzago		3	3.2	2.1	5	75	Lambro, IT	[51]
Olen		3	4.3	10	5	360	Albert Canal, BE	[51,56]
Ham		3	4.3	10	5	360	Albert Canal, BE	[51,56]
Hasselt		3	5	10	5	400	Albert Canal, BE	[11,27]
Rosko		6		1.7	4.5	60	Noteć, PL	[51,57]
Steinsau		6		1.4	3	30	Ill, FR	[51]
Marengo		6	3	1.6	3.7	51	Goito, IT	[58,59]

2.3. Case Studies

Analysis of current ASG power plants could help in finding common patterns, governing rules and important design points to make guidelines. Table 2 details current hydro power plants using more than one Archimedes screw installed in parallel.

2.4. Evaluation Criteria

In this study, to compare and evaluate the developed equations' results (i.e., estimations) with the case studies represented in Table 2 (i.e., observations), a combination of visualizations and statistical tests were conducted. Correlations were evaluated by Pearson correlation (R), and the relative differences were calculated by the percentage error (PE) and mean absolute percentage error (MAPE). In the following equations, O_i is the observed value, E_i is the estimated value, $\bar{O}$ is the average of the observed data, $\bar{E}$ is the average of the estimations, and n is the number of data points in the dataset [40].

In statistics, correlation refers to any statistically significant relationship between two variables. The Pearson correlation coefficient measures the linear relationship between two random variables [60] and describes it in a range between -1 to $+1$. Values close to $+1$ indicate a good and direct correlation, while values closer to -1 refer to a good but inverse relation between datasets. Values near zero indicate a lack of correlation [61]. These ranges could be represented in percent by multiplying the Pearson correlation to 100. The Pearson correlation is defined as [62].

$$R = \frac{\sum_{i=1}^{n}\left(E_i - \overline{E}\right)\left(O_i - \overline{O}\right)}{\sqrt{\sum_{i=1}^{n}\left(E_i - \overline{E}\right)^2}\sqrt{\sum_{i=1}^{n}\left(O_i - O\right)^2}} \tag{2}$$

The percentage (percent) error (PE) is a dimensionless error measure defined as the difference between the model estimations and the experimentally measured value:

$$PE = \frac{E_i - O_i}{O_i} \times 100 \tag{3}$$

The mean absolute percentage error (MAPE) is the average of absolute percentage errors and one of the most common accuracy measures [63] that is recommended in many textbooks (e.g., [64,65]). MAPE considers errors regardless of their sign, so positive and negative errors cannot cancel each other. MAPE is calculated as:

$$MAPE = \frac{100}{n} \sum_{i=1}^{n} \left| \frac{E_i - O_i}{O_i} \right| \tag{4}$$

3. Results and Discussion

3.1. Volumetric Flow Rate and Diameter of Archimedes Screws

According to Table 2, current multi-ASG hydropower plants were identified that are designed for the minimum, maximum, and average total flow rates of 4, 28.3 and 14.5 (m^3/s), respectively, for a wide range of hydraulic heads between 1.4 m to 10 m. Also, the minimum, maximum and average diameter of installed Archimedes screws in these plants ranges from 2.3 m to 5 m, with an average of about 3.58 m. This is a wide range of diameters per cubic meter of flow rate (between 0.3 m/m^3/s to 1.4 m/m^3/s and 0.71 m/m^3/s on average). Studies showed that in addition to the screw diameter, the inlet depth and rotation speed are also important factors in the volumetric flow rate of ASGs [40].

Figure 3 compares the volumetric flow rate and diameter of installed screws in hydropower plants listed in Table 2. The same size screws are installed in all these hydropower plants except at Linton Lock, where the two screws are different sizes. It is worth mentioning that, as discussed, ASGs could be used to upgrade or retrofit the current developments. The Linton Lock power plant is an example of a second ASG installed later as an upgrade.

According to Figure 3, many multi-ASG designs follow a relatively similar trend. For the current multi-ASG hydropower designs, the relationship between the flow rate (Q) and diameter of the screw (D$_O$) can be described linearly as follows. In this relationship, the coefficients are optimized using genetic algorithm optimization:

$$D_O = 0.2\,Q + 2.2 \tag{5}$$

where D_O must be in meters, and Q must be in m^3/s. This equation relates the Archimedes screw outer diameter (D_O) and passing volumetric flow rate (Q) with reasonable accuracy ($R = 69.81\%$, MAPE $= 10.59\%$).

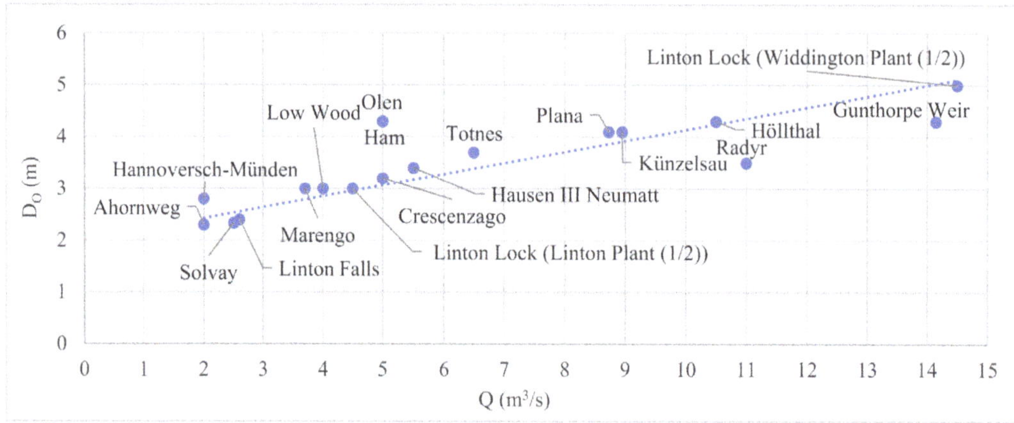

Figure 3. Comparison of the volumetric flow rate and diameter of installed screws in hydropower plants listed in Table 2. All values above are for a single screw.

3.2. Power and Diameter of Archimedes Screws

The available hydraulic power could be calculated by the following equation:

$$P_H = \gamma HQ \tag{6}$$

Figure 4 compares the potential available hydraulic power that is calculated using equations with the power that manufacturers estimated that their Archimedes screw designs could generate (Table 2). In this figure, all values are for a single screw, which are determined by dividing plant power by the number of screws, except at Linton Lock, where the two screws are different sizes.

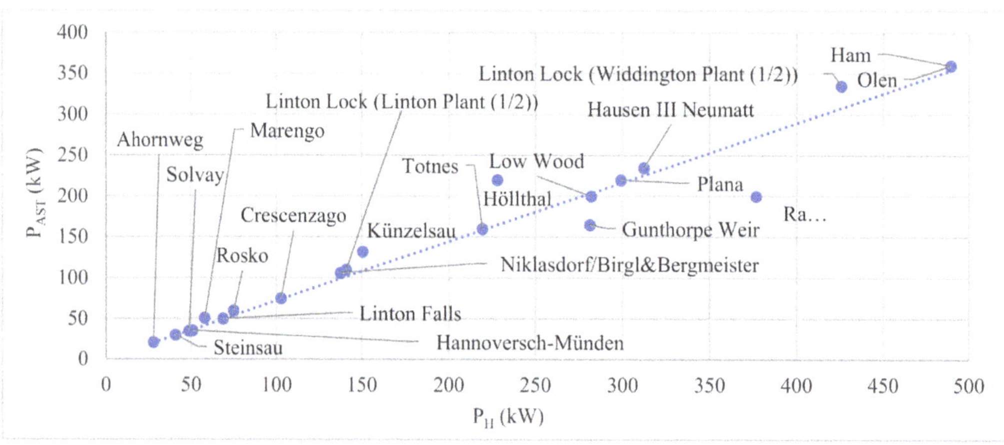

Figure 4. The theoretical available power vs. the power of installed screws published by manufacturers in current Archimedes screws hydropower plants represented in Table 2.

The efficiency of ASGs depends on many variables, including but not limited to the screw's design, installation, operation condition, fill height, rotation speed, etc. [6]. Typically, ASG water-to-wire efficiency is reported between 60% to 80% [66]. However, some studies reported hydraulic efficiencies of more than 80% in full-load and as high as

94% in partial-load situations [28]. Nonetheless, with reference to Figure 4, many of the studied designs follow a relatively similar trend that could be described as:

$$P = \eta\gamma HQ \tag{7}$$

where γ is the specific weight of water and H is the available hydraulic head. Using the generalized reduced gradient (GRG) algorithm [67,68] and the represented case studies in Table 2 to optimize the η as a coefficient indicates that for $\eta = 0.736$ the Equation (7) results are in good agreement with the case studies (R = 97.78%, MAPE = 7.90%). Since the available hydraulic power at a site is the product of head and flow (Equation), the coefficient $\eta = 0.736$ in Equation (7) suggests that the average efficiency of the plants based on the manufacturer's specifications in Table 2 is 73.6%, which is consistent with expectations from the literature. These results of estimated power by Equation (7) in comparison with the reported power by the manufacture that also visualized in Figure 5. In this figure, all estimated values are computed for a single screw, and the actual power is determined by dividing plant power by the number of screws except at Linton Lock, where the two screws are in different sizes.

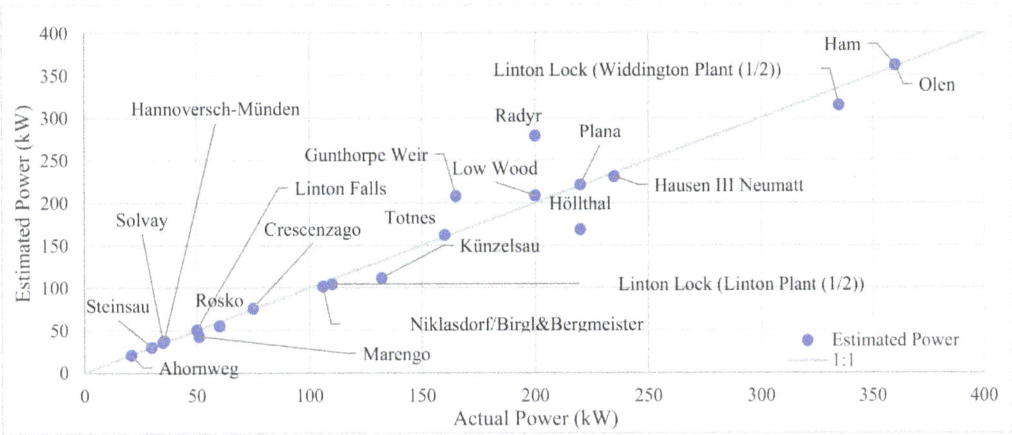

Figure 5. Comparison of the estimated power of each screw vs. the power of each installed screws in ASG hydropower plants.

Figure 6 compares the theoretical available hydraulic power (P_H) and the diameter of the Archimedes screw designs represented in Table 2. Since the size of screws is the same in almost all power plants, each point represents an ASG hydropower plant, and the number of installed screws is represented in the parentheses in front of each power plant. The only exception is for Linton Lock, where the size of two installed screws is different. Therefore, these screws are distinguished by marking them as (1 of 2).

According to Figure 6, the relationship between the available power and diameter of these Archimedes screws could be defined in the form:

$$D_O = \lambda(\gamma HQ)^{\psi} \tag{8}$$

By using the genetic algorithm (GA) and Table 2 to optimize the constants λ and ψ results in values of $\lambda = 0.213$ and $\psi = 0.232$ providing reasonable accuracy for these cases (R = 83.28%, MAPE = 10.64% and MPE = 1.52%).

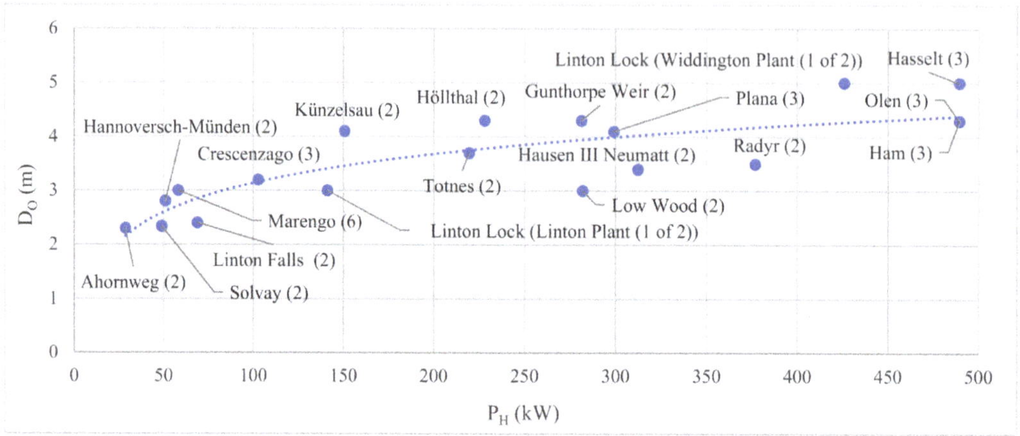

Figure 6. The theoretical available power vs. the diameter and number of installed screws in current hydropower plants represented in Table 2.

3.3. Analytical Equation

YoosefDoost and Lubitz (2021) developed a concept to define the maximum available area of the screw entrance for any inlet water level, which was called the effective area (A_E) [40]. The required parameters to define A_E are represented in Figure 7. Using the concept of effective inlet cross-section area [40], YoosefDoost and Lubitz (2021) offered an analytical equation to estimate the overall (outer) diameter of the screw (D_O) based on the volumetric flow rate (Q) for the desired rotation speed (ω) [33]:

$$D_O = \sqrt[3]{\frac{16\pi}{\sigma\omega\left(2\theta_O - \sin 2\theta_O - \delta^2(2\theta_i - \sin 2\theta_i)\right)}Q} \tag{9}$$

$$\theta_O = \pi - \arccos\left(\frac{y_O}{r_O} - 1\right) \tag{10}$$

$$\theta_i = \pi - \arccos\left(\frac{y_i}{r_i} - 1\right) \tag{11}$$

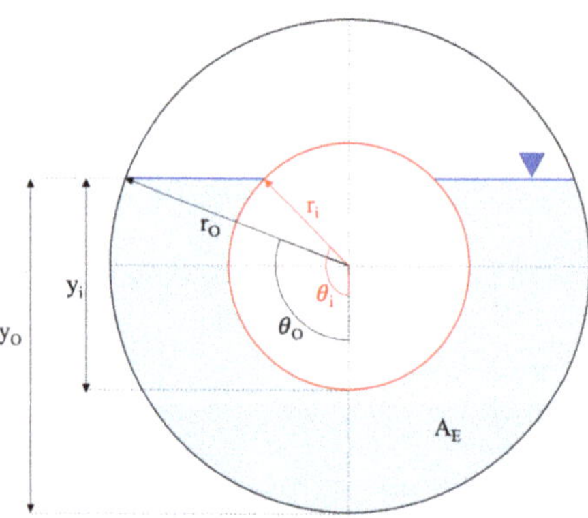

Figure 7. Parameters to define A_E.

The simplicity, efficiency and lower cost per watt of single-speed screws make them advantageous in a steady flow [6]. Single-speed Archimedes screws can operate in a partially full condition even if the flow rate is not sufficient to fill the screw at its operation speed [31]. The variable-speed screws are recommended to increase the power generation when the flow varies, as well as if hydropower is the sole source of electricity or for off-grid power plants [6]. Lashofer et al. studies [10] showed that many current industrial ASGs are designed with a rotation speed close to the maximum rotation speed recommended by Muysken (ω_M) [69]:

$$\omega_M = \frac{5\pi}{3D_o^{2/3}} \tag{12}$$

Therefore, for full-scale Archimedes screws running at a fixed rotation speed near ω_M, Equation (9) could be simplified as:

$$D_O = \left(\frac{48}{5\sigma\left(2\theta_O - \sin 2\theta_O - \delta^2(2\theta_i - \sin 2\theta_i)\right)} Q \right)^{3/7} \tag{13}$$

Equation (13) could be represented in the form of a power function such as $D_O = \lambda\, Q^\psi$. Studies showed that for most ASG power plants $\delta = 0.5$ and $\sigma = 1$ [10,33,46]. By defining $\Theta = 5\sigma\left(2\theta_O - \sin 2\theta_O - \delta^2(2\theta_i - \sin 2\theta_i)\right)/48$ the values of λ and ψ will be equal to $\Theta^{-3/7}$ and $3/7$ respectively and Equation (13) could be represented in a much simpler form:

$$D_O = \Theta^{-3/7}Q^{3/7} \tag{14}$$

$$Q = \Theta D_O^{7/3} \tag{15}$$

The Θ values for different δ and σ for a full range of dimensionless inlet depths (Ξ) could be determined by using Figure 8. Studies indicates that Equation (14) has the minimum relative difference with the current operating industrial ASG installations ($\delta = 0.5$ and $\sigma = 1$ [10,33,46]) for $\Xi = 69\%$ [33] where $\Theta = 0.32918$.

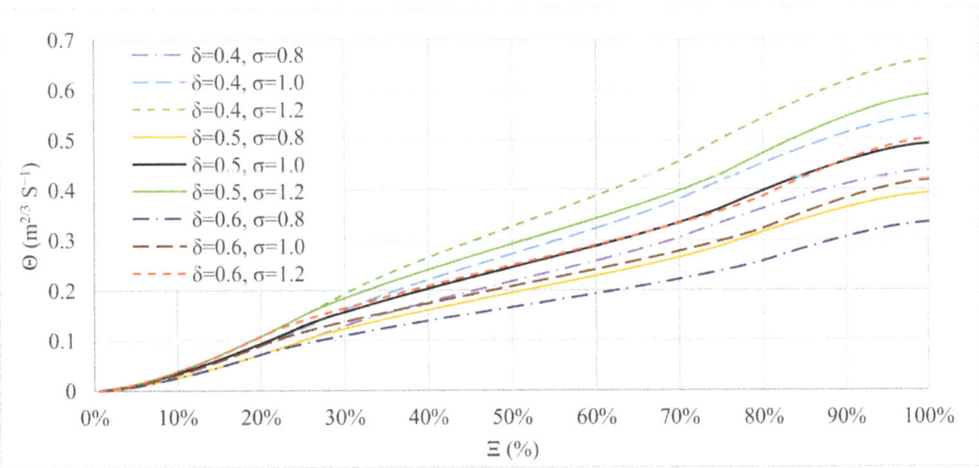

Figure 8. The corresponding Θ of different δ and σ for a full range of Ξ.

3.4. Evaluation of the Developed Equations

Equations (8) and (14) could be represented in a simplified form by applying the represented values of λ and ψ constants in Equation (8) and Θ in the analytical Equation (14).

The simplified version of the developed empirical Equations (5) and (8), as well as the analytical Equation (14), besides the evaluation results for Archimedes screws based on the evaluation criteria, are summarized in Table 3. Results of each equation for these cases are also visualized in Figure 9.

Table 3. Evaluation of equations based on the evaluation criteria.

Equation No.	Equation	R (%)	MAPE (%)
(5)	$D_O = 0.2\,Q + 2.2$	69.81	10.59
(8)	$D_O \approx 1.796(HQ)^{0.232}$	83.28	10.64
(14)	$D_O \approx 1.61\,Q^{3/7}$	74.38	9.69

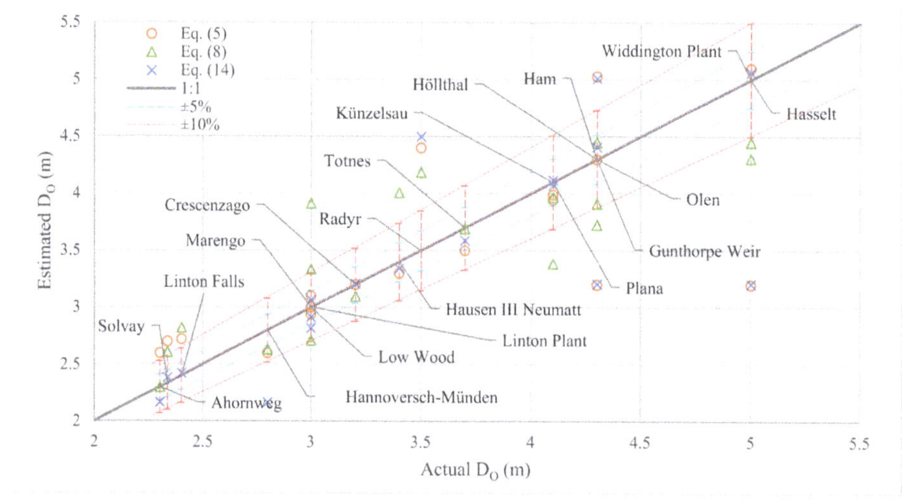

Figure 9. Graphical comparison of Equations (5), (8) and (14) estimations vs. actual overall diameter of ASG installations.

According to Table 3, Considering the small number of variables required for these equations, all of them indicate arguably reasonable results. The empirical Equation (8) shows the highest correlation with the case studies. Although both empirical Equations (5) and (8) are optimized for the case studies, analytical Equation (14) shows the highest agreement with the current ASG installations in terms of the relative difference. Moreover, the analytical Equation (14)'s strong theoretical basis makes it presumably more reliable and also general.

According to Table 3, Equation (14) estimations have the lowest relative difference with the case studies represented in Table 2. In addition, Figure 9 indicates many of these estimations are in the range of ±5% of the relative difference. However, in several cases, the estimations fall beyond ±10% of the relative difference: according to this figure, the relative difference of Equation (14)'s estimation for Hannoversch Münden multi-ASG hydropower plant seems considerable. More investigations indicate that this ASG has a unique flood protection design that enables it to have a 0° to 28° adjustable inclination angle to operate under variable tailwater levels. Therefore, it is probable that the published data for this ASG is not under a conventional operation condition [33,49]. For the Radyr, the manufacturer published details are roughly [50]. For Gunthorpe Weir Hydro Scheme, Q_{95} is reported as 28.33, which is according to the published data used in Table 2. However, reviewing the Fisheries and Geomorphology Assessment documents indicates that each screw is designed for 10 m^3/s [53]. Using this flow rate and Equation (7) results in much less relative difference for this power plant (PE = 0.44%).

The unique design could be considered as one of the important reasons for the high relative differences between Equation (14) estimations for the Ham and Olen hydropower plants. These screws are installed on the Albert Channel in Ham and Olen in Belgium. These screw designs are unique not only because of their length (21.6 m) but also because they are compact, switchable Archimedes screws. These screws have a unique design to operate as ASPs (780 kW pumping operation) and ASGs (360 kW power generating operation) [56].

The highest relative difference of Equation (14) estimations is for the longest hybrid Archimedean screws in the world, installed on the Albert Canal in the Hasselt hydropower plant. With the remarkable length of 30 m, these ASGs have the most unique designs in the case studies. A hydraulically operated jacket valve is used to switch from pump and turbine mode. In ASP mode, each of these hybrid Archimedes screws can pump $5 \text{ m}^3/\text{s}$ over a head of more than 11 m (installed power 800 MW) during dry seasons and drain $5 \text{ m}^3/\text{s}$ over a head of 10 m to generate 400 MW of energy in wet seasons [27].

4. A Quick Design Guideline for Archimedes Screw Power Plants

In developing hydropower plants, the site and connection assessment expenses, geotechnical, electrical, and civil engineering costs, as well as licensing or approval fees, are among the fixed or semi-fixed soft costs. Many of these activities are carried out just as larger projects. Therefore, regardless of the amount of power that could be generated, these costs are almost the same and so become disproportionate for small projects. For example, these costs are almost the same for 7.2 kW and 150 kW ASG hydropower power plants [70]. However, there are situations that some or several costs could be reduced, at least for initial investigations. For example, when a small dam or weir exists (i.e., upgrade retrofit). Moreover, the results of the analysis of the current multi-ASG hydropower plants help to make some rough estimations. Here are some guidelines for rough estimations of Archimedes screw hydropower plants:

4.1. Determination of the ASG Configuration

- For the known available average flow rate and head, Equation (7) could be used for rough estimations about the possible amount of power that could be generated.
- If the estimated power is less than the requirements, site properties could be checked to evaluate the possibilities of the series configuration of ASGs. e.g., to take advantage of low flow rates but reasonable available heads, ASGs could be installed in series to deal with technical considerations of a very long ASG.
- To take advantage of high flow rates, instead of a very large (in diameter) and heavy ASG, it is possible to install ASGs in parallel. This approach could help to reduce the challenges of technical limitations and offers several advantages.
- For very low flow rates and heads, using industrial pico-ASGs available in the market may facilitate the process or even save some costs. For example, currently, pico-ASGs can generate up to 500 W with a flow rate as low as $0.1 \text{ m}^3/\text{s}$ and 0.7 m of the head (more information is available in [6]). Obviously, several units of such screws could be used in parallel or in series to take advantage of higher flow rates or available heads. For higher flow rates and heads, custom ASGs designs could be more efficient options. For example, Fletcher's Horse World Archimedes Screw can generate up to 7.2 kW using a design flow rate, head and outer diameter of $0.536 \text{ m}^3/\text{s}$, 1.7 m [71,72] and 1.39 m [73], respectively.

4.2. Estimation of Archimedes Screws Design Properties

1. Determine the site properties: the river's historical data or hydrograph for the volumetric flow rate (Q) and the site geometry to find the appropriate head (H) and the Archimedes screw inclination angle (β). Studies show that many ASGs are installed at $\beta = 22°$ [10].

2. Determine the maximum and minimum overall diameter of the screw (D_{max}, D_{min}) based on the site properties and the Archimedes screw hydro power plants design assessments proposed in Section 5-1 of [6].

3. Use Equation (1) to determine the screw(s) length.

4. Use the historical dataset of the river's flow rate to determine the flow duration curve (FDC). The probability that a system will take on a particular value or collection of values could be described by a mathematical expression which is known as a distribution function. The cumulative distribution function (CDF) of a variable for a value is the probability that this variable will take values less than or equal to this value [74]. Therefore, for a time series with n items, for item i with m items equal or bigger than it:

$$P_i = \frac{m}{n} \times 100 \tag{16}$$

In an FDC, the horizontal axis p_i represents the percentage of the period that the flow rate is more than or equal to its corresponding flow rate represented in the vertical axis. Usually, the flow exceeding for 95% of the period (Q_{95}) is considered as minimum river flow rate. It is important to note that considering the regulations, hydropower plants may need to bypass a portion of flow for aesthetic, ecological, environmental, or other purposes. This flow is called reserved, residual, compensation, prescribed or hands-off flow and should be deducted from the available flow rates.

5. Use the volume of flow rate that is provided on the flow duration curve (e.g., Q_{95}) and use Equation (14) to estimate the corresponding diameter of the screw for this flow rate (D_O).

6. Check the estimated diameter:
   - If $D_{min} \leq D_O \leq D_{MAX}$, go to step 7.
   - If $D_O < D_{min}$ use a higher volume of flow rate (e.g., Q_{90}) and repeat step 6.
   - If $D_O > D_{max}$, several approaches could be considered:
     - **Identical screws**: divide the volume of flow rate by i = 2 and follow the process from step 6. If it ends to $D_O > D_{MAX}$ again, repeat it for i = i + 1 until the condition passes. Then use the analytical Archimedes screw design method that is offered in [33] to design the screw. In this approach, "i" Archimedes screw generators with the same geometry will be designed to handle this flow rate. The advantage of this approach is that similar screws are easier to build, operate and maintain.
     - **Design based on the maximum diameter**: design the screw for $D_O = D_{max}$ and use the analytical Archimedes screw design method that is offered in [33] to design the Archimedes screw generator. Then use Equation (15) to estimate the volumetric flow that passes through this screw and design the next Archimedes screw by following the process from step 6 for the remaining volume of flow rate. This approach could lead to reducing the number of Archimedes screws.
     - **Trial and error**: consider a higher probability that means a lower flow rate (e.g., $Q_{97.5}$ instead of Q_{95}) and do and follow the process from step 6 and perform a trial and error. This approach could lead to an increase in the design of Archimedes screw turbines with higher reliability in generating power.

7. To utilize the remaining available volumetric flow rate, more Archimedes screws could be designed (parallel ASG power plants). The next flow rate to design the screw could be selected from the FDC based on the desired step size (Δ). For example, for the previous design flow rate of Q_n choose the next flow rate $Q_{n-\Delta}$ which is the flow exceeding for n $-$ Δ % of the period and continue the design starting from step 5.

8. The design process should be halted based on logical constraints. For example, for economic reasons designing screws for volumetric flow rates less than the certain probability (e.g., Q_{limit}) is not reasonable. Or, due to site limitations, there may be some restrictions such as the total area of the power plant (the minimum required area to install the Archimedes screw generators is equal to the sum of the diameter times to the length of each screw. If it goes beyond the installation site limitations, some of the screws designed for flow rates with lower probabilities could be cancelled. In such conditions, an optimized larger screw for the flow rates with the highest probabilities or using variable speed screws could be considered as alternative solutions to utilize more flow rate).

9. Use Equation (7) to estimate the possible amount of power that each ASG can generate. Then, estimate the ideal overall power that the Archimedes screw hydropower plant could generate.

$$P_{Overall} = \sum p_i P_i \tag{17}$$

Figure 10 summarizes this guideline as a flow chart.

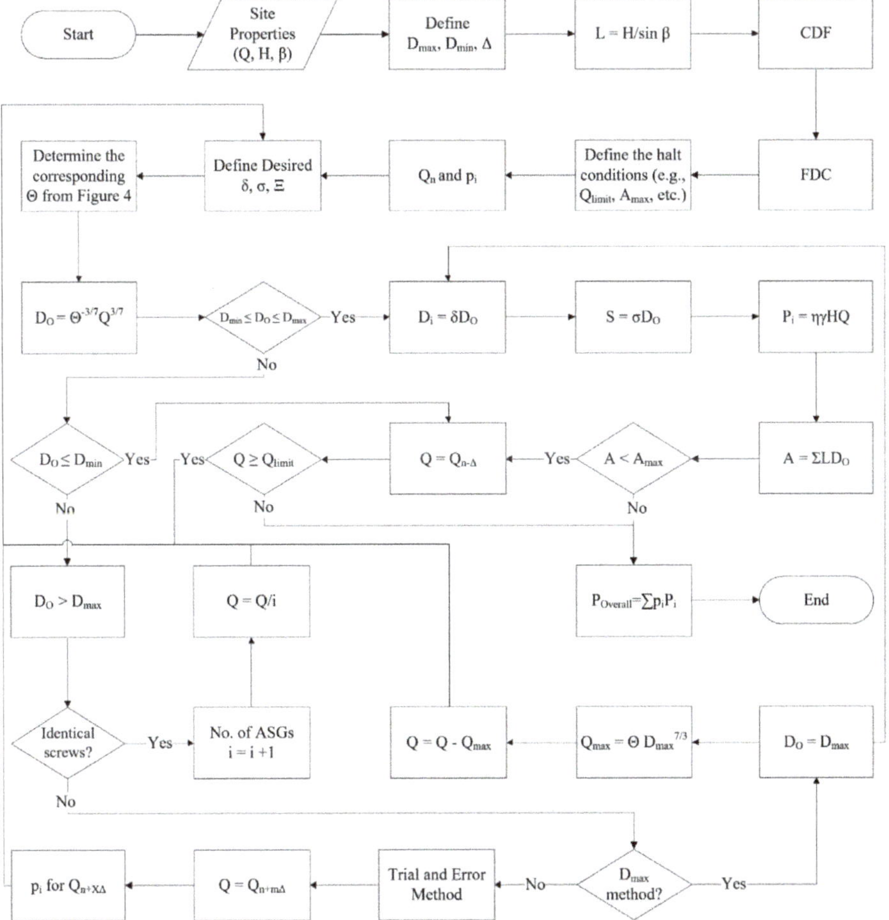

Figure 10. The guideline for quick estimations of Archimedes screw power plants design.

5. Conclusions

This study focuses on offering a simple method to address the important lack of a general and easy-to-use guideline for designing Archimedes screw power plants using more than one ASG. Then, a detailed design study should be undertaken on the ASGs suggested by this process's estimations. In addition, this study helps to address one of the significant burdens of small projects, the unscalable initial investigation costs.

Moreover, this study updates the records of the currently installed Archimedes screw generators and provides a better understanding of the advancements and the current possibilities in this technology. It also provides a list of currently installed and operating industrial multi-Archimedes screw hydropower plants in order to review and explore the common design properties between different manufacturers. Therefore, several empirical equations are proposed and optimized using the genetic algorithm and generalized reduced gradient. Moreover, this study provides an updated version of a new analytical method for designing Archimedes screws and offers a new graph that not only supports standard designs but also accelerates and simplifies custom designs. The evaluation of these equations with the list of industrial ASG installations provided indicates they have reasonable accuracy.

The proposed method enables everyone to evaluate the possibilities of green and renewable Archimedes screw hydropower generation where a flow is available and helps to make many potential sites feasible to study for this.

Author Contributions: Conceptualization, A.Y. and W.D.L.; methodology, A.Y. and W.D.L.; software, A.Y.; validation, A.Y. and W.D.L.; formal analysis, A.Y. and W.D.L.; investigation, A.Y. and W.D.L.; resources, A.Y.; data curation, A.Y.; writing—original draft preparation, A.Y.; writing—review and editing, A.Y. and W.D.L.; visualization, A.Y.; supervision, W.D.L.; project administration, W.D.L.; funding acquisition, W.D.L. All authors have read and agreed to the published version of the manuscript.

Funding: This work is part of a larger long-term research program that has been financially supported by the Natural Sciences and Engineering Research Council (NSERC) of Canada, Collaborative Research and Development (CRD) program (grant CRDPJ 513923-17) and Greenbug Energy Inc. (Delhi, ON, Canada).

Data Availability Statement: All data used in this study is reported and referenced in this article.

Conflicts of Interest: The authors declare no conflict of interest.

Abbreviations

The following symbols are used in this paper:

Notation

A_E	Effective cross-sectional area at the screw's inlet	(m^2)
D_i	The inner diameter of the Archimedes screw	(m)
D_O	The outer diameter of the Archimedes screw	(m)
E_i	The estimated value	
$\bar{E}$	The average of the estimations	
h_u	Upper (inlet) water level of the screw	(m)
h_L	Lower (outlet) water level of the screw	(m)
H	The available head	(m)
G_w	Gap width (The gap between the trough and screw)	(m)
L	The total length of the screw	(m)
$MAPE$	The mean absolute percentage error	$(\%)$
n	The number of data points in the dataset	
N	Number of helical planed surfaces	$(-)$
O_i	The observed value	
$\bar{O}$	The average of the observed data	

PE	The percentage (percent) error	(%)
Q	Total flow rate passing through the screw	(m^3/s)
r	Radius	(m)
R	Pearson correlation	(%)
S	Pitch of the screw (Distance along the screw axis for one complete helical plane turn)	(m)
y_i	The fill height of the inner diameter of the screw at the inlet	(m)
y_O	The fill height of the screw at the inlet	(m)
Z_u	The free surface elevations at the upstream	(m)
Z_L	The free surface elevations at the downstream	(m)
β	The inclination angle of the screw	(rad)
γ	The specific weight of water	(N/m^3)
δ	The screw's pitch to outer diameter ratio (S/D$_O$)	(-)
η	The average efficiency of the ASGs based on manufacturer's specifications in Table 2 ($\eta = 73.6\%$)	
Δ	Step size	(%)
Θ	A constant accounting for screw geometry and fill level	(m$^{2/3}$s^{-1})
θ	Angle of sector	(rad)
λ	The constant value in the power function form	
σ	The screw's inner to outer diameter ratio (D$_i$/D$_O$)	(-)
Ξ	The dimensionless inlet depth of the screw	(-)
ψ	The value of power in the power function form of diameter equation	(-)
ω	The rotation speed of the screw	(rad/s)
ω_M	The maximum rotation speed of the screw (Muysken limit)	(rad/s)
Subscripts		
i	inner	
min	minimum	
Max	Maximum	
O	Outer	

References

1. Williamson, S.J.J.; Stark, B.H.H.; Booker, J.D.D. Low head pico hydro turbine selection using a multi-criteria analysis. *RENE* **2014**, *61*, 43–50. [CrossRef]
2. Date, A.; Akbarzadeh, A. Design and cost analysis of low head simple reaction hydro turbine for remote area power supply. *Renew. Energy* **2009**, *34*, 409–415. [CrossRef]
3. Casini, M. Harvesting energy from in-pipe hydro systems at urban and building scale. *Int. J. Smart Grid Clean Energy* **2015**, *4*, 316–327. [CrossRef]
4. IRENA. *Renewable Energy Techlogies: Cost Analysis Series, Hydropower*; IRENA Innovation and Technology Centrer: Bonn, Germany, 2012; Volume 1.
5. Suh, S.H.; Kim, K.Y.; Kim, B.H.; Kim, Y.T.; Kim, T.G.; Roh, H.W.; Yoo, Y.I.; Park, N.H.; Park, J.M.; Shin, C.S.; et al. *Theory and Applications of Hydraulic Turbines*, 1st ed.; Dong Myeong Publishers: Paju, Korea, 2014.
6. YoosefDoost, A.; Lubitz, W.D. Archimedes screw turbines: A sustainable development solution for green and renewable energy generation-a review of potential and design procedures. *Sustainability* **2020**, *12*, 7352. [CrossRef]
7. Muller, G.; Senior, J. Simplified theory of Archimedean screws. *J. Hydraul. Res.* **2009**, *47*, 666–669. [CrossRef]
8. Simmons, S.; Lubitz, W. Archimedes screw generators for sustainable energy development. In Proceedings of the 2017 IEEE Canada International Humanitarian Technology Conference (IHTC), Toronto, ON, Canada, 21–22 July 2017; pp. 144–148. [CrossRef]
9. Moerscher, W. Water-power system. U.S. Patent US1434138A, 31 October 1922.
10. Lashofer, A.; Hawle, W.; Pelikan, B. State of technology and design guidelines for the Archimedes screw turbine. In Proceedings of the Hydro 2012—Innovative Approaches to Global Challenges, Bilbao, Spain, 29–31 October 2012.
11. Vandezande BVBA. Lock of Hasselt—Hybrid hydropower screw/screw pump. Youtube. 2018. Available online: https://youtu.be/FUlYjkzAIs8 (accessed on 8 August 2021).
12. Kibel, P. Fish Monitoring and Live Fish Trials. Archimedes Screw Turbine, River Dart. In *Phase 1 Report: Live Fish Trials, Smolts, Leading Edge Assessment, Disorientation Study, Outflow Monitoring*; Moretonhampstead: Devon, UK, 2007.
13. Boys, C.A.; Pflugrath, B.D.; Mueller, M.; Pander, J.; Deng, Z.D.; Geist, J. Physical and hydraulic forces experienced by fish passing through three different low-head hydropower turbines. *Mar. Freshw. Res.* **2018**, *69*, 1934–1944. [CrossRef]

14. McNabb, C.D.; Liston, C.R.; Borthwick, S.M. Passage of Juvenile Chinook Salmon and other Fish Species through Archimedes Lifts and a Hidrostal Pump at Red Bluff, California. *Trans. Am. Fish. Soc.* **2003**, *132*, 326–334. [CrossRef]

15. Kibel, P.; Pike, R.; Coe, T. Archimedes Screw Turbine Fisheries Assessment. In *Phase II: Eels and Kelts*; Moretonhampstead: Devon, UK, 2008.

16. Kibel, P.; Pike, R.; Coe, T. *The Archimedes Screw Turbine: Assessment of Three Leading Edge Profiles*; Fishtek Consulting Ltd.: Moretonhampstead, UK, 2009.

17. United Kingdom Environment Agency. Hydropower Good Practice Guidelines Screening requirements. *J. Hydraul. Res.* **2012**, *4*, 1–16.

18. Piper, A.T.; Rosewarne, P.J.; Wright, R.M.; Kemp, P.S. The impact of an Archimedes screw hydropower turbine on fish migration in a lowland river. *Ecol. Eng.* **2018**, *118*, 31–42. [CrossRef]

19. Pauwels, I.S.; Baeyens, R.; Toming, G.; Schneider, M.; Buysse, D.; Coeck, J.; Tuhtan, J.A. Multi-species assessment of injury, mortality, and physical conditions during downstream passage through a large archimedes hydrodynamic screw (Albert canal, Belgium). *Sustainability* **2020**, *12*, 8722. [CrossRef]

20. Durrani, A.M.; Mujahid, O.; Uzair, M. Micro Hydro Power Plant using Sewage Water of Hayatabad Peshawar. In Proceedings of the 2019 15th International Conference on Emerging Technologies (ICET), Peshawar, Pakistan, 2–3 December 2019.

21. Spaans Babcock Ltd. Hydro Power Using Waste Water at Esholt WwTW. 2009. Available online: https://cms.esi.info/Media/documents/54053_1316689157666.pdf (accessed on 29 July 2020).

22. Kozyn, A.; Ash, S.; Lubitz, W.D. Assessment of Archimedes Screw Power Generation Potential in Ontario. In Proceedings of the 4th Climate Change Technology Conference, Montreal, QC, Canada, 25–27 May 2015; pp. 1–11. Available online: https://www.cctc2015.ca/TECHNICALPAPERS/1570095585.pdf (accessed on 3 June 2019).

23. Shahverdi, K.; Loni, R.; Maestre, J.M.; Najafi, G. CFD numerical simulation of Archimedes screw turbine with power output analysis. *Ocean. Eng.* **2021**, *231*, 108718. [CrossRef]

24. UNIDO. *Renewable Energy: Micro Hydraulic Power Unit (Spiral Type Pico-Hydro Unit 'PicoPica10', 'PicoPica500')*; United Nations Industrial Development Organization: Tokyo, Japan, 2020. Available online: http://www.unido.or.jp/en/technology_db/5276/ (accessed on 7 July 2020).

25. Sumino, M. Ultra-Small Water Power Generator. 14 May 2019. Available online: https://youtu.be/XjEgFlngZ04 (accessed on 23 June 2020).

26. Sumino Co. Ltd. Japan. *Spiral Hydraulic Power Unit 'PicoPica'*; Ena: Gifu, Japan, 2020. Available online: http://www.unido.or.jp/files/sites/2/PICOPICA_Specification201127.pdf (accessed on 28 July 2021).

27. Vandezande Diksmuide. PS/WKC Lock Hasselt. Available online: https://web.archive.org/web/20210405220637/https://www.vandezande.com/en/projects/ps-wkc-lock-hasselt (accessed on 9 August 2021).

28. Rorres, C. *Archimedes in the 21st Century: Proceedings of a World Conference at the Courant Institute of Mathematical Sciences*; Springer: Cham, Switzerland, 2017.

29. Brada, K.; Radlik, K.-A. Wasserkraftschnecke: Eigenschaften und Verwendung. In Proceedings of the Heat Exchange and Renewable Energy Sources International Symposium, Szczecin, Poland, 17–19 June 1996; pp. 43–52.

30. OMNR. *Potential Waterpower Generation Sites (WPPOTSTE); Ontario Ministry of Natural Resources*. 2004. Available online: https://www.ontario.ca/data/potential-waterpower-generation-sites-wppotste (accessed on 10 January 2015).

31. Lubitz, W.D.; Lyons, M.; Simmons, S. Performance Model of Archimedes Screw Hydro Turbines with Variable Fill Level. *J. Hydraul. Eng.* **2014**, *140*, 04014050. [CrossRef]

32. Dragomirescu, A. Design considerations for an Archimedean screw hydro turbine. In Proceedings of the IOP Conference Series: Earth and Environmental Science, Bucharest, Romania, 21–24 October 2021; Volume 664, p. 12034. [CrossRef]

33. Yoosefdoost, A.; Lubitz, W.D. Archimedes Screw Design: An Analytical Model for Rapid Estimation of Archimedes Screw Geometry. *Energies* **2021**, *14*, 7812. [CrossRef]

34. Nuernbergk, D.M.; Rorres, C. Analytical Model for Water Inflow of an Archimedes Screw Used in Hydropower Generation. *J. Hydraul. Eng.* **2013**, *139*, 213–220. [CrossRef]

35. Aigner, D. *Current Research in Hydraulic Engineering 1993–2008*; Institut für Wasserbau und Technisch Hydromechanik der TU; Association: Dresden, Germany, 2008.

36. Schmalz, W. Studies on fish migration and control of possible fish loss caused by the hydrodynamic screw and hydropower plant. In *Fischo—Kologische und Limnol. Untersuchungsstelle Sudthurign, Rep.*; Thüringer Landesanstalt für Umwelt und Geol: Jena, Germany, 2010.

37. Lashofer, A.; Kaltenberger, F.; Pelikan, B. Does the Archimedean screw turbine stand the test? (Wie gut bewährt sich die Wasserkraftschnecke in der Praxis?). *WasserWirtschaft* **2011**, *101*, 76–81. [CrossRef]

38. Nuernbergk, D.M. *Wasserkraftschnecken—Berechnung und Optimaler Entwurf von Archimedischen Schnecken als Wasserkraftmaschine (Hydro-Power Screws—Calculation and Design of Archimedes Screws)*, 2nd ed.; Verlag Moritz Schäfer: Detmold, Germany, 2020.

39. Rorres, C. The Turn of the Screw: Optimal Design of an Archimedes Screw. *J. Hydraul. Eng.* **2000**, *126*, 72–80. [CrossRef]

40. YoosefDoost, A.; Lubitz, W.D. Development of an Equation for the Volume of Flow Passing Through an Archimedes Screw Turbine. In *Sustaining Tomorrow*; Ting, D.S.-K., Vasel-Be-Hagh, A., Eds.; Springer: Cham, Switzerland, 2021; pp. 17–37.

41. Dellinger, G.; Simmons, S.; Lubitz, W.D.; Garambois, P.A.; Dellinger, N. Effect of slope and number of blades on Archimedes screw generator power output. *Renew. Energy* **2019**, *136*, 896–908. [CrossRef]

42. Simmons, S.C.; Elliott, C.; Ford, M.; Clayton, A.; Lubitz, W.D. Archimedes screw generator powerplant assessment and field measurement campaign. *Energy Sustain. Dev.* **2021**, *65*, 144–161. [CrossRef]
43. Lyons, M. Lab Testing and Modeling of Archimedes Screw Turbines. Master's Thesis, University of Guelph, Guelph, ON, Canada, 2014.
44. Rosly, C.Z.; Jamaludin, U.K.; Azahari, N.S.; Mu'tasim, M.A.N.; Oumer, A.N.; Rao, N.T. Parametric study on efficiency of archimedes screw turbine. *ARPN J. Eng. Appl. Sci.* **2016**, *11*, 10904–10908.
45. Songin, K. Experimental Analysis of Archimedes Screw Turbines. Master's Thesis, University of Guelph, Guelph, Canada, 2017.
46. Nagel, G. *Archimedean Screw Pump Handbook*; RITZ-Pumpenfabrik OHG: Schwabisch Gmund, Germany, 1968.
47. Rose, R. Linton Falls and Low Wood Hydropower Schemes. No. UK Water Projects 2011. River Coast. 2011, pp. 197–202. Available online: https://web.archive.org/web/20190621110557/http://www.waterprojectsonline.com/case_studies/2011/Hydropower_Linton_Fall_2011s.pdf (accessed on 1 November 2021).
48. Landustrie Sneek, B.V. Totnes Weir (UK). Landustrie Worldwide Water Technology. 2015. Available online: https://web.archive.org/web/20210803231005/https://www.landustrie.nl/fileadmin/user_upload/Totnes_Times_November_2015.pdf (accessed on 3 August 2021).
49. Rehart Power. Hannoversch Münden CS. Available online: https://web.archive.org/web/20211115031954/https://www.rehart-power.de/en/reference-projects/hydropower-screw-type-cs/hannoversch-muenden-cs.html (accessed on 14 November 2021).
50. RenewablesFirst. Radyr Weir Hydro Turbines. 2015. Available online: https://web.archive.org/web/20210804003006/https://www.renewablesfirst.co.uk/project-blog/radyr-weir-hydro-scheme/ (accessed on 5 July 2020).
51. Ingenieurbüro Lashofer. Hydropower screws in Europe. Google Maps. Available online: Efort.info/AST-Map (accessed on 3 August 2021).
52. Vandezande Diksmuide. Hydropower Screws Höllthal. Vandezande. Available online: https://web.archive.org/web/20210809061648/https://www.vandezande.com/en/projects/hydropower-screws-höllthal (accessed on 9 August 2021).
53. Gratton, P.; Meadows, T.; Brook, T. *Gunthorpe Weir Hydropower Scheme: Fisheries and Geomorphology Assessment*; The Mill: Stroud, UK, 2019.
54. SinFin. *Solvay Industrial Plant*; SinFin Energy: Gijón, Spain, 2019. Available online: http://www.sinfinenergy.com/en/projects/solvay/ (accessed on 8 July 2020).
55. Landustrie. Linton Lock. Landustrie Sneek BV. 2017. Available online: https://web.archive.org/web/20210804020631/https://www.landustrie.nl/en/products/hydropower/projects/linton-lock.html (accessed on 29 July 2020).
56. Vandezande BVBA. Vandezande Specialist in Mechanics. Vandezande.com. Zeepziederijstraat, Brugge, Belgium. Available online: https://web.archive.org/web/20210809031945/https://www.vandezande.com/sites/default/files/vandezande-folder2017-engLR04.pdf (accessed on 8 August 2021).
57. Melbud, S.A. A Small Hydropower Plant–Rosko. Available online: https://web.archive.org/web/20210804011025/http://www.melbud.pl/language/pl/mala-elektrownia-wodna-rosko/ (accessed on 3 August 2021).
58. Fergnani, N. Hydroelectric Plants Energy Efficiency. Hydrosmart Srl. 2020. Available online: https://www.hydrosmart.it/energia-rinnovabile (accessed on 2 August 2020).
59. Sto98. Marengo Hydropower Plant-Goito [Centrale Idroelettrica Marengo—Goito]. YouTube. 2015. Available online: https://youtu.be/19px1EKa--4 (accessed on 19 July 2020).
60. Rodgers, J.L.; Nicewander, W.A. Thirteen Ways to Look at the Correlation Coefficient. *Am. Stat.* **1988**, *42*, 59–66. [CrossRef]
61. Yoosefdoost, A.; Yoosefdoost, I.; Asghari, H.; Sadeghian, M.S. Comparison of HadCM3, CSIRO Mk3 and GFDL CM2. 1 in Prediction the Climate Change in Taleghan River Basin. *Am. J. Civ. Eng. Archit.* **2018**, *6*, 93–100. [CrossRef]
62. Adler, J.; Parmryd, I. Quantifying colocalization by correlation: The pearson correlation coefficient is superior to the Mander's overlap coefficient. *Cytom. Part A* **2010**, *77*, 733–742. [CrossRef]
63. Kim, S.; Kim, H. A new metric of absolute percentage error for intermittent demand forecasts. *Int. J. Forecast.* **2016**, *32*, 669–679. [CrossRef]
64. Hanke, J.E.; Wichern, D. *Business Forecasting*, 9th ed.; Prentice Hall: London, UK, 2009.
65. Bowerman, B.L.; O'Connell, R.T.; Koehler, A.B. *Forecasting, Time Series, and Regression: An Applied Approach*, 4th ed.; Thomson Brooks/Cole: Belmont, CA, USA, 2005.
66. Hawle, W.; Lashofer, A.; Pelikan, B. Lab Testing of the Archimedean Screw. In Proceedings of the Hidroenergia Conference, Wroclaw, Poland, 23–26 May 2012.
67. Lasdon, L.S.; Waren, A.D.; Jain, A.; Ratner, M. Design and Testing of a Generalized Reduced Gradient Code for Nonlinear Programming. *ACM Trans. Math. Softw.* **1978**, *4*, 34–50. [CrossRef]
68. Waleed, D.; Alrabadi, H. Portfolio optimization using the generalized reduced gradient nonlinear algorithm An application to Amman Stock Exchange. *Int. J. Islam. Middle East. Financ. Manag.* **2016**, *9*, 570–582. [CrossRef]
69. Muysken, J. Calculation of the Effectiveness of the Auger. *De Ingenieur* **1932**, *21*, 77–91.
70. GreenBug Energy. Archimedes Screw Generators. GreenBug Energy. Available online: https://web.archive.org/web/20210128050704/https://greenbugenergy.com/how-we-do-it/archimedes-screw-generators (accessed on 6 August 2021).
71. Heron, L. Fletcher's Horse World—Nanticoke Creek—Archimedes Screw—7.2 kW. Ontario Rivers Alliance. 2014. Available online: https://web.archive.org/web/20210806171348/https://www.ontarioriversalliance.ca/fletchers-horse-world-nanticoke-creek-archimedes-screw-7-2-kw/ (accessed on 6 August 2021).

72. GreenBug Energy. *Fletchers Horse World*; GreenBug Energy Inc.: Delhi, ON, Canada. Available online: https://greenbugenergy.com/projects/fletchers-horse-world (accessed on 6 August 2021).
73. Passamonti, A. *Investigation of Energy Losses in Laboratory and Full-Scale Archimedes Screw Generators Advisor*; Politecnico Di Milano: Milan, Italy, 2017.
74. Deisenroth, M.P.; Faisal, A.A.; Ong, C.S. *Mathematics for Machine Learning*; Cambridge University Press: Cambridge, UK, 2020.

 *processes*

Article

Numerical Investigation of the Performance, Hydrodynamics, and Free-Surface Effects in Unsteady Flow of a Horizontal Axis Hydrokinetic Turbines

Aldo Benavides-Morán [1],*, Luis Rodríguez-Jaime [1] and Santiago Laín [2]

[1] Grupo de Modelado y Métodos Numéricos en Ingeniería, Departamento de Ingeniería Mecánica y Mecatrónica, Facultad de Ingeniería, Universidad Nacional de Colombia, Sede Bogotá, Carrera 30 No 45A-03, Edificio 453, Bogotá 111321, Colombia; lerodriguezj@unal.edu.co
[2] PAI+Research Group, Department of Energy and Mechanics, Universidad Autonoma de Occidente, Cali 760030, Colombia; slain@uao.edu.co
* Correspondence: agbenavidesm@unal.edu.co

Abstract: This paper presents computational fluid dynamics (CFD) simulations of the flow around a horizontal axis hydrokinetic turbine (HAHT) found in the literature. The volume of fluid (VOF) model implemented in a commercial CFD package (ANSYS-Fluent) is used to track the air-water interface. The URANS SST $k\text{-}\omega$ and the four-equation Transition SST turbulence models are employed to compute the unsteady three-dimensional flow field. The sliding mesh technique is used to rotate the subdomain that includes the turbine rotor. The effect of grid resolution, time-step size, and turbulence model on the computed performance coefficients is analyzed in detail, and the results are compared against experimental data at various tip speed ratios (TSRs). Simulation results at the analyzed rotor immersions confirm that the power and thrust coefficients decrease when the rotor is closer to the free surface. The combined effect of rotor and support structure on the free surface evolution and downstream velocities is also studied. The results show that a maximum velocity deficit is found in the near wake region above the rotor centerline. A slow wake recovery is also observed at the shallow rotor immersion due to the free-surface proximity, which in turn reduces the power extraction.

Keywords: computational fluid dynamics; volume of fluid; transition SST $k\text{-}\omega$ turbulence model; sliding mesh; wake

Citation: Benavides-Morán, A.; Rodríguez-Jaime, L.; Laín, S. Numerical Investigation of the Performance, Hydrodynamics, and Free-Surface Effects in Unsteady Flow of a Horizontal Axis Hydrokinetic Turbines. *Processes* **2022**, *10*, 69. https://doi.org/10.3390/pr10010069

Academic Editor: Alfredo Iranzo

Received: 29 November 2021
Accepted: 14 December 2021
Published: 30 December 2021

Publisher's Note: MDPI stays neutral with regard to jurisdictional claims in published maps and institutional affiliations.

1. Introduction

Hydraulic energy sources around the world are primarily used on large hydroelectric plants which provide electricity to densely populated areas. Colombia is rich in water resources with large hydropower schemes covering roughly 79% of the energy generation (54,532 GWh, data of 2019), while the remaining 21% mainly comes from coal and gas sources which both come at a high environmental cost [1]. The hydropower schemes are generally conventional hydroelectric dams which use a reservoir and dam set-up, and also run-of-the-river hydroelectricity which have no reservoir, similar to the world-renowned Hoover dam on the border between the states of Nevada and Arizona, USA [2]. The construction of conventional hydropower schemes poses a risk to the environment and nearby communities. The under-construction 2.4 GW Ituango hydroelectric project will be the largest hydropower plant in Colombia. One of Ituango's auxiliary diversion tunnels collapsed in 2018; the water had to be diverted to prevent the overflow of the dam and spillway. The resulting high-speed flow in the adduction tunnels produced rock instabilities that compromised the integrity of the whole structure [3]. The population of Colombia is expected to grow by 12% in 2050 [4], which will further increase the challenges in supplying sustainable renewable energy to the entire country and in reducing greenhouse gas emissions.

In Colombia, energy supply issues predominately affect isolated regions located away from the main electricity supply network. Alternative methods of energy generation from renewable resources offer a promising solution to tackle social inequality as well as reducing environmental impacts, such as air and water pollution and climate change due to the carbon emissions, and they would also reduce the local dependency on fossil fuels. The full potential of all types of renewable resources needs to be fully explored and integrated with novel solutions, such as hydrokinetic turbines. This technology does not require the use of large areas of land, contrary to conventional hydropower, while also being a significantly cheaper alternative due to the portability of hydrokinetic turbines as single assembled units, which facilitates deployment and maintenance [5]. To date, horizontal axis turbines have been preferred for large rivers due to their mature development owing to the wind industry. However, these devices are restricted to high-velocity regions and deep waters which are not always available in a riverine environment. It is therefore necessary to study the effect of rotor depth on the overall performance of horizontal-axis hydrokinetic turbines (HAHT), both experimentally and computationally. The work presented in this paper investigates the free-surface-turbine-wake interaction using computational fluid dynamics (CFD).

The most common numerical approaches to predict the rotor performance of HAHTs are borrowed from wind energy engineering. The BEM (blade-element momentum) theory, also referred to as strip theory, combines the balance of linear and angular momentum with the aerodynamic forces exerted on a blade section. BEM theory requires the knowledge of aerodynamic coefficients and the calculations of the axial and angular induction factors at the blade section. This information is used to integrate the balance equations along the blade length to estimate the total power from the rotor at the specified angular and flow velocities [6,7]. BEM theory can be used to estimate the performance of a HAHT when the rotor is well submerged into the water, so the rotor and free surface interaction can be neglected. Batten et al. [8] compared BEM results of thrust and power coefficients at various tip-speed ratios (TSR) with the experimental data of a laboratory-scale horizontal axis tidal turbine tested in a cavitation tunnel. Simulation results showed good agreement for power coefficient in the TSR range from 3 to 7; BEM underpredicted the thrust coefficient for the analyzed pitch angles. Similarly, Danao et al. [9] performed calculations with the open-source blade element momentum solver Qblade; and a good agreement was found for the power coefficient.

The actuator disk method (ADM) and the actuator line method (ALM) are two novel simulation techniques commonly used to simulate horizontal axis wind turbine rotors. In the ADM, the rotor is substituted by a non-rotating disk with the same area swept by the blades. A body-force term in the disk region is defined to account for the extraction of fluid momentum by the rotor [10]. Contrary to ALM, ADM is not able to reproduce root and tip vortices. i.e. the actuator disk just accounts for large-scale effects on the flow. Silva dos Santos et al. [11] used the ADM implemented in the commercial CFD software ANSYS-Fluent to simulate the wake characteristics of a hydrokinetic farm on a portion of the Brazilian Amazon river. In ALM, the complex geometry of rotor blades is simplified to a set of rotating lines. Distributed point forces are defined along those lines (actuator lines) to iteratively compute the loadings using BEM and available airfoil data with submodels for dynamic stall situations [12,13]. Next, the forces are projected onto the computational domain to account for the effect of rotating blades on the flow field. ALM avoids the computation of the blade boundary layer and the implementation of rotating meshes. The ALM technique has been successfully used to estimate the rotor performance and the wake behind wind turbines [14]. ALM combined with LES (large-eddy simulation) was used by Baba-Ahmadi and Dong [15] to simulate the flow through a horizontal axis tidal stream turbine. Axial velocity and turbulence intensity profiles behind the turbine at the rotor axis height showed good agreement with experimental results, although the velocity deficit was slightly overpredicted in the far wake. This study did not consider free-surface effects, i.e., it was a single-phase flow computation.

The number of investigations seeking to clarify the effects of the free surface on HAHT performance and wake dynamics is still limited. A complete CFD model of HAHT must include a fully resolved blade geometry and the definition of a rotating region that encloses the rotor in the computational domain. Tracking the interface requires the use of a multiphase framework, which further increases the computational cost. Disregarding the free surface in a computational approach can lead to significant differences in the predicted turbine performance. For a Darrieus turbine, Hocine et al. [16] reported an increment of 42.4% in power output and a 26.6% higher thrust when the free surface was neglected.

Previous computational and experimental studies have shown a decrease in thrust and power output by bringing the turbine closer to the free surface. Bahaj et al. [17] experimentally reported a reduction in power and thrust of approximately 8.8% and 4.5%, respectively, by taking the rotor to a blade tip immersion from $0.55D$ to $0.19D$ (D is the rotor diameter). Yan et al. [18] implemented a computational free-surface flow framework, with a novel level-set re-distancing technique and sliding interface, to perform three-dimensional, time-dependent simulations of a horizontal-axis tidal turbine. A turbulence model was not incorporated in their numerical approach. Results of power of thrust coefficients suggested the existence of a minimum immersion depth for optimal turbine operation. Conversely, Kolekar et al. [19] reported an increase in the power coefficient by reducing the rotor immersion, indicating a favorable depth region for the performance near a $0.27D$ immersion of the blade tip; they also observed significant changes in the wake dynamics, especially at shallow immersions in which the wake becomes asymmetric. Nishi et al. [20] showed an unusual increase in performance of up to 2.8 times when implementing a multiphase approach (water-air), compared to single-phase flow results (water only); the turbine output from the multiphase flow simulation is approximately 5% higher than the experimental data. Bai et al. [21] used the immersed boundary method and free surface method implemented in an in-house CFD code to simulate the turbine rotor of Bahaj et al. [17]. The computed power coefficient lay below the experimental data; a finer grid was shown to reduce the discrepancy at the TSR for maximum power output. Adamski [22] reported the appearance of large ripples (deformations) on the free surface when the distance to the rotor is decreased; the vertical deformation is approximately twice as much when the blade tip immersion is reduced from 1D to 0.75D.

Rodriguez et al. [23] performed two-phase flow simulations of the HAHT rotor described in Bahaj et al. [17], for two blade tip immersions and three TSRs. The computational approach incorporated a transition four-equation turbulence model, the Volume of Fluid (VOF) method for interface tracking, and the sliding-mesh technique to account for blade rotation. There was a good qualitative agreement between the simulation results of torque and thrust coefficients and the experimental data taken from Bahaj et al., at the analyzed TSRs. Their conclusions highlighted, however, the need to further assess the impact of model parameters such as mesh resolution and domain size, and the incorporation of the turbine's support structure.

The present study aims to extend the current understanding of the interaction between the free surface and a HAHT, including the rotor and support structure. The numerical approach adopts the four equation Transition SST turbulence model and the Volume of Fluid (VOF) method to compute the unsteady three-dimensional flow field. This paper analyzes the combined effect of rotor and support structure on the performance, the free surface evolution, and wake dynamics for deep and shallow rotor immersions.

The rest of the paper is organized as follows. Section 2 provides details about the HAHT computational model and its implementation, a description of the numerical approach, and the simulations performed. The simulation results are presented in Section 3 along with an analysis on their agreement with performance coefficients reported in the literature; the combined effects of the rotor and support structure on the performance and flow field are thoroughly discussed. Finally, Section 4 concludes the paper.

2. Methods

2.1. Governing Equations

The water flow around a HAHT is modeled using RANS (Reynolds-Averaged Navier-Stokes) equations for viscous, isothermal, unsteady, turbulent, three-dimensional, and incompressible fluid flow in rectangular coordinates. In addition, the well-established multiphase flow model known as VOF is used to track the air-water interface. VOF introduces an additional conservation property in the RANS equations, known as volume fraction, α. This conservation property represents how much of a computational cell is occupied by each phase [24]. In the case of a two-phase flow,

$$\alpha_w + \alpha_a = 1 \tag{1}$$

where the subscript w and a stand for water and air, respectively. In this work, water is arbitrarily chosen as the primary phase.

ANSYS-Fluent 19.0 uses the finite volume method to solve the fluid flow equation set. The continuity and momentum equations written in tensor notation are, respectively

$$\frac{\partial}{\partial t}(\alpha_w) + \frac{\partial}{\partial x_j}(\alpha_w \bar{v}_j) = 0, \tag{2}$$

$$\frac{\partial}{\partial t}(\rho \bar{v}_i) + \frac{\partial}{\partial x_j}(\rho \bar{v}_i \bar{v}_j) = \frac{\partial P}{\partial x_i} + \frac{\partial}{\partial x_j}\left[\mu\left(\frac{\partial \bar{v}_i}{\partial x_j} + \frac{\partial \bar{v}_j}{\partial x_i}\right) - \rho\overline{v'_i v'_j}\right] + \rho g_i, \tag{3}$$

where $\bar{v}$ denotes the mean velocity shared by both phases, P is the mean pressure, and g is the gravitational acceleration. The Reynolds stresses $\rho\overline{v'_i v'_j}$ are computed by means of the Boussinesq assumption for isotropic turbulence. More details on the turbulence model are given below. The viscosity μ and density ρ in Equation (3) are given by,

$$\mu = \alpha_w \mu_w + \alpha_a \mu_a, \tag{4}$$

$$\rho = \alpha_w \rho_w + \alpha_a \rho_a. \tag{5}$$

The well-established shear stress transport $k\text{-}\omega$ (SST) turbulence model [25] was initially adopted to approximate the turbulent flow. This two-equation eddy viscosity model combines the accuracy and robustness of the $k\text{-}\omega$ model in the near-wall region and the free-stream insensitivity of the $k\text{-}\epsilon$ model. The switch between turbulence models is accomplished by a blending function. The $k\text{-}\omega$ SST model is appropriate to compute flows with adverse pressure gradients and boundary layer separation which are flow phenomena commonly found in turbomachinery applications. Contreras et al. [26] showed that the $k\text{-}\omega$ SST turbulence model performs better than the $k\text{-}\epsilon$ model to predict the power coefficient of hydrokinetic turbines. Hocine et al. [16] used the $k\text{-}\omega$ SST and VOF approach to estimate the performance of a Darrieus hydrokinetic turbine; the computed power coefficient showed the highest discrepancy with experimental data (13%) near its maximum value. In the present work, the $k\text{-}\omega$ SST turbulence model is used to (i) compute the flow field in the absence of a free surface; (ii) initialize the flow field for the rotor and full turbine (with the support structure) simulation cases.

Laminar separation, transition to turbulence, flow reattachment, and turbulent separation are likely to occur along a hydrokinetic turbine blade. Such unsteady flow phenomena strongly depend on the local Reynolds number so they are not properly accounted for by the standard $k\text{-}\omega$ SST model. The four equation Transition SST (TSST) turbulence model couples two additional transport equations with the $k\text{-}\omega$ SST model to predict the laminar-turbulent boundary layer transition [27]; the first equation computes the intermittency (it triggers transition locally) and the second one makes use of the momentum thickness Reynolds number to determine the onset of transition (non-local effects).

The TSST turbulence model has shown its capability to predict the natural, separation-induced, and bypass transitions in wall-bounded flows [28]. Rezaeiha et al. [29] performed detailed CFD simulations of vertical axis wind turbines with seven different RANS turbulence models. They found that the TSST led to the lowest deviation of the computed power coefficient (18.6%) from the experimental data. Simulations of a Darrieus type vertical axis wind turbine showed that the calculated torque coefficient using the TSST turbulence model was in better agreement with experimental data [30]. The dynamic stall on vertical axis wind turbine blades was also investigated with the standard and transition k-ω SST models [31]. The drag coefficient computed with the standard k-ω SST and four-equation TSST turbulence models were significantly different; laminar separation bubbles were only observed when the TSST was used, and the TSST predicted the dynamic stall earlier than the k-ω SST. Contreras et al. [26] highlighted the transitional behavior of the boundary layer over a blade section resolved by the TSST turbulence model; a laminar flow envelope around the profile was also observed. Regarding vertical axis water turbines, the standard and transition versions of the k-ω SST model have been applied, among others by [32,33]. In order to properly capture the laminar-turbulent transition in the boundary layer of the hydrokinetic turbine blades, the four-equation TSST turbulence model is employed in the present investigation.

The performance of a HAHT is governed by a set of non-dimensional parameters. The tip-speed-ratio (λ) relates the rotor angular velocity (Ω) with the incoming flow velocity (V_{in}),

$$\lambda = \frac{\Omega D}{2V_{in}}, \tag{6}$$

where D is the rotor diameter. Three TSR values are analyzed in this work, namely $\lambda = 4$, 6 and 8.

The power coefficient (C_P) is the ratio between the power produced by the rotor (W) and the power from the fluid flow,

$$C_P = \frac{W}{\frac{1}{2}\rho_w A V_{in}^3}, \tag{7}$$

where A is the area swept by the rotor. The normal component (F_n) of the hydrodynamic force acting on the turbine (rotor and support structure) is used to compute the thrust coefficient (C_T) defined as,

$$C_T = \frac{F_n}{\frac{1}{2}\rho_w A V_{in}^2}. \tag{8}$$

Previous investigations highlight the effect of the turbine's support structure on the near wake flow field [34]. The HAHT wake recovery can be defined in terms of velocity deficit,

$$V_d = 1 - \frac{V_x}{V_{in}}, \tag{9}$$

where V_x stands for the streamwise velocity component in the wake.

2.2. Geometry and Computational Domain

The HAHT geometry is taken from Bahaj et al. [17]. The turbine consists of a three-blade rotor with a diameter of 0.8 m. The blade geometry is based on the airfoil series NACA 63-8XX; the software DesignFOIL R6.46 is used to generate the profiles along the blade span. Chord and twist distributions are taken from [17]. The reconstructed turbine geometry is shown in Figure 1. The blades are split into three sub-surfaces to ease the mesh generation, and the tower (vertical part of the structure) is divided into various surfaces to adjust the turbine immersion in the computational domain. The simulated scenarios reported in this work are summarized in Table 1.

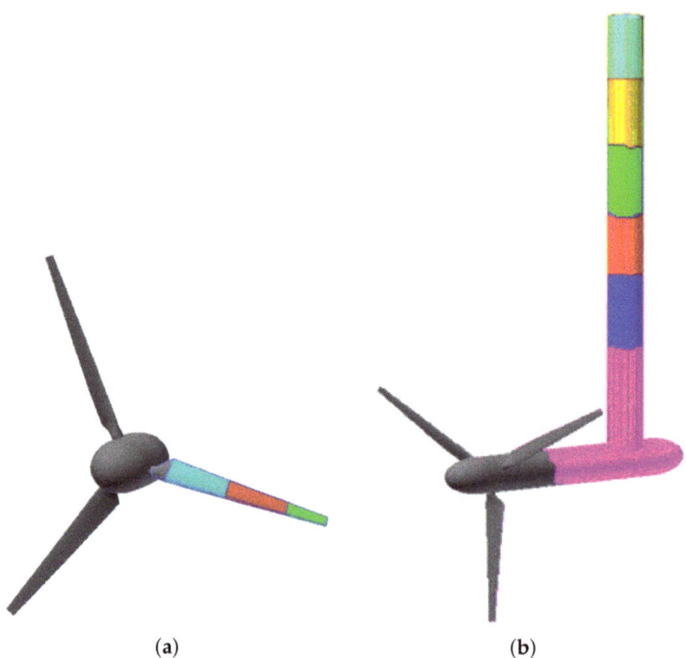

Figure 1. HAHT geometry (**a**) rotor (blades and hub) and (**b**) support structure.

Table 1. Summary of HAHT simulations.

Case	Modeling Approach	Geometry	Blade-Tip Immersion
1	Single-phase flow	Rotor only	-
2	Two-phase flow	Rotor only	$0.19D$ and $0.55D$
3	Two-phase flow	Rotor and support structure	$0.19D$ and $0.55D$

The dimensions of the computational domain are given in terms of the rotor diameter (box of $3.5D \times 3.5D \times 10D$) as is shown in Figure 2. The rotor is located at a distance of $3D$ relative to the inlet boundary. The computational domain is divided into two subdomains, namely, an inner rotating subdomain, which encloses the rotor, and an outer stationary subdomain available to the fluid flow. The rotating and stationary subdomains share cylindrical sliding non-conformal interfaces which account for flux continuity and transient interaction effects between the two subdomains.The sliding-mesh technique implemented in the commercial CFD software ANSYS-Fluent 19.0 is used to account for the relative motion between subdomains. As a result, all cells and boundaries in the rotating subdomain revolve at the prescribed angular velocity, Ω.

Boundary conditions for the second and third simulated cases are shown in Figure 2a,b, respectively. At the boundary named velocity inlet, a constant velocity ($V_{in} = 1.5$ m/s) is specified on the water portion of the inlet while zero velocity is applied elsewhere. Zero gauge pressure is set at the pressure-outlet boundary; the bottom boundary is set as a stationary non-slip wall. Surfaces of the support structure are also specified as non-slip walls when the full turbine is simulated. The top and side borders are defined as zero shear slip-walls which is accomplished by a symmetry type boundary condition. Because the turbulence quantities at the inlet boundary are not known, it was decided to prescribe a moderate turbulence intensity equal to 5%. When the free surface was not considered (the first simulated case) the top, bottom, and side boundaries of the domain were set as zero shear walls moving with the same velocity as that prescribed at the inlet boundary.

The rotating subdomain is a cylinder of diameter 1.15D; as it can be seen in Figure 2, the cylinder's length is slightly increased from 0.3D to 0.4D when the support is included. The shaded areas in Figure 2 indicate mesh refinement regions. In particular, a thin region of width 0.25D is refined to better capture the free surface evolution.

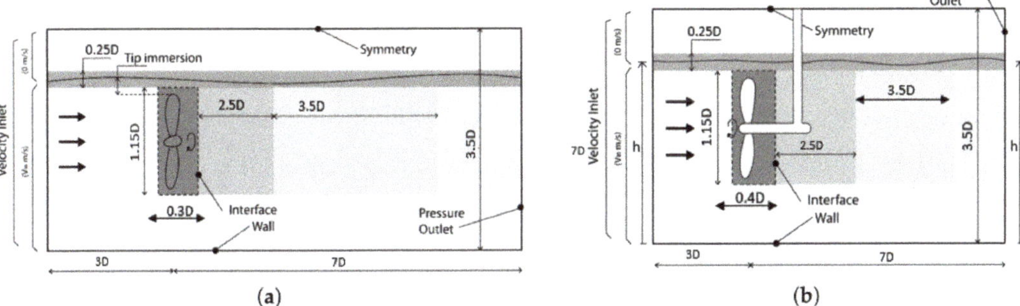

Figure 2. Computational domain and boundary conditions considering (**a**) the rotor only and (**b**) the full turbine. Shaded areas represent mesh refinement regions.

2.3. Spatial and Temporal Discretization

The SIMPLE segregated solver is used together with second-order discretization schemes. The least-squares cell-based method is employed for spatial discretization of gradients. The transient formulation is set to second-order implicit. The HRIC scheme is used to discretize the convective term in Equation (2). The HRIC formulation works with an explicit upwind difference scheme to satisfy the convective boundedness criterion, so the scheme introduces some numerical diffusion due to its first-order accuracy [35].

The computational domain meshed in ANSYS-Meshing 19.0 consists primarily of tetrahedral cells. A planar cut of the three-dimensional mesh is shown in Figure 3a, it corresponds to the mesh used for the second case (see Table 1). Details of the mesh around the blade are shown in Figure 3b; twenty prismatic layers are set around each blade to guarantee that the non-dimensional wall-normal coordinate lies in the viscous sublayer ($y^+ < 1$) and the boundary layer development is properly determined. A tetrahedral grid of a similar aspect ratio grows just outside the prismatic layer; about 85% of the computational cells are found in the rotating subdomain. A finer grid resolution is used downstream of the rotor to better capture the wake evolution. Figure 3c shows a close-up of the computational mesh around the rotor hub and support structure (case 3, Table 1). Prismatic layers are also used to capture the boundary layer effects on all structural components of the turbine.

A second-order implicit scheme is used for temporal discretization. The time step is chosen so that it corresponds to a blade rotation of approximately 1°; this choice is a compromise between computational cost and accuracy. The choice of time-step size is based on a sensitivity analysis performed for case one (single-phase flow). In all cases, the calculations start with the steady MRF model to obtain a preliminary flow field, which is used as an initial condition. When the transient computation is started, first-order schemes are employed during a few time-steps before switching to second-order schemes. A maximum of 50 iterations per time step is necessary to lower the residuals below 10^{-6}.

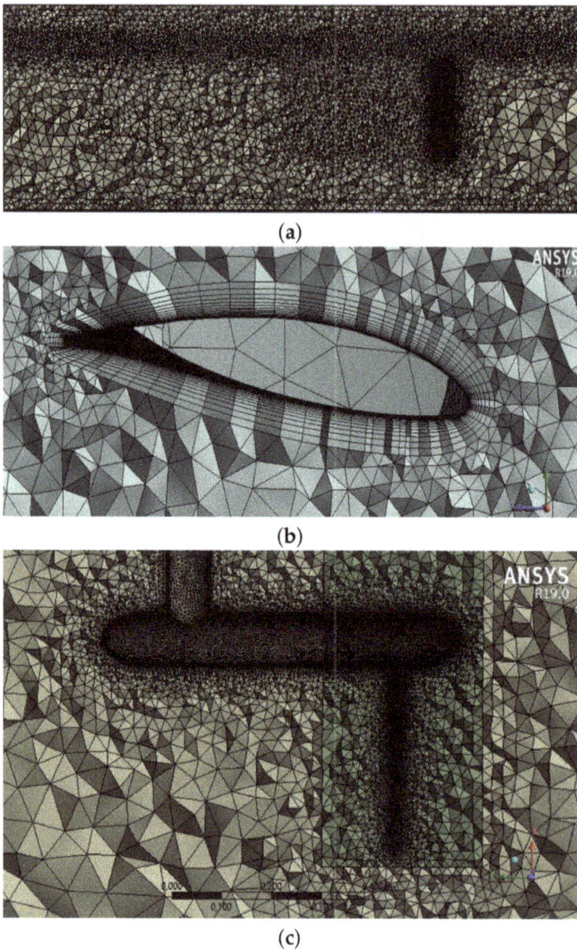

Figure 3. Two-dimensional cut of the mesh used in (**a**) case 2; (**b**) mesh distribution around the blades; (**c**) details of the mesh around the support structure (case 3).

3. Results and Discussion

3.1. Single-Phase Flow Simulations

Single-phase flow simulations were performed to assess the sensitivity to grid resolution, time-step size, and turbulence models. Simulation results presented in this section correspond to case 1 in Table 1.

The average power coefficient obtained with the standard SST k-ω and the four-equation TSST models at $\lambda = 6$ is presented in Figure 4a. The SST k-ω turbulence model predicts a lower C_P than the four-equation TSST model even when the grid is refined. When the domain is discretized with 8.5 million cells or more, the C_P calculated with the TSST turbulence model remains approximately 5% higher than the result obtained with the SST k-ω model. The difference can be attributed to the ability of the TSST model to capture boundary layer transition effects which are directly related to the hydrodynamic forces acting on the rotor blades [32]. Predictions of C_P obtained with the four-equation TSST turbulence model are compared with experimental data at various TSRs in Figure 4b. Simulation results are based on a grid resolution of 8.8 million cells and a time step of 5×10^{-4} s. As it can be noticed, the numerical results slightly underestimate the experimental points and also, the simulated C_P curve shows a small displacement toward the right regarding the measurements. However, in the range of tip speed ratios between 5 and 8, the maximum difference between the numerical and experimental points is about 6%, which at this moment constitutes a good enough agreement.

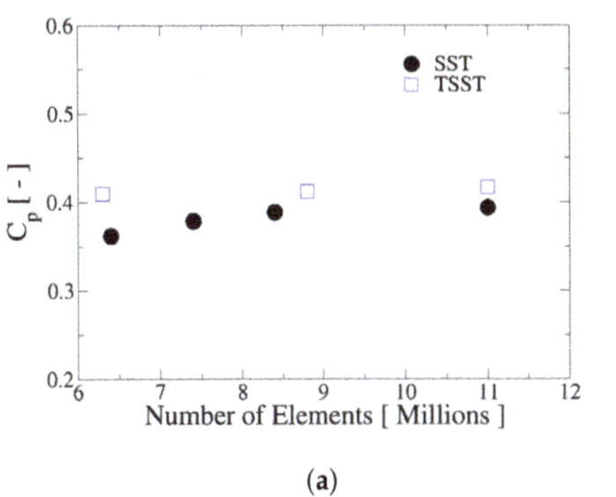

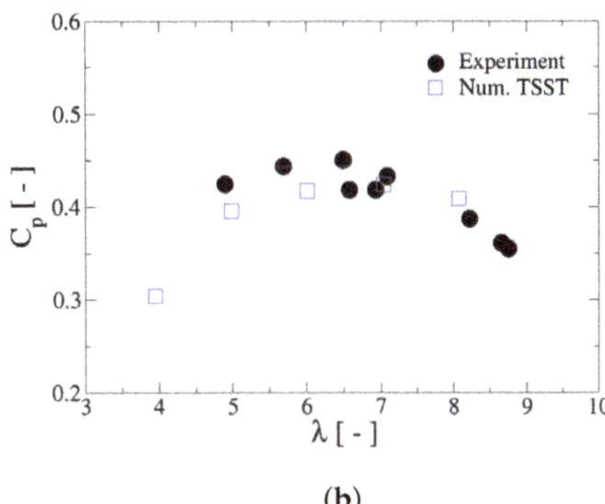

Figure 4. Average power coefficient obtained from single-phase flow simulations. (**a**) Sensitivity to turbulence model at $\lambda = 6$; (**b**) C_P calculated with the TSST model for various TSRs.

Table 2. Effect of time-step size on C_P at $\lambda = 8$.

Time Step (s)	Blade Rotation (°)	C_P
1×10^{-3}	1.72	0.3982
5×10^{-4}	0.83	0.4015
2.5×10^{-4}	0.34	0.4034

The effect of time-step size on the predicted power coefficient is investigated at $\lambda = 8$. The instantaneous values of C_P in time can be seen in Figure 5 whereas Table 2 shows the relationship between time-step, blade rotation, and the average C_P. It is found that reducing the time-step size in half results in a C_P increment of just 0.5%. Therefore, a time step of 5×10^{-4} s has been chosen as a compromise between accuracy of results and simulation time; this time step corresponds to a rotation angle of 0.83° which is below the value of 1° recommended previously as a rule of thumb [26,36].

3.2. Effect of Free Surface and Support Structure on Performance

This section describes the results obtained when free surface effects and the support structure are included in the computational model. Although the support consists of a simple geometry, it increases the number of cells in the stationary subdomain by 11%. Simulation results presented in this section were computed with the four-equation TSST turbulence model.

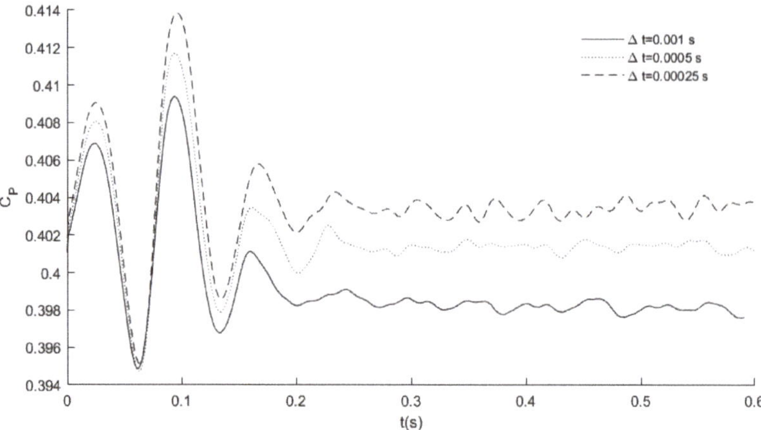

Figure 5. Effect of time-step size on the computed C_P at $\lambda = 8$.

It is a fact that the presence of the free surface and its dynamics affects the performance of the turbine as well as wake development and its recovery [36]. For instance, in Kolekar et al. [19] it is concluded that the free surface compresses the wake and that it produces a blocking effect which increases for larger rotational velocities and lower tip immersions. Figures 6 and 7 show the time histories of C_P and C_T, respectively, experienced by the blades for both, shallow and deep tip immersion cases at a TSR of 6. They are compared with the results without a support structure. In first place, it is seen that power and thrust coefficients are smaller in the shallow immersion than in the deeper configuration, a result that agrees with the observations of [18,37]. Such reduction of the coefficients is related with a stronger deformation of the free surface in the case of shallow immersion as it can be seen in Figures 8 and 9 below. The interaction between this deformation and the flow behind the HAHT causes a slower wake recovery than in the case of deep immersion, which is responsible for the reduction of the turbine performance. Additionally, the presence of the support decreases further both coefficients in a similar percentage for both turbine immersions (around 3% for C_P and 1.6% for C_T). Such decrease is attributed to a small flow velocity reduction caused by the presence of the supporting rod; this effect is similar to the effect of the tower on C_P in case of horizontal wind turbines [38]. In the configuration of shallow tip immersion, the time evolution of the two coefficients display clear ripples, either considering or not the supporting structure; such ripples are due to the periodic blades passage close to the free surface and their frequency equals to that of the blades rotational frequency. On the other hand, for the deeper immersion conditions the ripples are still present but with very much attenuated amplitude, reflecting that the turbine in such configuration is close to free stream conditions. Moreover, a long period undulation in the values of the C_P and C_T coefficients can be appreciated in Figures 6 and 7, which is present for both immersion conditions. This behavior was not seen in Yan et al. [18], where the coefficients monotonically reached a statistically steady state. The reason behind this long period coefficient undulation is at the moment not clear and deserves future investigation.

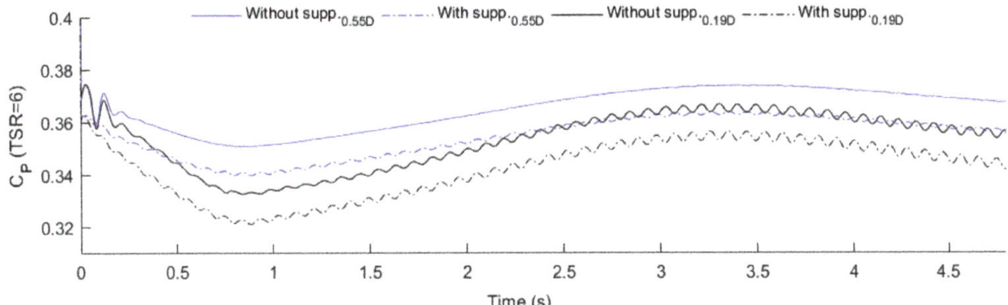

Figure 6. Time history of power coefficient for deep and shallow tip immersions, with and without support structure. Results are for $\lambda = 6$.

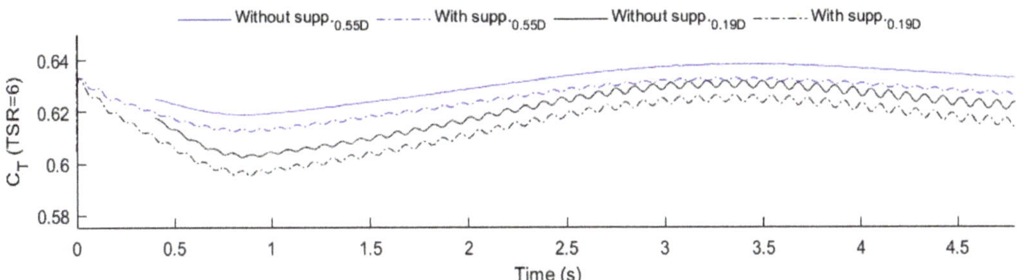

Figure 7. Time history of thrust coefficient for deep and shallow tip immersions, with and without support structure. Results are for $\lambda = 6$.

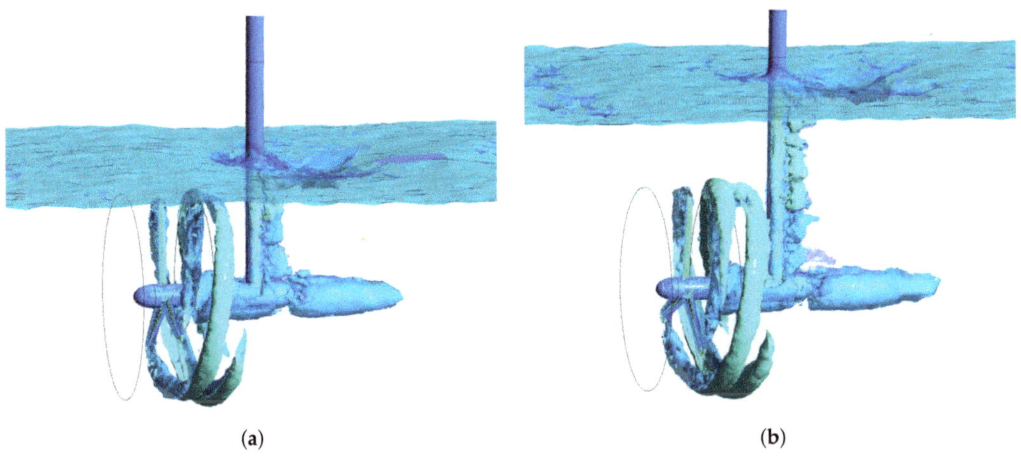

Figure 8. Snapshots of air-water interface and vorticity isosurface of 20 Hz at (**a**) $0.19D$ and (**b**) $0.55D$ tip immersions.

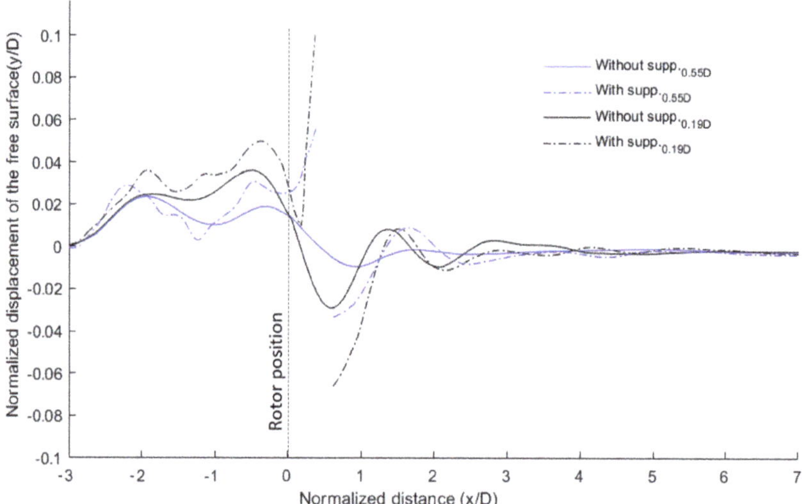

Figure 9. Vertical displacement of the free surface at the channel mid plane. Results for $\lambda = 6$ and two immersions.

The individual contribution of each blade to the power coefficient over one revolution taking into account the effect of the support structure is shown in Figure 10. In that figure, the azimuthal angle θ is set at 0° when blade 1 (identified by red color) is located upwards directly pointing towards the free surface. Results correspond to a TSR of 6. The reduction in the value of C_P for the full assembly is noted in comparison to the results when the support structure is not included in the computational model. Each blade provides the lowest C_P when it is placed vertically upwards, aligned with the vertical bar of the support structure, and the highest C_P when it is located vertically downwards. The difference between these two extreme values is very much reduced for the deep immersion case, regarding the shallow configuration, although it is noticeable even when the support structure is not included in the computation. Moreover, it can be observed for the case of 0.19D immersion without support that the minimum value of C_P attained is at the position where the blade tip is closest to the free surface, a fact that illustrates that the proximity of the blades to the air-water interface greatly decrease the turbine performance, which in this case is of the same order than the presence of the supporting structure. Interestingly, the maximum value of the instantaneous power coefficient, reached when the azimuthal angle is 180°, happens for the shallow immersion configuration, regardless of the supporting rod presence.

The fact that the variation of C_P along one revolution increases as the blade tip is closer to the free surface can be understood as follows. In the absence of a support structure, the flow accelerates by surrounding the blades more or less symmetrically; on the other hand, the flow through the rotor moves more slowly towards the free surface when the structure is included. This behavior is more pronounced at the shallow tip immersion. As can be seen in Figure 10, C_P reaches its maximum when blade 1 sweeps the high flow velocity region at 180°, i.e. the blade tip is farthest from the free surface. Peak values of C_P for blades 2 and 3 are shifted 120° and 240°, respectively. The support structure lowers the contribution of blade 1 to C_P when the blade is pointing upwards (at 0°) and the difference between the C_P curves with and without support structure is the highest. It resembles the tower shadow effect experienced by horizontal axis wind turbine blades, which is known to be the main cause of power fluctuation [38].

3.3. Wake and Free Surface Interactions

Figure 8 shows the vorticity isosurface (20 Hz) and free surface for both deep and shallow tip immersion simulations in the case which includes the supporting structure. Significant deformation of the air-water interface can be observed at both tip immersions, especially in the vicinity of the support rod. Figure 8 shows the wake past the support structure, revealed by the vorticity isosurface, which could not be detected in the previous study of Yan et al. [18] but is present in the recent work of [37]. As a comment, blade root vortices are observed in the results of [18], which can be attributed to the presence of a gap between the rotor and the support in the employed computational model. That space is not considered in this work, so the horizontal part of the support hinders the development of root vortices in the present case.

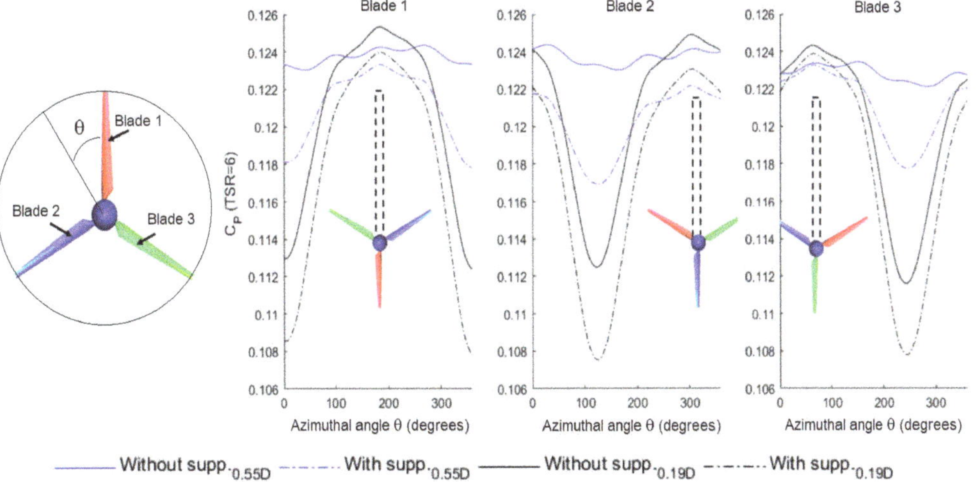

Figure 10. Blade contribution to the power coefficient along one revolution with and without the support structure. Results are for $\lambda = 6$.

The vertical displacement of the free surface at the domain mid-plane is presented in Figure 9. The displacement is normalized by the rotor diameter for both the shallow and deep tip immersions. Time-averaged results were obtained in the last three revolutions for $\lambda = 6$. The configuration without support, solid lines in Figure 9, induces an elevation of the water level just upstream the rotor and a gentle drop of it downstream. This effect is of course due to the obstruction generated by the turbine in the flow and it is more pronounced in the shallow immersion. On the other hand, the inclusion of the support structure causes a discontinuity immediately downstream of the rotor position; the surface rapidly elevates in front of the pole (vertical rod) while a sort of dimple forms just downstream. Two peaks are observed upstream regardless of the presence of the structure. Moreover, the free surface deformation upstream is increased for both immersions when the structure is included in the model. For the deep tip immersion, the free surface exhibits little disturbance downstream from the rotor in the absence of the structure which suggests that the free surface effect on the rotor performance is marginal when the rotor is well below it, in agreement with the conclusions of [18,37]. The highest depression downstream on the free surface occurs for the 0.19D immersion when the structure is present, followed by a peak that is approximately equal for both deep and shallow immersions with the support structure. In order to provide a quantitative estimation, the normalized maximum free-surface displacement (Y_m/D) and its location downstream relative to the rotor position (X_m/D) are summarized in Table 3. The results are compared with interpolated data taken from the study of Adamski [22]. Froude number (*Fr*) is based on the rotor diameter and d_r represents the rotor depth-to-diameter ratio. As it can be seen, the computed maximum displacement agrees reasonably well with the results reported by [22] for both tip immersions.

Table 3. Normalized maximum displacement of the free surface (Y_m/D) and its location downstream the rotor (X_m/D).

Tip Immersion	*Fr*	d_r	X_m/D	Y_m/D	X_m/D Interp. from [22]	Y_m/D Interp. from [22]
0.55D	0.536	1.05	0.8745	0.02954	0.8843	0.02265
0.19D	0.536	0.69	0.5901	0.05764	0.7874	0.05126

On the other hand, the study of wake development is of importance to determine the optimal spacing between HAHTs in a hydrokinetic farm [11,39–41]. Time-averaged velocity deficit profiles downstream of the simulated HAHT are shown in Figure 11a and Figure 11b for the deep tip and shallow tip immersions, respectively. The profiles are taken from the vertical center plane so $y/D = 0$ corresponds to the rotor centerline. The dot-dashed horizontal line represents the level of the undisturbed free surface. Profiles are shown at four positions downstream of the HAHT; a comparison is drawn between the combined effect of the rotor and support structure on the wake and in the absence of a support structure (case 2 in Table 1). Significant differences between the profiles can be appreciated, especially in the near wake region (at a distance less than $4D$ from the rotor). At $x = 1D$, the maximum velocity deficit is found off the centerline both for the deep-tip and shallow tip immersions when the support structure is included. The corresponding curves in this case show two unequal peaks instead of the two symmetric maximums in the case without supporting structure. Simulation results suggest that the support shifts the maximum velocity deficit towards the free surface line, a fact clearly observed in Figure 11b for the shallow immersion configuration. It can be also noticed that the combined effect of rotor and support structure reduces the velocity in the wake above the rotor centerline for the analyzed positions and tip immersions. Due to the free surface proximity, the wake recovers somewhat slower when the tip immersion is 0.19D. In this case, the wake expansion is quickly constrained by the free surface, which in turn reduces the power and thrust coefficients at shallow tip immersions.

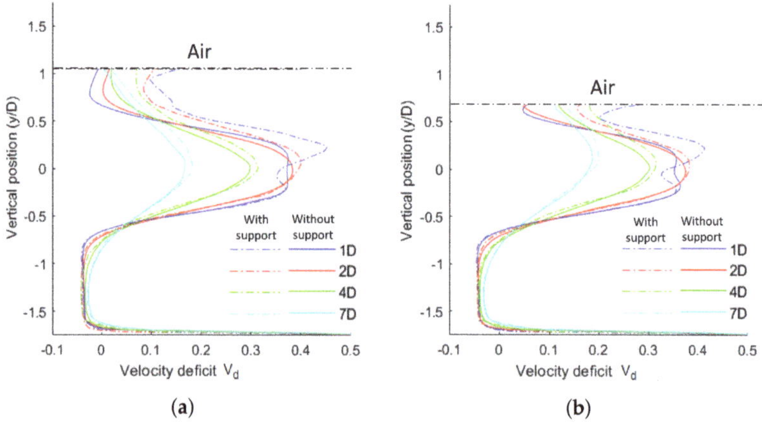

Figure 11. Comparison of time-averaged velocity deficit profiles at four downstream positions of the HAHT at $\lambda = 6$. (**a**) Deep tip immersion (0.55D); (**b**) Shallow tip immersion (0.19D).

4. Conclusions

A computational study of the influence of the free surface and supporting structure on the flow around a HAHT was conducted using the VOF method combined with the sliding mesh technique and RANS turbulence models. The commercial CFD software ANSYS-Fluent 19.0 was used to implement the computational model. First, the HAHT performance was analyzed in the absence of the free surface and without the support structure. Computations showed that the four-equation TSST turbulence model led to predictions of C_P which are in better agreement with the experimental results of Bahaj et al. [17] for TSRs between 4 and 8 than those obtained with the standard SST model. A thorough sensitivity analysis showed that a mesh with about eight million cells and a time-step of about 5×10^{-4} s are required to estimate the power and thrust coefficients within a reasonable computing time. Then, the performance was analyzed by taking into account the free surface in the computational domain. The rotor was placed at two tip immersions showing that a shallow immersion results in lower power extraction in agreement with experimental results. In addition, predictions of power and thrust coefficients were found to be affected by the support structure. Also, the dynamics of the wake behind the turbine was shown to be affected by both the free surface and the presence of the support structure. In the shallow immersion case, the development of the wake is constrained by the free surface and supporting structure leading to a slower wake recovering and to a decrease of the turbine power coefficient.

Future work will explore the overset mesh technique to replace the sliding nonconformal interface currently used for the interpolation between stationary and rotating subdomains. The simulation of very shallow turbine immersions is also planned, where the blades emerge above the free surface, inclusive of fluid-structure interaction and the possible appearance of cavitation effects.

Author Contributions: Conceptualization, A.B.-M. and S.L.; methodology, A.B.-M. and S.L.; validation, L.R.-J.; formal analysis, L.R.-J., A.B.-M. and S.L.; investigation, L.R.-J.; resources, S.L.; data curation, L.R.-J.; writing—original draft preparation, A.B.-M.; writing—review and editing, S.L. and L.R.-J.; visualization, L.R.-J.; supervision, A.B.-M. and S.L. All authors have read and agreed to the published version of the manuscript.

Funding: This research received no external funding.

Institutional Review Board Statement: Not applicable.

Informed Consent Statement: Not applicable.

Data Availability Statement: The data presented in this study are available on request from the corresponding author.

Acknowledgments: The authors thank the financial support provided by the Research Vice-rectory of Universidad Autónoma de Occidente. We acknowledge the use of the ANSYS license and computer time provided by the Laboratorio de Bioprocesos y Flujos Reactivos (BIOFRUN) at the Facultad de Minas of the Universidad Nacional de Colombia.

Conflicts of Interest: The authors declare no conflict of interest.

Abbreviations

The following abbreviations are used in this manuscript:

HAHT	Horizontal Axis Hydrokinetic Turbine
HRIC	High Resolution Interface Capturing
SIMPLE	Semi-Implicit Method for Pressure-Linked Equations
MRF	Moving Reference Frame
TSR	Tip Speed Ratio
TSST	Transition Shear Stress Transport
URANS	Unsteady Reynolds Average Navier-Stokes

References

1. International Energy Agency. Available online: https://www.iea.org (accessed on 30 August 2021).
2. Rogers, J.D. Celebrating Hoover Dam: The Majesty of Hoover Dam. *Civ. Eng. Mag.* **2010**, *80*, 52–65. [CrossRef]
3. Suárez, B.; Vera Rodríguez, J.D.; Botero, F.; Suárez Agudelo, B.H.; Giraldo Jiménez, W. Hidroituango Intake Gate Closure—Emergency Conditions. *Revista Facultad de Ingeniería Universidad de Antioquia* **2021**. Available online: https://revistas.udea.edu.co/index.php/ingenieria/article/view/344363 (accessed on 10 September 2021).
4. Organization for Economic Co-operation and Development. Available online: https://stats.oecd.org (accessed on 28 August 2021).
5. Anyi, M.; Kirke, B. Evaluation of small axial flow hydrokinetic turbines for remote communities. *Energy Sustain. Dev.* **2010**, *14*, 110–116. [CrossRef]
6. Batten, W.; Bahaj, A.; Molland, A.; Chaplin, J. The prediction of the hydrodynamic performance of marine current turbines. *Renew. Energy* **2008**, *33*, 1085–1096. [CrossRef]
7. Lee, J.H.; Park, S.; Kim, D.H.; Rhee, S.H.; Kim, M.C. Computational methods for performance analysis of horizontal axis tidal stream turbines. *Appl. Energy* **2012**, *98*, 512–523. [CrossRef]
8. Batten, W.; Bahaj, A.; Molland, A.; Chaplin, J. Experimentally validated numerical method for the hydrodynamic design of horizontal axis tidal turbines. *Ocean. Eng.* **2007**, *34*, 1013–1020. [CrossRef]
9. Danao, L.A.; Abuan, B.; Howell, R. Design Analysis of a Horizontal Axis Tidal Turbine. In Proceedings of the 3rd Asian Wave and Tidal Conference-AWTEC, Singapore, 24–28 October 2016.
10. Bowman, J.; Bhushan, S.; Thompson, D.S.; O'Doherty, D.; O'Doherty, T.; Mason-Jones, A. A physics-based actuator disk model for hydrokinetic turbines. In Proceedings of the Fluid Dynamics Conference, Atlanta, GA, USA, 25–29 June 2018.
11. Silva dos Santos, I.F.; Camacho, R.G.R.; Filho, G.L.T. Study of the wake characteristics and turbines configuration of a hydrokinetic farm in an Amazonian river using experimental data and CFD tools. *J. Clean. Prod.* **2021**, *229*, 126881.
12. Mikkelsen, R. Actuator Disc Methods Applied to Wind Turbines. Ph.D. Thesis, Technical University of Denmark, Lyngby, Denmark, 2003.
13. Sørensen, J.N.; Shen, W.Z. Numerical Modeling of Wind Turbine Wakes. *ASME J. Fluids Eng.* **2002**, *124*, 393–399. [CrossRef]
14. Henao Garcia, S.; Benavides-Morán, A.; Lopez Mejia, O.D. Wake and Performance Predictions of Two- and Three-Bladed Wind Turbines Based on the Actuator Line Model. *ASME J. Fluids Eng.* **2021**, *143*, 051206. [CrossRef]
15. Baba-Ahmadi, M.H.; Dong, P. Validation of the actuator line method for simulating flow through a horizontal axis tidal stream turbine by comparison with measurements. *Renew. Energy* **2017**, *113*, 420–427. [CrossRef]
16. Hocine, A.E.B.L.; Jay, R.W.; Poncet, S. Multiphase modeling of the free surface flow through a Darrieus horizontal axis shallow-water turbine. *Renew. Energy* **2019**, *143*, 1890–1901. [CrossRef]
17. Bahaj, A.; Molland, A.F.; Chaplin, J.R.; Batten, W.M.J. Power and thrust measurements of marine current turbines under various hydrodynamic flow condition in a cavitation tunnel and a towing tank. *Renew. Energy* **2007**, *32*, 407–423. [CrossRef]
18. Yan, J.; Deng, X.; Korobenko, A.; Bazilevs, Y. Free-surface flow modeling and simulation of horizontal-axis tidal-stream turbines. *Comput. Fluids* **2017**, *158*, 157–166. [CrossRef]
19. Kolekar, N.; Vinod, A.; Banerjee, A. On Blockage Effects for a Tidal Turbine in Free Surface Proximity. *Energies* **2019**, *12*, 3325. [CrossRef]
20. Nishi, Y.; Sato, G.; Shiohara, D.; Inagaki, T.; Kikuchi, N. Performance characteristics of axial flow hydraulic turbine with a collection device in free surface flow field. *Renew. Energy* **2017**, *112*, 53–62. [CrossRef]
21. Bai, X.; Avital, E.J.; Munjiza, A.; Williams, J.J.R. Numerical simulation of a marine current turbine in free surface flow. *Renew. Energy* **2014**, *63*, 715–723. [CrossRef]
22. Adamski, S.J. Numerical Modeling of the Effects of a Free Surface on the Operating Characteristics of Marine Hydrokinetic Turbines. Master's Thesis, University of Washington, Seatle, GA, WA, 2013.
23. Rodríguez, L.; Benavides-Moran, A.; Laín, S. Three-Bladed Horizontal Axis Water Turbine Simulations with Free Surface Effects. *Int. J. Appl. Mech. Eng.* **2021**, *26*, 187–197. [CrossRef]
24. Hirt, C.W.; Nichols, B.D. Volume of fluid (VOF) method for the dynamics of free boundaries. *J. Comput. Phys.* **1981**, *39*, 201–225. [CrossRef]
25. Menter, F.R. Zonal two equation k-turbulence models for aerodynamic flows. In Proceedings of the 23rd Fluid Dynamics, Plasmadynamics and Lasers Conference, Orlando, FL, USA, 6–9 July 1993.
26. Contreras, L.T.; López, O.D.; Laín, S. Computational fluid dynamics modelling and simulation of an inclined horizontal axis hydrokinetic turbine. *Energies* **2018**, *11*, 3151. [CrossRef]
27. Langtry, R.B.; Menter, F.R. Transition Modeling for General CFD Applications in Aeronautics. In Proceedings of the 43rd AIAA Aerospace Sciences Meeting and Exhibit, Reno, NV, USA, 10–13 January 2005.
28. Langtry, R.B.; Menter, F.R.; Likki, S.R.; Suzen, Y.B.; Huang, P.G.; Völker, S. A Correlation-Based Transition Model Using Local Variables—Part II: Test Cases and Industrial Applications. *ASME J. Turbomach.* **2006**, *128*, 423–434. [CrossRef]
29. Rezaeiha, A.; Montazeri, H.; Blocken, B. On the accuracy of turbulence models for CFD simulations of vertical axis wind turbines. *Energy* **2019**, *180*, 838–857. [CrossRef]
30. Arab, A.; Javadi, M.; Anbarsooz, M.; Moghiman, M. A numerical study on the aerodynamic performance and the self-starting characteristics of a Darrieus wind turbine considering its moment of inertia. *Renew. Energy* **2017**, *107*, 298–311. [CrossRef]

31. Almohammadi, K.M.; Ingham, D.B.; Ma, L.; Pourkashanian, M. Modeling dynamic stall of a straight blade vertical axis wind turbine. *J. Fluids Struct.* **2015**, *57*, 144–158. [CrossRef]
32. Laín, S.; Taborda, M.A.; López, O.D. Numerical study of the effect of winglets on the performance of a straight blade Darrieus water turbine. *Energies* **2018**, *11*, 297. [CrossRef]
33. López, O.; Meneses, D.; Quintero, B.; Laín, S. Computational study of transient flow around Darrieus type Cross Flow Water Turbines. *J. Renew. Sustain. Energy* **2016**, *8*, 014501. [CrossRef]
34. Myers, L.; Bahaj, A.S. Near wake properties of horizontal axis marine current turbines. In Proceedings of the 8th European Wave and Tidal Energy Conference-EWTEC, Upsala, Sweden, 7–10 September 2009.
35. Waclawczyk, T.; Koronowicz, T. Comparison of CICSAM and HRIC high-resolution schemes for interface capturing. *J. Theor. Appl. Mech.* **2008**, *46*, 325–345.
36. El Fajri, O.; Bowman, J.; Bhushan, S.; Thompson, D.; O'Doherty, T. Numerical study of the effect of tip-speed ratio on hydrokinetic turbine wake recovery. *Renew. Energy* **2022**, *182*, 725–750. [CrossRef]
37. Tian, W.; Ni, X.; Mao, Z.; Zhang, T. Influence of surface waves on the hydrodynamic performance of a horizontal axis ocean current turbine. *Renew. Energy* **2020**, *158*, 37–48. [CrossRef]
38. Wen, B.; Wei, S.; Wei, K.; Yang, W.; Peng, Z.; Chu, F. Power fluctuation and power loss of wind turbines due to wind shear and tower shadow. *Front. Mech. Eng.* **2017**, *12*, 321–332. [CrossRef]
39. Müller, S.; Muhawenimana, V.; Wilson, C.A.M.E.; Ouro, P. Experimental investigation of the wake characteristics behind twin vertical axis turbines. *Energy Convers. Manag.* **2021**, *247*, 114768. [CrossRef]
40. Myers, L.; Bahaj, A.S. Experimental analysis of the flow field around horizontal axis tidal turbines by use of scale mesh disk rotor simulators. *Ocean. Eng.* **2010**, *37*, 218–227. [CrossRef]
41. Silva, P.A.S.F.; De Oliveira, T.F.; Junior, A.C.P.B.; Vaz, J.R.P. Numerical Study of Wake Characteristics in a Horizontal-Axis Hydrokinetic Turbine. *Ann. Braz. Acad. Sci.* **2016**, *88*, 2441–2456. [CrossRef] [PubMed]

processes

Article

Power Regulation and Fault Diagnostics of a Three-Pond Run-of-River Hydropower Plant

Ahmad Saeed [1], Adnan Umar Khan [1], Muhammad Iqbal [1,2], Fahad R. Albogamy [3], Sadia Murawwat [4], Ebrahim Shahzad [1], Athar Waseem [1] and Ghulam Hafeez [5,*]

[1] Department of Electrical Engineering, International Islamic University, Islamabad 44000, Pakistan; ahmedsaeed771@gmail.com or ahmad.phdee85@iiu.edu.pk (A.S.); adnan.umar@iiu.edu.pk (A.U.K.); muhammad.iqbal@iiu.edu.pk (M.I.); ebrahimawan@gmail.com or ebrahim.phdee84@iiu.edu.pk (E.S.); athar.waseem@iiu.edu.pk (A.W.)

[2] KIOS Research and Innovation Center of Excellence, Department of Electrical and Computer Engineering, University of Cyprus, Nicosia 1678, Cyprus

[3] Computer Sciences Program, Turabah University College, Taif University, P.O. Box 11099, Taif 21944, Saudi Arabia; f.alhammdani@tu.edu.sa

[4] Department of Electrical Engineering, Lahore College for Women University, Lahore 51000, Pakistan; sadia.murawwat@lcwu.edu.pk

[5] Department of Electrical Engineering, University of Engineering and Technology, Mardan 23200, Pakistan

[*] Correspondence: ghulamhafeez393@gmail.com; Tel.:+92-3005003574

Abstract: Hydropower generation is one of the most prominent renewable sources of power. Run-of-river hydropower is like traditional hydropower but has significantly less environmental impact. Faults in industrial processes are a cause for large amounts of losses in monetary value and off times in industrial processes and consumer utilities. It is more efficient for the system to identify the occurring faults and, if possible, to have the processes running without interruption with the occurrence of a fault. This work uses a model previously proposed—the three-pond hydraulic run-of-river system and integrates it with a turbine and regulated power generation. After integration of the hydraulic system with the turbine and power generation, we then design a diagnostic system for commonly occurring faults within the system. Mathematical models of the faults are formulated and residues are calculated. Fault detection and identification is achieved by analyzing the residues and then a fault-tolerant control is proposed. The Fault Diagnostic Module can correctly detect the faults present and offers sufficient fault compensation to make the system run nearly normally in the event of fault occurrence. With the emergence of distributed power generation smart grids and renewable energy, this fault diagnostic is able to reliably offer uninterrupted power to the grid and thus to consumers.

Keywords: fault diagnostics; model-based fault detection; fault tolerance; fuzzy control

Citation: Saeed, A.; Khan, A.U.; Iqbal, M.; Albogamy, F.R.; Murawwat, S.; Shahzad, E.; Waseem, A.; Hafeez, G. Power Regulation and Fault Diagnostics of a Three-Pond Run-of-River Hydropower Plant. *Processes* **2022**, *10*, 392. https://doi.org/10.3390/pr10020392

Academic Editors: Santiago Lain and Omar Dario Lopez Mejia

Received: 30 December 2021
Accepted: 11 February 2022
Published: 17 February 2022

Publisher's Note: MDPI stays neutral with regard to jurisdictional claims in published maps and institutional affiliations.

1. Introduction

Hydropower plants are one of the most common and widely used renewable energy sources. These plants provide up to 80% of the total renewable energy. Traditional hydropower plants consist of a rather large water reservoir filled by constructing a dam in front of a river or a water way and flooding it to make the aforementioned reservoir. Hydropower is generated by allowing the water from the pond to flow through the turbine. The turbine and the generator are located in the power house. One of the main disadvantages of a traditional run-of-river hydropower plant is the substantial environmental impact and sometimes socio-economic impact on the population due to the need to flood the reservoir and displace the existing population. Due to these reasons, run-of-river hydro plants are an alternative to traditional hydropower plants. A wall is built on the river bed in the absence of a reservoir. This wall is known as a weir, to gain some degree of level regulation. Another essential part of the industrial process and power generation is fault diagnostics. During industrial processes, many known and unknown faults or errors can occur, resulting in undesirable outputs or off times, which can result in monetary loss and service on the consumer side. Many times, it is not desired to shut down the system for maintenance at the occurrence of every fault, as an unexpected system shutdown may cause monetary loss and inconvenience. The types of faults that require active maintenance are more severe, like loss of power or severe physical damage to the system due to any natural or manmade disaster. Otherwise, the maintenance is scheduled. Therefore, it is required for the control system to identify the occurring faults and make the system operate in such a way that minimizes the loss and inconvenience. A fault diagnostic system has two main factions: firstly, to identify and isolate any fault that has occurred, and secondly, to compensate the system in such a way that the effects of the fault are minimized.

In this work, the authors consider the three-pond hydraulic system from [1] and add a Francis turbine to it from [2]. Next, the authors propose a three-level control system to integrate the two parts and regulate the power at the desired level. After achieving power regulation, fault models are introduced, and their mathematical model is developed. These faults are introduced into the system. The introduced faults are mathematically modeled, and a fault model is discussed in detail as an example. After achieving power regulation and fault modeling, fault diagnostics and fault-tolerant control are targeted using model-based fault-tolerant control. Model-based fault-tolerant control is achieved by first obtaining the residue as the difference in the output of the system and system model. Next, the residue is analyzed to identify any fault; after fault identification, proper fault compensation is fed to the system if required. The fault diagnostic process operates in three modes for fault identification and tolerance. These three fault diagnostic modes are residue generation, fault identification, and fault compensation.

Both fault diagnostics and run-of-river hydropower plants have been topics of interest for research. Especially fault tolerance and diagnostics is the new hot field, especially in the application of remote power plants such as wind turbines and off shore wave power plants, which are harder to access for service and maintenance personnel. Directly relevant to this work, [2] proposed a fault-tolerant control of a simulated hydropower plant. The faults of a hydropower plant discussed in [2] are the actuator fault, i.e., the response time of the actuator increases, the turbine flow fault, and the turbine speed sensor fault. The authors in [3] discuss different faults that occur in Modular Multilevel Converters (MMCs) and present various fault diagnostics and fault-tolerant schemes. The faults of a 132 kV power distribution system are detected using stationary wavelet transform to extract the features. Then these features were used with artificial neural networks to detect and classify faults [4,5]. Another researcher [6] presents fault diagnostics using a Markov model and regression vectors for the faults. Another interesting study of fault diagnostics is by [7], regarding fault detection and diagnostics of ventilation units in a building and using multiple readings from adjacent sensors and detecting the deviation in sensor readings. The sensors considered are temperature, air, and fan-speed sensors. The authors in [8] proposed a contrastive learning algorithm for software and data-based fault diagnostics and a fault-detection system. In [9], fault detection and diagnostics of a vehicle's internal combustion engine and mechanical parts was achieved using chaos analysis and signal processing on the sound of the running vehicle. Similar to [9,10], they used spectrum analysis for fault diagnostics in rotating machines. In [11], for the fault diagnostics of a hydroelectric generator, they used two sensors: a vibration sensor with both horizontal and vertical movement and a pendulum and bond phase sensor. These sensors are attached to the generating unit. The faults of the generating unit are diagnosed by analyzing the sensor data over the data communication line. The authors of [12] discuss the fault diagnostics of the hydro turbine governing system. First, a simplified non-linear model of a hydro turbine non-linear system is taken and analyzed using Volterra models in the frequency domain. In addition to these, many theses, such as [13–15], have been done on fault detection and diagnostics using observers and residues.

2. System Model

The system considered here is a redirected run-of-river hydropower plant, which is discussed in [1]. The hydraulic part of the plant consists of three head ponds, pond 1 (T_1), 2 (T_2), and 0 (T_0). Water from the river enters ponds 1 and 2. Ponds 1 and 2 are connected at the bottom to the head pond 0 through tunnels. Both tunnels from pond 1 to pond 0 and pond 2 to pond 0 have a controllable valve, s_1 and s_2, respectively. The head pond 0 is connected from the bottom to the surge tank T_s through the long headrace tunnel. The level of water of the ith pond is denoted by x_i. Figure 1 shows the construction of the system.

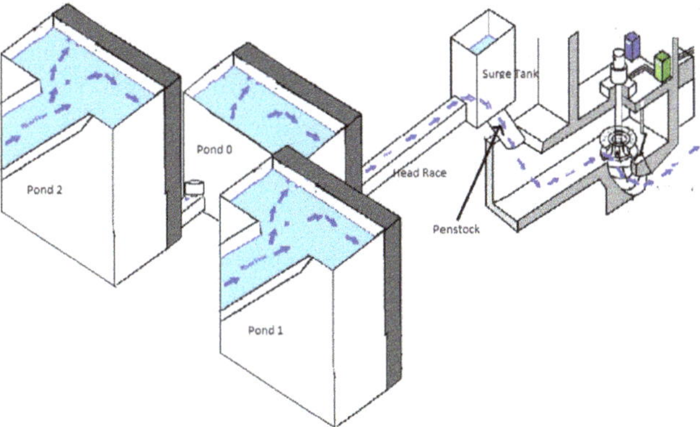

Figure 1. Three-pond hydraulic system.

The equations describing the hydraulic part will be discussed according to [1]. A modification in the model will be proposed, and its interaction with a Francis turbine will be discussed. The head ponds are described by the rate of change of their water levels. The rate of change of the water level is a function of its area and the net difference of the outflows and inflows. For pond 1 the equation is given by [1].

$$\frac{d}{dt}x_1 = -\frac{n\,a}{A}s_1\sqrt{2g|x_1 - x_0|}sgn(x_1 - x_0) + \frac{U_1}{A} \tag{1}$$

where x_1 is the water level in pond 1, and A is the cross-sectional area of the pond. The duct connecting ponds 1 and 0 have a cross-sectional area, a. The controllable valve s_1 is between ponds 1 and 0. The constants are n, the dimensionless water flow constant of value 0.98 [16], and g, the gravitational constant of 9.8 m/s^2.

In Equation (1), $sgn(z)$ is defined as

$$sgn(z) = \begin{cases} 1 & z > 0 \\ 0 & z = 0 \\ -1 & z < 0 \end{cases} \tag{2}$$

and $|z|$ is

$$|z| = \begin{cases} z & z \geq 0 \\ -z & z < 0 \end{cases} \tag{3}$$

Furthermore, note that the flow between two tanks or ponds depends on the difference in the levels between them and the cross-sectional area of the duct connecting them.

Similarly, the equation describing the rate of change in the water level in pond 2 is

$$\frac{d}{dt}x_2 = -\frac{n\,a}{A}s_2\sqrt{2g|x_2 - x_0|}sgn(x_2 - x_0) + \frac{U_2}{A} \tag{4}$$

In ponds 1 and 2, the river's inflow is assumed controllable, whereas the outflow is to pond 0. The dynamics of pond 0 are described by the following equation.

$$\frac{d}{dt}x_0 = \frac{n_a}{A}\left[s_1\sqrt{2g|x_1 - x_0|}\,sgn(x_1 - x_0) + s_2\sqrt{2g|x_2 - x_0|}\,sgn(x_2 - x_0)\right] - \frac{Q_t}{A} \tag{5}$$

The equation describing the dynamics of the headrace is the same as the rate of flow in a tube, which depends on the difference in the water levels at both ends, its cross-sectional area, and the tube's length. The flow frictional losses are also accounted for by subtracting them from the expression. Therefore, the rate of flow of water in the headrace is given by

$$\frac{d}{dt}Q_t = \frac{gA_t}{L_t}(x_0 - x_s) - C_t Q_t |Q_t| \tag{6}$$

where Q_t is the flow of water through the headrace, A_t is the headrace's cross-sectional area, and L_t is the length of the headrace.

For the surge tank, the rate of change in its level x_s is given by

$$\frac{d}{dt}x_s = \frac{Q_t}{A_s} - \frac{n_a}{A_s}\sqrt{2gx_s} \tag{7}$$

where A_s is the cross-sectional area of the surge tank. The above Equations (1) and (4)–(7) describe the system from [1]. Note also that the control variables are U_i, where $i = 1, 2$. U_i is the controllable flow of river into the pond i and is varied between a maximum value and zero. The other control variables are s_j, where $j = 1, 2$. s_j is the controllable valve between ponds. s_1 is the valve between pond 1 and 0 while s_2 is the valve between pond 2 and 0. The valve s_j can be varied between zero and unity.

We have proposed two modifications to the above mentioned system. The first one is a modification to the flow of water entering ponds 1 and 2 and the second is to couple it with a turbine.

In [1], it is assumed that the inflow of river in ponds 1 and 2 are completely controllable; however, this is not practically achievable. The flow of the river can be controlled by adding a gate in its path. The river flow can have some short and long-term variations and some base flow rates. For simplicity, the river's flow is taken as a combination of the constant base flow rate and short-term variable flow rate and is modeled by a sinusoidal function. In contrast, the long-term variation is considered constant and is included in the base flow rate. Therefore, the flow rate in cubic meter per second of the river is given by

$$l = m + p\,\sin\omega t \tag{8}$$

where l is the total flow rate, m is the baseline constant flow rate, and $p\,\sin\omega t$ is the short-term variation in the flow rate. It is assumed that a controllable sluice gate w is present at the mouth of the pond from the river. Now, U is the controllable flow rate from [1], and is given by

$$U = wl \tag{9}$$

where w ranges from 0 to 1. Now, as a turbine with penstock is added to the hydraulic system, Equation (7) of the surge tank is modified. The out flow from the surge tank is changed to be a single variable instead of the expression. The outflow from the surge tank and inflow to the penstock and turbine is denoted by Q as

$$na\sqrt{2gx_s} = Q \tag{10}$$

Now one substitutes the river and turbine flow values in the model and rewrite the equations. After substituting Equation (9) in Equations (1) and (4), the equations for pond 1 and 2 become

$$\frac{d}{dt}x_1 = -\frac{n_*a}{A}s_1\sqrt{2g|x_1 - x_0|}sgn(x_1 - x_0) + \frac{w_1l_1}{A} \tag{11}$$

$$\frac{d}{dt}x_2 = -\frac{n_*a}{A}s_2\sqrt{2g|x_2 - x_0|}sgn(x_2 - x_0) + \frac{w_2l_2}{A} \tag{12}$$

Note that U_1 and U_2 are replaced by w_1l_1 and w_2l_2, which is the product of the river flow and the sluice gate opening into ponds 1 and 2. The equations for pond 0 and the headrace are unchanged.

$$\frac{d}{dt}x_0 = \frac{n_*a}{A}\left[s_1\sqrt{2g|x_1 - x_0|}sgn(x_1 - x_0) + s_2\sqrt{2g|x_2 - x_0|}sgn(x_2 - x_0)\right] - \frac{Q_t}{A} \tag{13}$$

$$\frac{d}{dt}Q_t = \frac{gA_t}{L_t}(x_0 - x_s) - C_tQ_t|Q_t| \tag{14}$$

Finally, the equation of the surge tank is modified by plugging in Equation (10) in Equation (7) as

$$\frac{d}{dt}x_s = \frac{Q_t}{A_s} - \frac{Q}{A_s} \tag{15}$$

Therefore, the system equations from the [1] are modified for the current system. The inputs of the system are

$$\vec{v} = \begin{bmatrix} U_1 & U_2 & s_1 & s_2 & Q \end{bmatrix}^T \tag{16}$$

where Q is the flow through the turbine and is dictated by the equations of the turbine, while the rest of the inputs of the system are the control variables which are given as

$$\vec{u} = \begin{bmatrix} U_1 & U_2 & s_1 & s_2 \end{bmatrix}^T \tag{17}$$

Note also that U_1, U_2 are given by the level controller, which is discussed in the next section and is equated using Equation (9).

Comparing the above Equations (16) and (17), the input of the system is

$$\vec{v} = \begin{bmatrix} \vec{u}^T & Q \end{bmatrix}^T \tag{18}$$

The states vector $\vec{x}$ consisting of the state variables is

$$\vec{x} = \begin{bmatrix} x_1 & x_2 & x_0 & Q_t & x_s \end{bmatrix}^T \tag{19}$$

The output of the hydraulic system is

$$\vec{y} = C\vec{x} \tag{20}$$

where the matrix C is given by

$$C = \begin{bmatrix} 0 & 0 & 1 & 0 & 0 \\ 1 & 0 & 0 & 0 & 0 \\ 0 & 1 & 0 & 0 & 0 \\ 0 & 0 & 0 & 0 & 1 \end{bmatrix} \tag{21}$$

Resultantly $\vec{y}$ becomes:

$$\vec{y} = \begin{bmatrix} x_0 & x_1 & x_2 & x_s \end{bmatrix}^T \tag{22}$$

After the formulation of the hydraulic system model, the model of the turbine and penstock is needed to complete the hydropower system. In this work, the three-pond hydraulic system has a Francis turbine and penstock taken from [2]. The Francis turbine is described as a function of its rotational speed n and gate opening G to achieve a flow rate Q. The equation for the flow of water from the turbine is given as

$$Q = h(G, n) \tag{23}$$

The characteristics of a Francis turbine are described by a set of points described by hill graphs [15]. For Hill graphs, a predetermined operating point of the turbine is set, and then the static values of G and n are used in the calculations. However, [2] describes the characteristics of the turbine in the form of a quadratic equation that is non dimensional. In addition to providing a dynamic model of the Francis turbine, this model is also completely scalable over different operating conditions and rated values. The equation for the Francis turbine from [2] is described as

$$\frac{Q}{Q_r} = G\left[a_1\frac{n^2}{n_r^2} + b_1\frac{n}{n_r} + c_1\right] \tag{24}$$

where n_r and Q_r are the rated speed and the rated flow rate of the turbine, and the constants a_1, b_1, c_1 are the constants within the above quadratic equation, as stated by [2]. The torque generated by the turbine is given by the following equation [2].

$$\frac{M}{M_r} = \frac{\frac{Q}{Q_r}\frac{H}{H_r}}{\frac{n}{n_r}} \tag{25}$$

where M and M_r are the torque and rated torque. H and H_r are the height or level of water and the rated height, respectively. In the case of our system, the level or height (H) of concern is x_0 and the rated height (H_r) is the height of the pond 0 so the rated height is H_0.

Now, rewriting Equation (25)for the generated torque of the turbine in our system, we get

$$\frac{M}{M_r} = \frac{Q}{Q_r}\frac{x_0}{H_0}\frac{n_r}{n} \tag{26}$$

The power generated at the turbine is given by the product of rotational speed and torque of the turbine as in [2]:

$$P = nM \tag{27}$$

The total output of the system is the output of the hydraulic system and power.

$$\vec{Y} = \begin{bmatrix} x_0 & x_1 & x_2 & x_s & P \end{bmatrix}^T \tag{28}$$

We shall be requiring the explicit values of $\vec{Y}$ in order to conveniently find the effects of the faults on the output. As the faults occur, we diagnose them by computing the values of the states by integrating the state equations rather than just taking the output value. The value of x_0 is computed by integrating Equation (13).

$$x_0 = \frac{n_a}{A}\int_0^t \left[s_1\sqrt{2g|x_1 - x_0|}sgn(x_1 - x_0) + s_2\sqrt{2g|x_2 - x_0|}sgn(x_2 - x_0)\right]dt - \int_0^t \frac{Q_t}{A}dt \tag{29}$$

Similarly, for x_1, x_2, and x_s, Equations (11), (12), and (15) are integrated.

$$x_1 = -\frac{n_a}{A}\int_0^t s_1\sqrt{2g|x_1 - x_0|}sgn(x_1 - x_0)dt + \int_0^t \frac{w_1l_1}{A}dt \tag{30}$$

$$x_2 = -\frac{n_a}{A}\int_0^t s_2\sqrt{2g|x_2 - x_0|}sgn(x_2 - x_0)dt + \int_0^t \frac{w_2l_2}{A}dt \tag{31}$$

$$x_s = \int_0^t \frac{Q_t}{A_s} dt - \int_0^t \frac{Q}{A_s} dt \tag{32}$$

Furthermore, power is calculated by Equation (27), and we plug the value of the torque from Equation (26) into Equation (27) after simplifying.

$$P = \frac{Q M_r x_0 n_r}{Q_r H_0} \tag{33}$$

Now, plugging in the value of Q from Equation (24) to Equation (33) and simplifying.

$$P = \frac{M_r x_0 n_r}{H_0} G \left[a_1 \frac{n^2}{n_r^2} + b_1 \frac{n}{n_r} + c_1 \right] \tag{34}$$

Although this step appears to be an unnecessary complication, it will help us find the residues easily in case of faults.

3. Control Design

The control system for the system is done in a pseudo hierarchal and nested style. There are three levels of control dealing with level control, level reference, and gate (power) control. Now, these three controls will be discussed separately. The simplified form of the control system with the three levels of control is given in Figure 2.

3.1. Level Controller

The primary level control governs the hydraulic part of the system. The primary level control is discussed in [1]. Level control has the controllable variables s_1, s_2, and U_1, U_2. s_1 is the valve between pond 1 and 0 while s_2 is the valve between pond 2 and 0. While U_1 and U_2 are the flow of river in the ponds 1 and 2 through the sluice gates. In this work's current scheme, the level control is essentially the same as in [1], with two differences. The first difference is discussed in the previous section regarding the controllable flow of river in the ponds 1 and 2. While the next change is just that the reference levels of the three ponds are set at the same level by the reference control, as opposed to [1], in which all the reference levels could be set at different levels. Combining Equations (1) and (4)–(7) of the hydraulic system gives us the nonlinear system of equations as

$$\frac{d}{dt} \vec{x} = \vec{f} \left(\vec{x}, \vec{u} \right) \tag{35}$$

where the state variables are

$$\vec{x} = \begin{bmatrix} x_1 & x_2 & x_0 & Q_t & x_s \end{bmatrix}^T \tag{36}$$

and the control variables are

$$\vec{u} = \begin{bmatrix} U_1 & U_2 & s_1 & s_2 \end{bmatrix}^T \tag{37}$$

Here, U_1, U_2 range from 0 to U_{Max} and s_1, s_2 range from 0 to 1. However, in the current system the flow U is equal to the product of the flow-rate of the river l and the opening of the sluice gate w, as shown in Equation (9). The controller gives U_1, U_2, the *required* flow-rate output to maintain the ponds' water levels at the desired reference level. According to Equation (9):

$$U_i = l_i w_i \tag{38}$$

where $i = 1, 2$ for ponds 1 and 2. We have control input U_i and river flow rate l_i from the river. We control the sluice gate by the expression

$$w_i = \begin{cases} \frac{U_i}{l_i} & \frac{U_i}{l_i} \leq 1 \\ 1 & \frac{U_i}{l_i} > 1 \end{cases} \tag{39}$$

This ensures that U_i cannot exceed l_i, the actual flow-rate in the river; the actual controlled system is slightly slower to reach its steady state as compared to [1], as it makes a more realistic approach towards the problem.

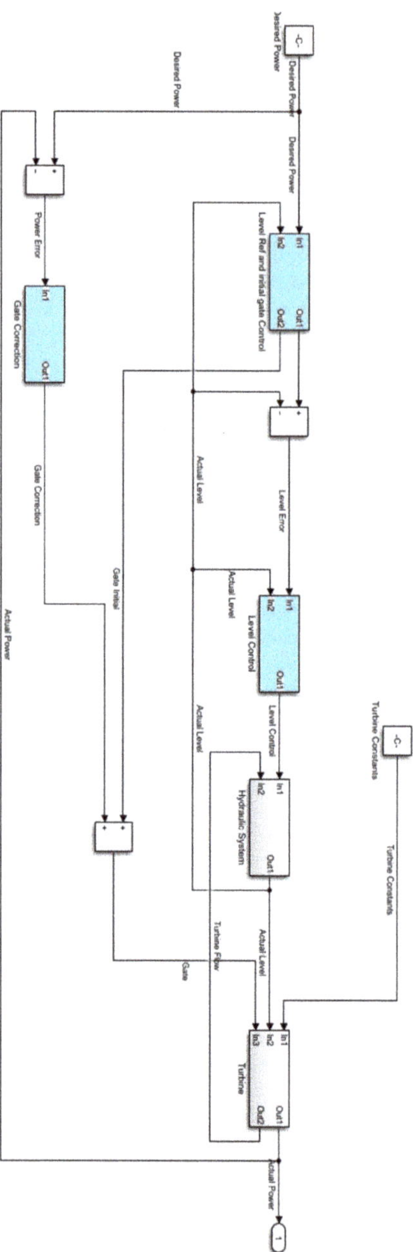

Figure 2. Control scheme for the system showing three different controllers (in color).

A fuzzy controller is used for level control due to its flexibility and ease of controlling different nonlinear and complex applications. Another advantage is its speed and higher accuracy compared to traditional PID controllers. A fuzzy controller is a linguistic and intuitive controller as it takes the input in the form of linguistic variables and gives output in the same way. A fuzzy system consists of a set of IF-THEN rules. In the first part, the input variable is a crisp value that is fuzzified to convert it into a linguistic variable. The system takes input in the form of linguistic variables. It then gives output in the form of a linguistic variable for the Mamdani type system or a linear combination of inputs for a Takagi Sugeno type system. Similarly, the output linguistic variable is defuzzified to convert it into the crisp output. For example, a two-input fuzzy system will have a rule of this form:

$$Rule1 : IF\ x_1\ is\ A_1^1\ and\ x_2\ is\ A_2^1\ THEN\ y\ is\ B^1 \tag{40}$$

As shown in Figure 2, level control is driven by the state errors and the states of the hydraulic system. The actual level control is the same as [1] and is done using Mamdani fuzzy control. The level reference is given by the level reference control, which is discussed next. The levels $x_1\ x_2\ x_0$ of the ponds 1, 2, and 0 are given the membership functions *Low, Medium,* and *High*. Similarly, the membership functions of the errors $E_1\ E_2\ E_0$ are given as *Negative, Near Zero,* and *Positive*. The outputs have five membership functions each, from L_0 to L_4, with L_0 indicating the complete closure of the valve or river water input, while L_4 means full throttle. The number of rules for the system depends on the number of inputs and the total number of membership functions. So, there are six membership functions for each of the three ponds. The total possible rules are 729, the same as the total possible combinations, which is rather computationally intensive. Therefore, the level control is divided into two parts. One control is for pond 1 and 0, while the second control is for ponds 2 and 0. Dividing the control in two parts reduces the number of rules to 81 for each controller. It may appear that a clash may theoretically exist between level control 1 and 2, but as the level reference of pond 0 is the same for both controls, no such clash exists. Although this level control can maintain the ponds at different levels, as discussed in [1], all the three references are set at the same level by the level reference controller. The controllable variables are the valve position s and the inflow U, which are nonzero numbers ranging from 0 (close) to full throttle. Therefore, the levels are adjusted by balancing the inflows and the outflows to increase, decrease, and maintain the ponds' levels. Depending on the required levels of the three ponds, the controller can achieve and maintain the water levels in the ponds. In the current case, the reference controller sets the reference levels of all three ponds at the same level. The controller currently does not use the level controller's ability to achieve and maintain the water levels at different levels.

We shall consider one of the two controllers for discussion as the second controller mirrors the first. The Mamdani controller works in three main modes, each for when the current levels are below, on, or higher than the required levels. When the actual level is lower than the required level in the ponds, the sluice gate (inflow) is set to maximum or high so that pond 1 reaches the required level. The valve s_1 is kept at full throttle, so that pond 0 also reaches the required level. When errors are near zero, both the inflow and the valve s_1 are kept at level 1 where inflows and outflows are balanced and the water level is retained in the ponds. Finally, if the actual water level exceeds the required level, the inflow is stopped, and valve s_1 is kept at full throttle so that the levels reduce and reach the required level. Another small addition to this is that the controller prioritizes the level of pond 0. That means if somehow the level of pond 1 is increased due to some disturbance or unknown noise, the disturbance will not appear in pond 0 as the controller prioritizes the level of pond 0. The controller does not allow pond 1 to decrease to the required level if it means to increase the level of pond 0 above its required level. An example of the rule when the system is in an equilibrium state is given below:

$$IF\ E1\ is\ ZERO\ and\ E0\ is\ ZERO\ and\ X1\ is\ HIGH\ and\ X0\ is\ HIGH$$
$$THEN\ U1\ is\ L1\ and\ S1\ is\ V1 \tag{41}$$

All of the 81 rules for each level controller 1 and 2 have a generalized form:

$$u(t) = \frac{\sum_{j=1}^{81} \overline{u}_j \left(\prod_i^4 \mu_{A_i^j}(k_i) \right)}{\sum_{j=1}^{81} \left(\prod_i^4 \mu_{A_i^j}(k_i) \right)} \tag{42}$$

where $u(t)$ is the controller's output for U_1 or s_1, k_i is the input value of x_0, x_1, E_0, and E_2. The linguistic variable for input, which can be $NEG, ZERO, POS$ or LOW, MED, HI, is A; u_A is the membership function of A; and $\overline{u}_j$ is the firing strength of the jth fuzzy rule.

3.2. Level Reference Controller

The second controller that controls both the hydraulic system and turbine is also a fuzzy controller. The reference controller takes the input in the form of required power and the level of the pond 0. This controller requires the exact value of the required power or at least a scaled version of it. Therefore, the hydraulic system parameters are taken from [1], and the parameters of the hydraulic turbine are taken arbitrarily to match with the hydraulic system. After that, the total theoretical power is calculated, which comes around 2 Mega Watts of power. Next, the minimum level of the pond is also decided arbitrarily as it is very rare for the head pond of a power plant to be empty. That minimum level and minimum power are arbitrarily decided as 20 m and 800 kilo watts.

The reference controller does not run the errors as opposed to the level controller but with the reference power and the feedback from the pond 0 (level). The head ponds' operating levels range from 20 m to a maximum of 35 m [1], and the operating power is between 800 kW and 2 MW.

The reference controller is in the form of the Takagi Sugeno Fuzzy Inference Engine instead of Mamdani Fuzzy Inference Engine for the previously discussed level controller. The Takagi Sugeno controller is different from the Mamdani controller in the form as its output is a constant or a linear combination of its inputs for each rule.

The function of the reference controller is to set a level reference for the ponds and give the initial base position to the gate of the turbine. As the power plant operates between 800 kW and 2 MW and from the 20 to 35 m pond level, the available power for the turbine is given by the following equation:

$$P = \eta \rho g H Q \tag{43}$$

where η is the turbine efficiency, ρ is the density of water, g is the gravitational constant, and H and Q are the level of water in pond and the flow rate through the turbine, respectively. As the efficiency of the turbine is not in our control, as the model taken from [2] assumes constant efficiency for the turbine, and the density of water and gravitational constants are fixed, the level of the water in head pond and the flow rate through the turbine can be adjusted. According to Equation (34), the turbine's rotational speed can also be used as a control parameter, but we kept it constant and is left for future work. The flow rate through the turbine is adjusted by adjusting the opening of the turbine wicket gate. The pond level is divided from 20 m to 35 m in increments of 1 m, and the operating power is adjusted from 800 kW to 2 MW in increments of 25 kW. The position of the turbine gate is estimated against the required power and water level while fixing the other parameters. This estimation is the basis for the rules of level reference and initial gate position of the turbine.

The inputs of the level reference control are the required power and the level of pond 0. These inputs are assigned membership functions as follows: The pond level can be varied from 20 to 35 m, so the level is assigned four membership functions. The first membership function deals with levels less than 20 m, while the rest three deal with from 20 to 35 m in increments of 5 m. The level membership functions are named as m_{20}, m_{25}, m_{30}, and m_{35}, respectively. All four level membership functions are trapezoidal, with the first membership function being non symmetric to describe all the levels less than 20 m. The required power level ($REQP$) is divided into increments of 25 kW between 800 kW and 1.95 MW. That means that 45+2 membership functions describe the required power. The 45 membership functions are between 800 kW and 1.95 MW, while the other two represent power levels less than 800 kW and greater than 1.95 MW. The inner 45 membership functions are triangular, while the boundary membership functions are trapezoidal. Therefore, the total number of rules for the level reference controller is 4 × 47 or 188. The controller requires the current level of the head pond to give the level reference; for when the required power level is reduced, and the current level of the pond is higher than the theoretical level, to stop the controller from quickly reducing the level of the pond. The other aspect to know about the current level is to set the initial position of the turbine gate according to the current level rather than the desired level. An example of a rule is given below.

$$IF\ REQP\ is\ 165P\ and\ LEVEL\ is\ m_{25}\quad THEN\ LEVREF\ is\ 30\ and\ G_{Fuzzy}\ is\ 0.92 \tag{44}$$

All of the 188 rules for the level reference controller have a generalized form:

$$u(t) = \frac{\sum_{j=1}^{188} \overline{u}_j \left(\prod_i^2 \mu_{A_i^j}(k_i) \right)}{\sum_{j=1}^{188} \left(\prod_i^2 \mu_{A_i^j}(k_i) \right)} \tag{45}$$

where $u(t)$ is the output of the controller for $LEVREF$ or G_{Fuzzy}, k_i is the input value of $REQP$ and $LEVE$. The linguistic variable for the input, which has been described above is, A; u_A is the membership function of A; and $\overline{u}_j$ is the firing strength of the jth fuzzy rule.

3.3. Turbine Gate Correction Control

The turbine gate correction control is the final fine adjustment control for the turbine gate. The level reference control discussed earlier gives the required level to the level control and gives the turbine gate's initial gate position. The final adjustment to the turbine gate is made using a traditional PID controller. The output of the PID controller is added to the previously assigned initial position of the gate by the level reference controller. Therefore, the final value of the turbine gate becomes

$$G_{FINAL} = G_{Fuzzy} + G_{PID} \tag{46}$$

The PID control is a traditional control scheme that is driven by the error between the required power and generated power, and the output of controller is given by

$$G_{PID} = K_P e(t) + K_I \int_0^t e(t)dt + K_D \frac{d}{dt} e(t) \tag{47}$$

where $e(t)$ is the power error, and K_P, K_I, and K_D are the proportional, integral, and derivative constants of the PID controller.

By the interaction of these three controllers, the required levels of the ponds are maintained, and the generated power comes within a hundred watts of the required power.

4. Fault Model

A few commonly occurring faults were modeled and added to the system to make fault diagnostics and a fault-tolerant control. These faults were added to the system model, detected, and identified by the fault diagnostic module, and then suggested fault-tolerant control action is taken. In this work, currently, only saturation and leakage faults are considered.

The saturation and leakage faults most commonly occur within the valves and gates due to age, obstruction, or a weakened or faulty actuator. Saturation fault occurs when the actuator cannot fully open the valve or gate; this reduces flow which is amplified by propagating through the system and affects the output by making it sluggish or having lower than the desired value. Similarly, a leakage fault is the same as but opposite to a saturation fault. During a leakage fault, the gate or valve cannot close or shut completely, resulting in a constant leakage, having the output creeping higher than the desired value, or having a sluggish response when braking or lowering the desired states of the system. During the normal mode of operation (i.e., from zero or lower initial conditions to higher desired states), the effects of saturation fault appear during the transient state but are not present during the steady state. In comparison, the case is vice-versa for leakage faults where their effects appear in the steady state but are invisible in the transient state. Although this rule is not strictly followed, and it will be seen later when these faults are added to the system. As a general rule of thumb, the faults whose effects appear in the steady state are more severe than those whose effects only appear during the transient state.

From a purely fault-tolerant point of view, the faults whose effects are only present during the transient state may not strictly require corrective action. As in this case, the system is merely somewhat slower than a normal system unless it is so slow that it becomes undesirable. In comparison, a fault whose effect is evident in the steady state has to be dealt with using corrective action to keep the undesirable effects of the faults to a minimum. However, detecting, identifying, and isolating both kinds of faults are very important for early warning and avoiding the system from failing.

If we consider a single equation for a generalized linear system, then

$$\dot{x} = Ax + Bu \tag{48}$$

where A is the state gain matrix and B is the input gain matrix. In case of the occurrence of a fault, the equation is modified as follows:

$$\dot{x} = Ax + Bu + Ef \tag{49}$$

where E is the fault gain and f is the fault. In case of a nonlinear system with a nonlinear fault, the system becomes

$$\dot{x} = g(x, u, f) \tag{50}$$

Now let us consider our system equations and see how they are affected by the faults. The types of faults being considered are saturation, leakage, and a combination of the two. The components affected by these faults are the sluice gates w_1 and w_2, which allow the river's flow in ponds 1 and 2, the controllable valves s_1 and s_2 between the ponds T_1, T_0 and T_1, T_0, and finally, the wicket gate G of the Francis turbine. Sometimes the nonlinear fault can be separated as an additive or a multiplicative function, which is sometimes not easily possible. First, let us describe the faults mathematically before finding their effect on the components. The saturation fault, nonlinear in nature, is described as

$$f^1(z) = \begin{cases} z & z < Sat \\ Sat & z \geq Sat \end{cases} \tag{51}$$

where z is an arbitrary faulty component, and Sat is the constant saturation limit of the effect of the fault. However, plugging this value of fault into the system is not feasible. Therefore, we transform this nonlinear fault as an additive fault to the system. The additive effect may be time varying or nonlinear, as is needed by the system. The fault f^1 is described in its additive form as

$$f^1(z) = \begin{cases} z & z < Sat \\ z + f_z^1 & z \geq Sat \end{cases} \tag{52}$$

In the term f_z^1, z is the faulty component and f_z^1 is the time varying term that satisfies the equation below.

$$z + f_z^1 = Sat \tag{53}$$

Similarly, the leakage fault is defined as

$$f^2(z) = \begin{cases} Leak & z < Leak \\ z & z \geq Leak \end{cases} \tag{54}$$

Here, *Leak* is the constant leakage effect of the fault. Now, one can describe the leakage fault in additive form as

$$f^2(z) = \begin{cases} z + f_z^2 & z < Leak \\ z & z \geq Leak \end{cases} \tag{55}$$

The time-varying term f_z^2 satisfies the following equation:

$$z + f_z^2 = Leak \tag{56}$$

Similarly, when these two types of faults occur at the same time, it is defined as f^3, which is defined as

$$f^3(z) = \begin{cases} Sat & z > Sat \\ z & Leak \leq z \leq Sat \\ Leak & z < Leak \end{cases} \tag{57}$$

To describe the f^3 type of fault, f_z^3 is not needed, as shown in the additive form of the fault below.

$$f^3(z) = \begin{cases} z + f_z^1 & z > Sat \\ z & Leak \leq z \leq Sat \\ z + f_z^2 & z < Leak \end{cases} \tag{58}$$

The above equations from Equation (51) to Equation (58) shall be referred to multiple times to define f_z^i, where the faulty component is z and $i = 1, 2$, depending on whether the component is in saturation or leakage fault. This will be used extensively to convert the fault equations into additive forms.

Now let us look at how these faults affect the components of the system. We shall examine the sluice gate, pond valve, and turbine wicket gate fault one by one. There are two sluice gates and pond valves; only one of each will be examined to avoid redundancy.

4.1. Sluice Gate Faults

The sluice gates are on the opening of the river to the ponds T_1 and T_2. Although, the level controller assumes that the flow of the river into the ponds is controllable explicitly, the sluice gates w_1 and w_2 control the inflow by monitoring the river flow using Equation (38). The equation for pond 1 T_1, Equation (11), is rewritten adding the effect of fault f^1 below.

$$\frac{d}{dt}x_1 = -\frac{n_a}{A}s_1\sqrt{2g|x_1 - x_0|}sgn(x_1 - x_0) + \frac{f^1(w_1)l_1}{A} \tag{59}$$

According to Equation (52), the effect of the sluice gate fault in the saturation region can be written as a sum with the $f_{w_1}^1$ term. When the effect of the f^1 fault appears for the sluice gate, Equation (57) is written as

$$\frac{d}{dt}x_1 = -\frac{n_a}{A}s_1\sqrt{2g|x_1 - x_0|}sgn(x_1 - x_0) + \frac{w_1 l_1}{A} + \frac{f_{w_1}^1 l_1}{A} \tag{60}$$

Note that the value of $f_{w_1}^1$ is negative in the above case. To find the fault's total effect, we integrate the above Equation (60).

$$x_1 = -\frac{n_a}{A}\int_0^t s_1\sqrt{2g|x_1 - x_0|}sgn(x_1 - x_0)dt + \int_0^t \frac{w_1 l_1}{A}dt + \int_0^t \frac{f_{w_1}^1 l_1}{A}dt \tag{61}$$

Similarly, for the case of a leakage fault, the equation becomes

$$\frac{d}{dt}x_1 = -\frac{n\,a}{A}s_1\sqrt{2g|x_1-x_0|}sgn(x_1-x_0) + \frac{f^2(w_1)l_1}{A} \tag{62}$$

Going through similar steps and defining $f^2_{w_1}$ the fault f^2 is converted to an additive form and the effect of the fault in leakage mode is written as

$$\frac{d}{dt}x_1 = -\frac{n\,a}{A}s_1\sqrt{2g|x_1-x_0|}sgn(x_1-x_0) + \frac{w_1l_1}{A} + \frac{f^2_{w_1}l_1}{A} \tag{63}$$

The total effect of the fault is given by integrating the above Equation (63) as before:

$$x_1 = -\frac{n\,a}{A}\int_0^t s_1\sqrt{2g|x_1-x_0|}sgn(x_1-x_0)dt + \int_0^t \frac{w_1l_1}{A}dt + \int_0^t \frac{f^2_{w_1}l_1}{A}dt \tag{64}$$

Using the similar reasoning the effect of third type of fault appears in the equation as

$$\frac{d}{dt}x_1 = -\frac{n\,a}{A}s_1\sqrt{2g|x_1-x_0|}sgn(x_1-x_0) + \frac{f^3(w_1)l_1}{A} \tag{65}$$

Since the f^3 fault is a combination of the saturation and leakage faults, the additive form of the fault f^3 on the sluice gate w_1 is described by Equations (61) and (64), depending on whether the fault is in the saturation or leakage region. The fault model for w_2 is the same as above and discussing it will be redundant.

4.2. Pond Valve Fault

In the case of a fault in the valve s_1, its effect appears in two equations. The equations are of pond 1 and pond 0. The equation for pond 1, Equation (11), with the occurrence of fault f^1, is modified as follows:

$$\frac{d}{dt}x_1 = -\frac{n\,a}{A}f^1(s_1)\sqrt{2g|x_1-x_0|}sgn(x_1-x_0) + \frac{w_1l_1}{A} \tag{66}$$

Similarly, the equation for the head pond 0 (13) is modified as follows

$$\frac{d}{dt}x_0 = \frac{n\,a}{A}\left[f^1(s_1)\sqrt{2g|x_1-x_0|}sgn(x_1-x_0) + s_2\sqrt{2g|x_2-x_0|}sgn(x_2-x_0)\right] - \frac{Q_t}{A} \tag{67}$$

Rewriting the faults in additive form according to Equation (50) and defining $f^1_{s_1}$, the equation for pond 1 in the saturation fault region is

$$\frac{d}{dt}x_1 = -\frac{n\,a}{A}s_1\sqrt{2g|x_1-x_0|}sgn(x_1-x_0) - \frac{n\,a}{A}f^1_{s_1}\sqrt{2g|x_1-x_0|}sgn(x_1-x_0) + \frac{w_1l_1}{A} \tag{68}$$

Likewise, the equation for pond 0 during the saturation region of the fault is

$$\frac{d}{dt}x_0 = \frac{n\,a}{A}\left[s_1\sqrt{2g|x_1-x_0|}sgn(x_1-x_0) + f^1_{s_1}\sqrt{2g|x_1-x_0|}sgn(x_1-x_0) + s_2\sqrt{2g|x_2-x_0|}sgn(x_2-x_0)\right] - \frac{Q_t}{A} \tag{69}$$

The total effect of the fault on pond 1 is given by integrating Equation (68).

$$x_1 = -\frac{n\,a}{A}\int_0^t s_1\sqrt{2g|x_1-x_0|}sgn(x_1-x_0)dt - \frac{n\,a}{A}\int_0^t f^1_{s_1}\sqrt{2g|x_1-x_0|}sgn(x_1-x_0)dt + \int_0^t \frac{w_1l_1}{A}dt \tag{70}$$

Similarly, the total effect of fault on pond 0 is given by integrating Equation (69).

$$x_0 = \frac{n\,a}{A}\int_0^t \left[s_1\sqrt{2g|x_1-x_0|}sgn(x_1-x_0) + f^1_{s_1}\sqrt{2g|x_1-x_0|}sgn(x_1-x_0) + s_2\sqrt{2g|x_2-x_0|}sgn(x_2-x_0)dt\right] - \int_0^t \frac{Q_t}{A}dt \tag{71}$$

We shall be delving more deeply into the effects of the saturation fault of the pond valve in the Appendix A. Using similar reasoning as above, the occurrence of fault f^2 on the valve s_1 modifies the equation of the pond 1 Equation (11) as follows:

$$\frac{d}{dt}x_1 = -\frac{n_v a}{A}f^2(s_1)\sqrt{2g|x_1 - x_0|}sgn(x_1 - x_0) + \frac{w_1 l_1}{A} \tag{72}$$

Similarly, the equation of the pond 0 (13) with the occurrence of fault is:

$$\frac{d}{dt}x_0 = \frac{n_v a}{A}\left[f^2(s_1)\sqrt{2g|x_1 - x_0|}sgn(x_1 - x_0) + s_2\sqrt{2g|x_2 - x_0|}sgn(x_2 - x_0)\right] - \frac{Q_t}{A} \tag{73}$$

Now, defining $f_{s_1}^2$ and rewriting the fault equations for pond 1 and pond 0, the equation for pond 1 during the effects of leakage fault is given as

$$\frac{d}{dt}x_1 = -\frac{n_v a}{A}s_1\sqrt{2g|x_1 - x_0|}sgn(x_1 - x_0) - \frac{n_v a}{A}f_{s_1}^2\sqrt{2g|x_1 - x_0|}sgn(x_1 - x_0) + \frac{w_1 l_1}{A} \tag{74}$$

Similarly, the equation for pond 0 during the effects of leakage fault is

$$\frac{d}{dt}x_0 = \frac{n_v a}{A}\left[s_1\sqrt{2g|x_1 - x_0|}sgn(x_1 - x_0) + f_{s_1}^2\sqrt{2g|x_1 - x_0|}sgn(x_1 - x_0) + s_2\sqrt{2g|x_2 - x_0|}sgn(x_2 - x_0)\right] - \frac{Q_t}{A} \tag{75}$$

The total effect of the fault on pond 1 is given by integrating Equation (74).

$$x_1 = -\frac{n_v a}{A}\int_0^t s_1\sqrt{2g|x_1 - x_0|}sgn(x_1 - x_0)dt - \frac{n_v a}{A}\int_0^t f_{s_1}^2\sqrt{2g|x_1 - x_0|}sgn(x_1 - x_0)dt + \int_0^t \frac{w_1 l_1}{A}dt \tag{76}$$

Similarly, the total effect of fault on pond 0 is given by integrating Equation (75).

$$x_0 = \frac{n_v a}{A}\int_0^t\left[s_1\sqrt{2g|x_1 - x_0|}sgn(x_1 - x_0) + f_{s_1}^2\sqrt{2g|x_1 - x_0|}sgn(x_1 - x_0) + s_2\sqrt{2g|x_2 - x_0|}sgn(x_2 - x_0)\right]dt - \int_0^t \frac{Q_t}{A}dt \tag{77}$$

Now when describing the fault combination f^3, which is actually a combination of the first two faults, the equation for pond 1 is

$$\frac{d}{dt}x_1 = -\frac{n_v a}{A}f^3(s_1)\sqrt{2g|x_1 - x_0|}sgn(x_1 - x_0) + \frac{w_1 l_1}{A} \tag{78}$$

and the fault equation for pond 0 is

$$\frac{d}{dt}x_0 = \frac{n_v a}{A}\left[f^3(s_1)\sqrt{2g|x_1 - x_0|}sgn(x_1 - x_0) + s_2\sqrt{2g|x_2 - x_0|}sgn(x_2 - x_0)\right] - \frac{Q_t}{A} \tag{79}$$

As the fault f^3 of the valve s_1 behaves like f^1 or f^2 depending on whether the fault is in the saturation or leakage region, the effects of f^3 are described by Equations (69,70,76,77), respectively.

4.3. Turbine Wicket Gate Fault

Now let, us look at the fault model for the turbine wicket gate. In the case of the occurrence of fault f^1 in the turbine wicket gate, the effect appears in the turbine Equation (24) as

$$\frac{Q}{Q_r} = f^1(G)\left[a_1\frac{n^2}{n_r^2} + b_1\frac{n}{n_r} + c_1\right] \tag{80}$$

Now, if we convert the fault to the additive form using Equation (52) and define f_G^1, the effect of the saturation fault for the turbine in the saturation region is given by the following equation:

$$\frac{Q}{Q_r} = G\left[a_1\frac{n^2}{n_r^2} + b_1\frac{n}{n_r} + c_1\right] + f_G^1\left[a_1\frac{n^2}{n_r^2} + b_1\frac{n}{n_r} + c_1\right] \tag{81}$$

We find the effect of turbine wicket gate fault on power by plugging in Equation (81) to Equation (34), so the faulty power becomes

$$P = \frac{M_r x_0 n_r}{H_0}G\left[a_1\frac{n^2}{n_r^2} + b_1\frac{n}{n_r} + c_1\right] + \frac{M_r x_0 n_r}{H_0}f_G^1\left[a_1\frac{n^2}{n_r^2} + b_1\frac{n}{n_r} + c_1\right] \tag{82}$$

In the case of leakage fault f^2 of the turbine gate, the equation of the turbine is

$$\frac{Q}{Q_r} = f^2(G)\left[a_1\frac{n^2}{n_r^2} + b_1\frac{n}{n_r} + c_1\right] \tag{83}$$

The effect of the leakage fault is written by converting the above Equation (83) to the additive form by defining f_G^2 using Equation (55). The additive form of the turbine wicket gate leakage fault in the faulty region is given by

$$\frac{Q}{Q_r} = G\left[a_1\frac{n^2}{n_r^2} + b_1\frac{n}{n_r} + c_1\right] + f_G^2\left[a_1\frac{n^2}{n_r^2} + b_1\frac{n}{n_r} + c_1\right] \tag{84}$$

Similarly, the effect of the fault on power is given by

$$P = \frac{M_r x_0 n_r}{H_0}G\left[a_1\frac{n^2}{n_r^2} + b_1\frac{n}{n_r} + c_1\right] + \frac{M_r x_0 n_r}{H_0}f_G^2\left[a_1\frac{n^2}{n_r^2} + b_1\frac{n}{n_r} + c_1\right] \tag{85}$$

Similarly, the case of the f^3 fault of the turbine wicket gate is given by

$$\frac{Q}{Q_r} = f^3(G)\left[a_1\frac{n^2}{n_r^2} + b_1\frac{n}{n_r} + c_1\right] \tag{86}$$

As it was done in earlier cases, the additive form of the turbine gate fault f^3 is given as Equation (81) or (84) for flow through the turbine. Similarly, Equation (80) or (83) gives the effect of the fault f^3 on power. These all depend on whether the turbine gate is in saturation or leakage fault mode. With the effects of the faults modeled, we shall find the effects of the faults on the residue in the next section.

5. Fault Diagnostics and Tolerance

The faults considered in this work are modeled in the previous section. Now there is the case of detection, identifying, and finally proposing a control action to mitigate the effect of the faults. In this work, fault diagnostics are done in three modes: residue generation, fault identification, and fault tolerance. The first step is residue generation and saving the said residue in a memory unit. In the second diagnostic and identification step, the fault type is identified, and a relevant fault code is generated for the fault-tolerant control. In the last step, the fault tolerant control reads the fault code and gives the corresponding corrective action to make the system fault tolerant.

A model-based fault diagnostic approach is used to detect and identify the faults and then suitable corrective action is taken to make the system fault tolerant. The first step in this direction is to generate residue for the faults.

5.1. Residue Generation

In the model-based fault-tolerant control, the residue is generated by the difference between the system's actual output and the output of the system model. As the system and its model are theoretically the same, the residues should be zero or near to it. However, there is almost always some disturbance or model uncertainties present in the residue. This makes the residue nonzero, but it is near to zero if there is minimum model mismatch and disturbance. Equations (87) and (88) give the generalized system with faults:

$$\dot{x} = Ax + Bu + Ed + Ff \tag{87}$$

$$y = Cx + Du + Hd + Jf \tag{88}$$

where E and F are the disturbance and fault distribution matrices for the state of the system, H and J are the output disturbance and fault distribution matrices, while d and f are the disturbance and fault, respectively. The system model is described by the equations

$$\dot{x} = A_m x + B_m u \tag{89}$$

$$y = C_m x + D_m u \tag{90}$$

The residue of the system is given as the difference between Equations (88) and (90).

$$r = Cx + Du + Hd + Jf - C_m x - D_m u \tag{91}$$

In the absence of the model mismatch or uncertainty, the residue will only contain faults and disturbances; therefore,

$$r = Hd + Jf \tag{92}$$

This residue is processed to identify and isolate the faults for fault diagnostics and tolerance. In our system, which consists of the hydraulic system and the turbine, the outputs are Equation (22) and Power. The combined output of the system is given by

$$\vec{Y} = [x_0 \quad x_1 \quad x_2 \quad x_s \quad P]^T \tag{93}$$

In the current case the residue is given by

$$\vec{r} = \vec{Y} - \vec{Y}_{Faulty} \tag{94}$$

Now the residue in the case of the occurrence of the faults will be examined for each fault. The explicit effects of the faults on the states are given in the previous section. Although the effects of the faults tend to propagate through the system, due to the layered structure of the control, the effects of the faults are compensated for at the next level. Even with that, the faults can affect the system, thus requiring some kind of fault identification and tolerance. The implicit effects of the faults on the system are not mathematically discussed, but they are graphically shown in the residue graphs in the results section.

5.1.1. Residue Generation of Sluice Gate Faults

In the case of the saturation fault f^1 at the sluice gate w_1, directly affected is pond 1 T_1. The residue for this fault is given as the difference between Equations (30) and (61).

$$r_1^{f^1(w_1)} = \int_0^t \frac{f_{w_1}^1 l_1}{A} dt \tag{95}$$

Here, $r_1^{f^1(w_1)}$ is the residue of pond 1 with the occurrence of fault f^1 at sluice gate w_1. Although the rest of the residues may or may not be non zero due to the implicit effects of the saturation fault, these effects are shown graphically rather than mathematically. The total residue is given as

$$
\overrightarrow{r^{f^1(w_1)}} =
\begin{bmatrix}
r_0^{f^1(w_1)} \\
\int_0^t \frac{f_{w_1}^1 l_1}{A} dt \\
r_2^{f^1(w_1)} \\
r_s^{f^1(w_1)} \\
r_P^{f^1(w_1)}
\end{bmatrix}
\tag{96}
$$

As this residue is due to the saturation fault of w_1, this residue will only be nonzero when the sluice gate w_1 is in the saturation region; this happens when the current levels of the ponds are less than the required level, and the sluice gate needs to open fully to allow the levels of the ponds to rise to the required level quickly. However, in the presence of the saturation fault in the sluice gate, the gate is not opened fully, thus restricting the water flow to the pond (pond 1 in this case). Due to this, the pond's level is less than the faultless state. Due to this pond level difference, the effect of the fault appears in the headpond 0 and is thus transmitted to power because at this level the turbine gate is also at its maximum opening, such as during normal operation, but the head pond level is lower; thus, there exists a power residue. However, once the required level is achieved, the sluice gate is no longer in saturation; thus, all residues drop to zero in steady-state mode.

In the case of the leakage fault f^2 at the sluice gate w_1, directly affected is pond 1. The residue for this fault is given as the difference between Equations (30) and (64).

$$
r_1^{f^2(w_1)} = \int_0^t \frac{f_{w_1}^2 l_1}{A} dt
\tag{97}
$$

Here, $r_1^{f^2(w_1)}$ is the residue of pond 1 with the occurrence of fault f^2 at the sluice gate w_1. The rest of the residues will also be non zero due to the implicit effects of the saturation fault; these effects are shown graphically rather than mathematically. The total residue is given as

$$
\overrightarrow{r^{f^2(w_1)}} =
\begin{bmatrix}
r_0^{f^2(w_1)} \\
\int_0^t \frac{f_{w_1}^2 l_1}{A} dt \\
r_2^{f^2(w_1)} \\
r_s^{f^2(w_1)} \\
r_P^{f^2(w_1)}
\end{bmatrix}
\tag{98}
$$

Unlike the saturation fault, the effects of the leakage fault are not present when the current levels of the ponds are below the required level as the sluice gate is in normal operating mode. However, once the system is in steady-state mode, the sluice gate enters the leakage mode, and the effects of the fault appear in the residue. The effects of this fault are constrained in pond 1 for a while until it propagates to head pond 0 and then later appears as an increase in power.

In the case of fault f^3 at the sluice gate w_1, this is a combination of the above two faults and the residue is given by either Equation (96) or Equation (98), depending on whether it is in transition or steady state.

5.1.2. Residue Generation of Pond Valve Faults

In the case of the saturation fault f^1 at the valve s_1, the primary affected are pond 1 and pond 0. The residue for pond 1 is the difference between Equations (30) and (70).

$$
r_1^{f^1(s_1)} = -\frac{n_v a}{A} \int_0^t f_{s_1}^1 \sqrt{2g|x_1 - x_0|} sgn(x_1 - x_0) dt
\tag{99}
$$

The residue for pond 0 is given by the difference between Equations (29) and (71).

$$r_0^{f^1(s_1)} = \frac{n_v a}{A} \int_0^t f_{s_1}^1 \sqrt{2g|x_1 - x_0|} sgn(x_1 - x_0) dt \tag{100}$$

The rest of the residues may or may not be nonzero (its mathematics is shown in the Appendix A) due to the implicit effects of the faults, and the total residue is given as

$$\xrightarrow{r^{f^1(s_1)}} = \begin{bmatrix} \frac{n_v a}{A} \int_0^t f_{s_1}^1 \sqrt{2g|x_1 - x_0|} sgn(x_1 - x_0) dt \\ -\frac{n_v a}{A} \int_0^t f_{s_1}^1 \sqrt{2g|x_1 - x_0|} sgn(x_1 - x_0) dt \\ r_2^{f^1(s_1)} \\ r_s^{f^1(s_1)} \\ r_P^{f^1(s_1)} \end{bmatrix} \tag{101}$$

This is the residue due to the saturation fault of the valve s_1. The effects of this fault occur as Equation (71), when the valve s_1 goes to the saturation region. The valve s_1 only goes in the saturation region when the current levels of the ponds are fairly below the required level and the valve s_1 has to be fully open to make the water levels quickly rise to the required level. Thus, the effects of this fault appear only in the transition state and show their effect by slightly increasing the level of pond 1 and decreasing the level of pond 0. The fault effects are transmitted to the other parts of the system, but they disappear as soon as the system reaches steady state.

In the case of leakage fault f^2 at the valve, s_1, the primary affected are pond 1 and pond 0. The residue for pond 1 is the difference between Equations (30) and (76).

$$r_1^{f^2(s_1)} = -\frac{n_v a}{A} \int_0^t f_{s_1}^2 \sqrt{2g|x_1 - x_0|} sgn(x_1 - x_0) dt \tag{102}$$

The residue for the pond 0 is given by the difference between Equations (29) and (77).

$$r_0^{f^2(s_1)} = \frac{n_v a}{A} \int_0^t f_{s_1}^2 \sqrt{2g|x_1 - x_0|} sgn(x_1 - x_0) dt \tag{103}$$

The rest of the residues may be nonzero due to the implicit effects of the faults, and the total residue is given as

$$\xrightarrow{r^{f^2(s_1)}} = \begin{bmatrix} \frac{n_v a}{A} \int_0^t f_{s_1}^2 \sqrt{2g|x_1 - x_0|} sgn(x_1 - x_0) dt \\ -\frac{n_v a}{A} \int_0^t f_{s_1}^2 \sqrt{2g|x_1 - x_0|} sgn(x_1 - x_0) dt \\ r_2^{f^1(s_1)} \\ r_s^{f^1(s_1)} \\ r_P^{f^1(s_1)} \end{bmatrix} \tag{104}$$

Unlike the above discussed leakage fault f^2 at the sluice gate w_1, the valve's fault only affects the system when the current levels of the system are higher than the required level, and the water level in the ponds need to be reduced to the lower level. Unless the leakage fault is of very high magnitude, the effects of the fault disappear when the system reaches steady state. This is because the valve s_1 is open significantly during the steady state to equalize the inflows and the outflows.

In the case of fault f^3 to the valve s_1, the above two faults are combined. The residue is given by either Equation (101) or Equation (104) whether the current pond level of the system is lower or higher than the required level.

5.1.3. Residue Generation of Turbine Wicket Gate Faults

In the case of the saturation fault f^1 at the turbine wicket gate G, the effected part of the system is the turbine power output. Although due to some back propagation, the other parts of the residue may or may not be zero. The residue of the turbine is given by the difference between Equations (34) and (82).

$$r_P^{f^1(G)} = \frac{M_r x_0 n_r}{H_0} f_G^1 \left[a_1 \frac{n^2}{n_r^2} + b_1 \frac{n}{n_r} + c_1 \right] \tag{105}$$

The rest of the residues, which may or may not be zero, are given by

$$\overrightarrow{r^{f^1(G)}} = \begin{bmatrix} r_0^{f^1(G)} \\ r_1^{f^1(G)} \\ r_2^{f^1(G)} \\ r_s^{f^1(G)} \\ \frac{M_r x_0 n_r}{H_0} f_G^1 \left[a_1 \frac{n^2}{n_r^2} + b_1 \frac{n}{n_r} + c_1 \right] \end{bmatrix} \tag{106}$$

The saturation fault f^1 of the turbine wicket gate is fairly critical as it directly affects the output power of the turbine. The effects of this fault occur whenever the turbine gate needs to be operated in a higher flow region, which is nearly all of the time other than when the system's power level needs to be reduced from a higher level. In the case of this fault, it is imperative to have some kind of fault-tolerant scheme for the system to run within acceptable levels of performance.

In the case of the leakage fault f^2 at the turbine wicket gate G, the primary affected part is the power output. The residue for the power output is given by the difference between Equations (34) and (85).

$$r_P^{f^2(G)} = \frac{M_r x_0 n_r}{H_0} f_G^2 \left[a_1 \frac{n^2}{n_r^2} + b_1 \frac{n}{n_r} + c_1 \right] \tag{107}$$

The rest of the residue, which may or may not be zero, are given by

$$\overrightarrow{r^{f^2(G)}} = \begin{bmatrix} r_0^{f^2(G)} \\ r_1^{f^2(G)} \\ r_2^{f^2(G)} \\ r_s^{f^2(G)} \\ \frac{M_r x_0 n_r}{H_0} f_G^2 \left[a_1 \frac{n^2}{n_r^2} + b_1 \frac{n}{n_r} + c_1 \right] \end{bmatrix} \tag{108}$$

Unlike the above saturation fault, the residue for this fault only appears if the turbine needs to be shutdown or the required power is suddenly reduced to a very low level when the system is under operation.

In the case of fault f^3 at the turbine wicket gate G, the residue is given by Equation (106) or Equation (108), whether the system is in normal operation mode or needs to be shutdown.

Although there is some disturbance present in the system's output or the residue, it is assumed that the disturbances present in the power output, head pond, and surge tank are negligible. In the current case, some white noise is added to ponds 1 and 2 of the system approximating the sensor noise due to the water's bubbling or turbulent effect as it falls from the river into ponds 1 and 2. Therefore, the disturbance vector that will appear in the residue is given as

$$\vec{d} = \begin{bmatrix} 0 \\ d_1 \\ d_2 \\ 0 \\ 0 \end{bmatrix} \tag{109}$$

So, the total residue in case of the occurrence of fault is given as

$$\overleftrightarrow{r^{fi}(z)} = \begin{bmatrix} r_0^{fi(z)} \\ r_1^{fi(z)} \\ r_2^{fi(z)} \\ r_s^{fi(z)} \\ r_P^{fi(z)} \end{bmatrix} + \begin{bmatrix} 0 \\ d_1 \\ d_2 \\ 0 \\ 0 \end{bmatrix} \tag{110}$$

where z is the faulty component and i is the type of fault, both of which have been discussed above. As seen from Equation (110), the sensor noise is clearly not needed as it will tend to give a nonzero residue even if the system is running in its normal state and no faults or unexpected disturbances have occurred. As the disturbances usually have relatively high frequency components, while the system's hydraulic and power dynamics are in a lower frequency region, it makes sense to filter or average the residue to attenuate the effects of disturbance. In this case, a moving average is used to reduce the effects of disturbance. After the averaging, the residue is multiplied by a high gain to make it more sensitive to the effects of the faults. This makes the residue sensitive and makes the effects of the fault more apparent when the effects of the faults appear in the system output. The residue given to the fault identification module is given as

$$\overleftrightarrow{r^{fi(z)}_{Final}} = g_r \frac{1}{T} \int_t^{t+T} \overleftrightarrow{r^{fi(z)}}\, dt \tag{111}$$

Here, $\vec{r^{fi(z)}}_{Final}$ is the residue vector for fault identification, T is the averaging time window, and g_r is the constant gain of the residue. In this way, all the faults' residues $\vec{r^{fi(z)}}_{Final}$ are found and saved in the database for further usage. The simplified form of the residue generation and the preprocessing scheme is given in Figure 3. Note also that the 'Total System' in Figure 3 represents the whole Figure 2 as the total system, which is the hydraulic system, turbine, and its control system.

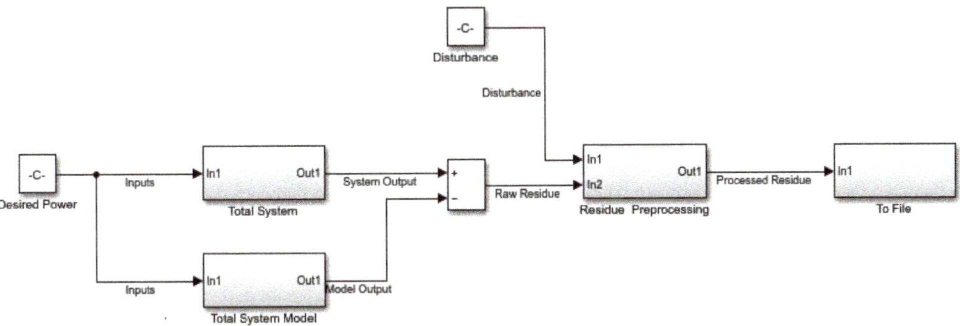

Figure 3. Residue generation and pre-processing.

5.2. Fault Identification

After the residue is generated, the next task is to identify and isolate the fault. As the residue is generated by the system and passed to the diagnostic module, it is processed to determine the system integrity and to identify any fault. The system residue is compared to all fault residues saved in the database beforehand to determine the presence or absence of a fault. The effects of the faults are more deterministic rather than probabilistic. According to assumptions, only one component can be faulty at a given time. It is more efficient to compare the generated residue with the known fault residues from the database than any other method to identify the fault. A simplified form of the fault identification process is shown in Figure 4.

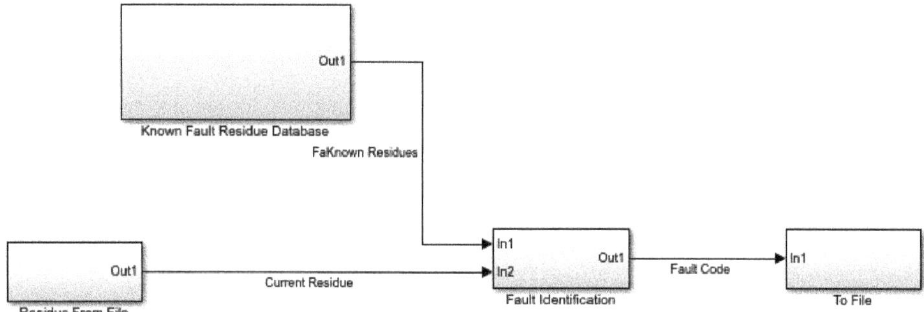

Figure 4. Fault identification module.

The residue comparison is given by taking the difference of the given residue with all the identified residues present in the database. The difference for a particular fault is given as

$$\overrightarrow{d_{diff}^{fi(z)}} = \overrightarrow{r_{Final}^{Unidentified}} - \overrightarrow{r_{Final}^{fi(z)}} \tag{112}$$

Here, $\overrightarrow{r_{Final}^{Unidentified}}$ is the unidentified residue given by the system and $\overrightarrow{r_{Final}^{fi(z)}}$ is one of the known residues of the system present in the database. Similarly, the given residue is compared with all of the residues present in the database and is prepared for the next step.

As we know that the residue is a set of five time series and so is the difference of the residues, we need to make the difference vector independent from time while retaining all the information present in the difference vector. We square all of the five parts of the difference vector first. Writing the difference vector in its expanded form:

$$\overrightarrow{d_{diff}^{fi(z)}} = \begin{bmatrix} d_0^{fi(z)} \\ d_1^{fi(z)} \\ d_2^{fi(z)} \\ d_s^{fi(z)} \\ d_P^{fi(z)} \end{bmatrix} \tag{113}$$

The vector containing all the squared time series of the difference vector is donated by $\overleftrightarrow{d_{diff}}^{fi(z)}$. We individually square all the elements of the difference vector as

$$\overleftrightarrow{D_{diff}}^{fi(z)} = \begin{bmatrix} \left(d_0^{fi(z)}\right)^2 \\ \left(d_1^{fi(z)}\right)^2 \\ \left(d_s^{fi(z)}\right)^2 \\ \left(d_s^{fi(z)}\right)^2 \\ \left(d_p^{fi(z)}\right)^2 \end{bmatrix} \tag{114}$$

This makes $\overleftrightarrow{d_{diff}}^{fi(z)}$ to retain the information in the difference vector by squaring the time series. Now, to turn the squared difference vector consisting of five time series, we integrate the squared difference vector over time to make it a time-independent vector consisting of five scalar values. The elimination of the time dependency is given as

$$\overleftrightarrow{I^{fi(z)}} = \int_0^t \overleftrightarrow{D_{diff}}^{fi(z)}\, dt \tag{115}$$

Here, $\overleftrightarrow{I^{fi(z)}}$ is the difference vector, consisting of five positive scalar elements each. Finally, the square of the magnitude of $\overleftrightarrow{I^{fi(z)}}$ is found by having a dot product with itself.

$$K^{fi(z)} = \left\| \overleftrightarrow{I^{fi(z)}} \right\|^2 = \overleftrightarrow{I^{fi(z)}} \cdot \overleftrightarrow{I^{fi(z)}} \tag{116}$$

The steps of Equation (112) to Equation (116) are repeated for each known fault residue in the database. This procedure generates a scalar for each known fault. After the scalar values are generated, the minimum of that is taken and compared to a predetermined threshold value. If the minimum value is lower than the threshold, that fault is determined to have occurred and is identified. Each fault is given a unique code that describes the fault, and then this code is passed onto the fault tolerant control.

5.3. Fault Tolerance

In the event of fault occurrence, it corrects or modifies the system output parameters for the system to behave more desirably. Fault-tolerant action may not be necessary in the case of the occurrence of every fault. In the case of the faults discussed above, there are two possible cases for the effects of faults. If the effects of the fault are only apparent during the transient state and disappear when the system enters steady state, there is no need to have dedicated fault tolerance in this case. The second case is when the faults affect the outputs when the system is in steady state, fault-tolerant action is required. However, if the faults do not affect the output critical to the system, dedicated fault tolerant action may not be required. If the fault affects the critical outputs in the steady state, fault tolerance is required to correct the critical output. In the case of this system, the critical output which needs to have fault tolerance is the output power, while the actual levels of the ponds are not important.

In this case, the faults discussed are all the actuator faults, and it is rather hard to give corrective action to an already malfunctioning actuator. For this reason, fault tolerance is achieved by giving corrective action to the reference controller. The level reference controller gives the required level to the hydraulic system. By adjusting the required level to a corrected level, the effects of the fault are minimized and suppressed, thus achieving fault tolerance. By adjusting the reference, the rest of the system will behave more desirably while still being faulty. The fault-tolerant or fault-compensation process is shown in Figure 5.

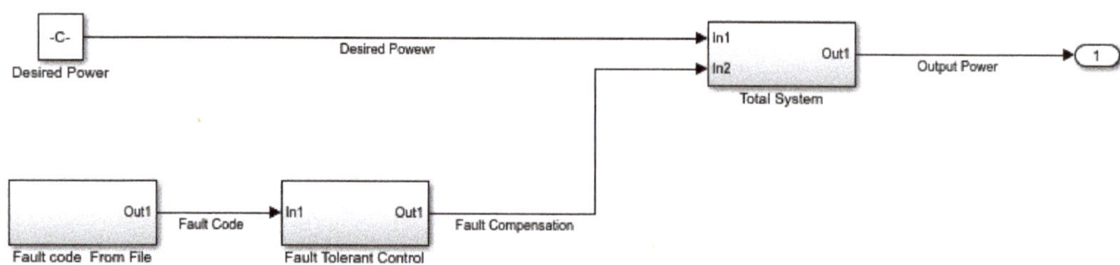

Figure 5. Fault tolerance and fault compensation process.

For each type of identified fault from the fault code given by the fault identification module, the actual name of the fault is given as output and whether it requires fault corrective action or not. In the case where fault corrective action is required, the fault corrective reference is added to the level reference as

$$Level_{RefAdj} = Level_{Ref} + Level_{RefCorr} \tag{117}$$

So, this $Level_{RefAdj}$ is given as the required level to the level controller. When the fault tolerance action is not needed, $Level_{RefCorr}$ is given as zero to the level reference. Therefore, by combining the three steps of fault diagnostics, the fault diagnostic process is of the type shown in Figure 6.

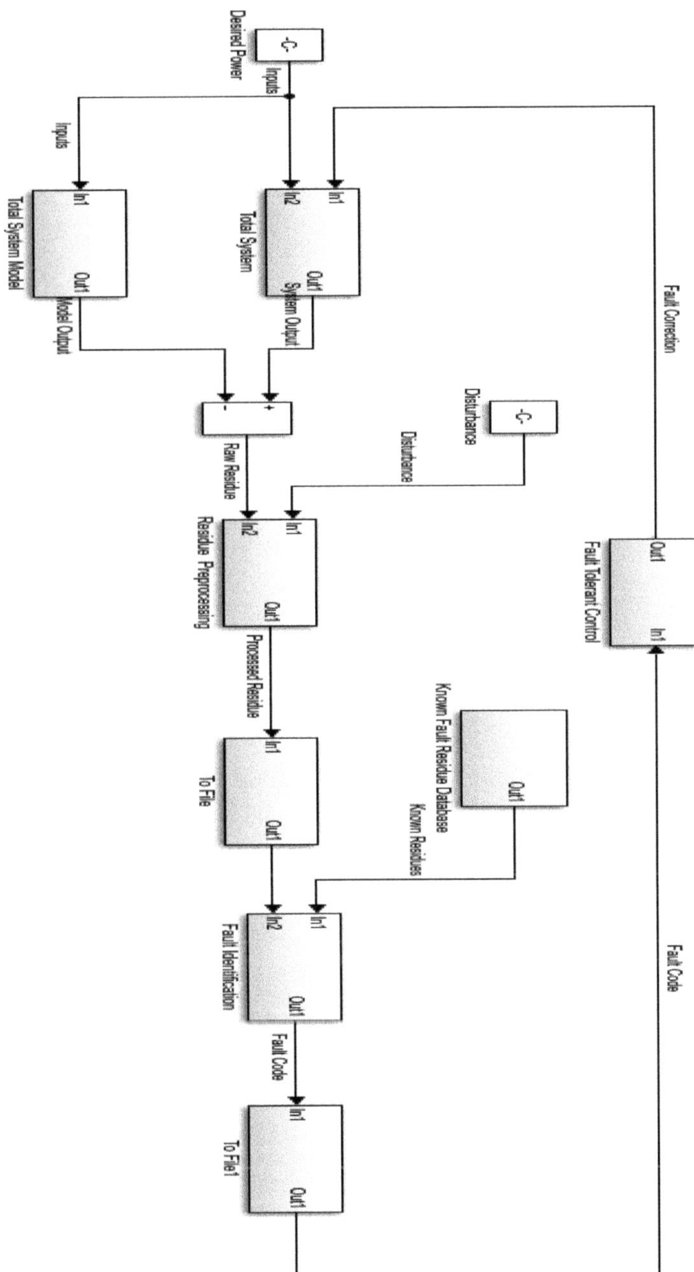

Figure 6. Fault diagnostic scheme.

6. Results and Discussion

The system was simulated in MATLAB and Simulink to analyze its validity. The basic three-pond hydraulic model is presented in [1] and was modified according to the requirements of this work. The Francis turbine model is taken from [2] and was implemented using MATLAB. First, the total system was simulated with the three-level control. After implementing faults and residue generation, the residue is saved in a memory unit. The saved unidentified residue is read by the fault identification module and identified. After identifying the fault or lack of it, a suitable error code is passed to the fault tolerant module. The system is simulated in fault-tolerant mode in case of a fault. The model parameters were taken arbitrarily but taken to be in line with the parameters of the hydraulic system [1], which was also taken arbitrarily. So, the parameters of the hydraulic system for simulation are as follows:

$H_0 = H_1 = H_2 = Height\ of\ pond\ 0, 1, 2 = 35$ m;

$H_s = Height\ of\ Surge\ Tank = 20$ m;

$C_t = Friction\ constant\ of\ headrace. = 0.98; a = Cross\ sectional\ area\ of\ the\ ducts\ between\ T_1\&T_1,\ T_2\&T_0\ = 6.25$ m^2;

$A = Area\ of\ the\ Ponds\ T_0, T_1, T_2 = 50$ m $\times\ 50$ m $= 2500$ m^2;

$A_s = Area\ of\ Surge\ Tank. = 10$ m $\times\ 10$ m $= 100$ m^2;

$L_t = Length\ of\ headrace. = 200$ m; $At = Cross-sectional\ area\ of\ headrace = 6.25$ m^2;

$U_{max} = Maximum\ controllable\ inflow\ of\ water\ in\ ponds\ T_1\ and\ T_2 = 100$ m^3/sec;

$m = Base\ inflow\ of\ water\ in\ ponds\ T_1\ and\ T_2\ = 75$ m^3/sec;

$p = The\ maximum\ transient\ inflow\ of\ water\ in\ ponds\ T_1\ and\ T_2\ = 25$ m^3/sec.

According to Equation (8) both the base and the transient flow rate will become equal at 50 m^3/sec

The parameters of the turbine penstock were arbitrarily set as

$Q_r = $ Rated flow-rate through the turbine $= 6.5$ m^2;
$n_r = $ Rated rotational speed of the turbine $= 10$ rev/$textsec$;
$n = $ Rotational speed at which the turbine is set $= 8.33$ rev/sec;
$H_r = $ Rated height of the turbine which is same as the Height of Pond $0 = 35$ m;
$M_r = $ Rated torque of the turbine $= 200{,}655$ Nm;
$P_r = $ Rated maximum power of the turbine $= 2{,}006{,}550$ Watts $\cong 2$ MegaWatts.

Although the turbine is rated at slightly more than 2 MW, the generated maximum power of the turbine maxes out at nearly 1.95 MW at the set conditions.

6.1. Power Generation

With all of the system parameters out of the way, the system operation was tested at some required power. The required power at which the system was tested was assumed to be the usual required power at which the system would operate most of the time. The required power was set at 1.65 MW, and the system's normal operation was tested. Although only the hydraulic system was tested in [1], the hydraulic system had zero initial conditions. In this work, as stated earlier, the minimum level of the head pond is 20 m. Therefore, the initial conditions for the ponds were set at 20 m. The total system was tested with the required power of 1.65 MW and the initial conditions described.

The concerned outputs of the system were the pond levels 0, 1, and 2, and the output power is shown graphically. Furthermore, due to the massive difference between the output power and pond levels, it made sense to show both graphs separately for all the cases, such as the system operation in normal or in faulty conditions; also, the level and power graphs for the residue are shown separately due to this reason.

For the case of the normal operation of the system, with a required power at 1.65 MW, the reference controller sets the required level to 30 m and the system is started from the initial conditions. The level and the power graphs are shown in Figures 7 and 8.

As shown in Figures 7 and 8, the system behaves normally and it reaches both the required levels and required power in some time with a minimum level of error for both the pond levels and the generated power in steady state. Another point to consider is that there are comparisons of the level control of this model with the traditional single-pond model in [1], which are not repeated here. Another comparison about the power regulation can be made with the results of [17], but the scale of power is different, and although [17] still uses a fuzzy controller as our work, their approach is different as [17] only deals with power regulation with a static head under different head conditions, such as a low, medium, and high head and changing from power requirement from 2 MW to 10 MW or from 10 MW to 40 MW by only controlling the flow rate through the turbine.

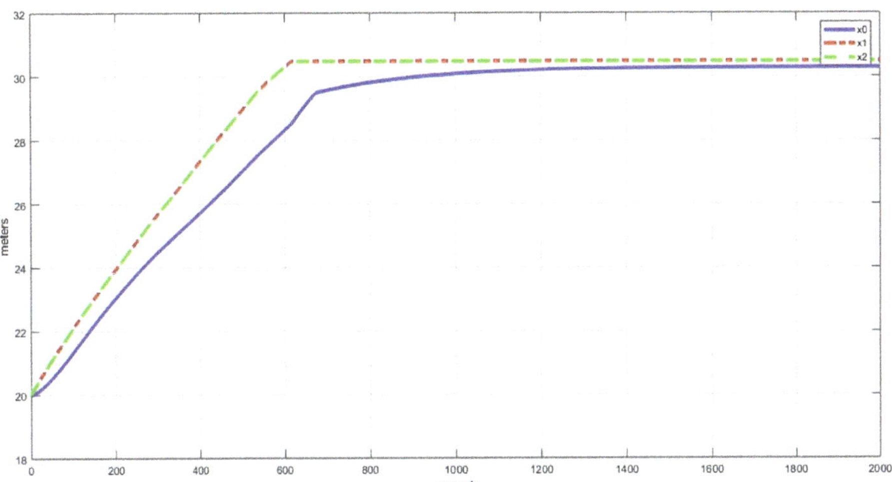

Figure 7. Pond levels (0, 1, and 2) in normal operation at the desired power of 1.65 MW.

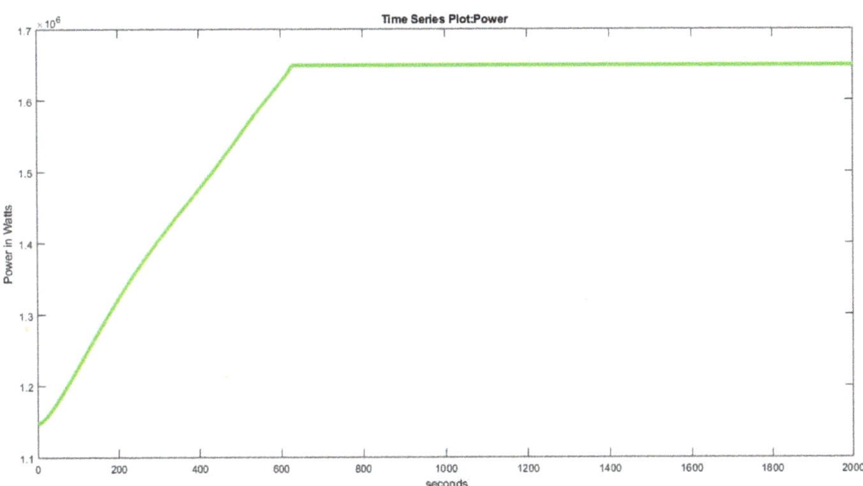

Figure 8. Power level in normal operation and desired power of 1.65 MW.

6.2. Faults and Residues

As the main crux of this work relates to residue generation and fault identification, we shall be plotting the graphs for the residues of the faults. The residue generation mode consists of a fixed time window with the standard initial conditions and standard required power, which is 1.65 MW. It is unnecessary to plot the residue for the normal operation as it will be zero with noise added to r_1 and r_2. The magnitudes of the saturation and leakage faults are arbitrarily taken at 10%; i.e., the faulty valve or gate will be in saturation mode if it is equal to or greater than 90% of its opening and it will be in leakage mode when it is equal to or less than 10% of its opening. The graphs of the residues are generated through the system sensors in MATLAB/Simulink, while to aid in calculations, the fault effects were taken mathematically. Now, let us look at the residues for each fault discussed above with the exception of those which do not have any effect on the residue in the normal mode of operation and also faults that are redundant after discussing these faults; i.e., faults of sluice gate w_2 and pond valve s_2.

6.2.1. Saturation Fault of Sluice Gate

As discussed earlier, Equation (96) is used to generate the residue of fault f^1 of the sluice gates w_1 or w_2; it also was noted earlier that the effects of the saturation fault are only visible in the residue when the sluice gate is in its saturation region. This occurs when the system is in a transient region, and the levels of the ponds are below the required levels. When the system reaches its steady state, the sluice gate is no longer in its saturation mode, and the system starts behaving in its normal mode. Therefore, the residue for this fault is only visible in the transient region of the system. Although the exact mathematical model of this fault was only given for the directly affected pond (in this case pond 1), it was also stated that the effects of this fault would be transmitted to the other ponds and the output power. The residue graph for the saturation fault of w_1 is given in Figure 9. If the same fault is in w_2, Residues 1 and 2 are swapped, while the rest of the residues will remain the same.

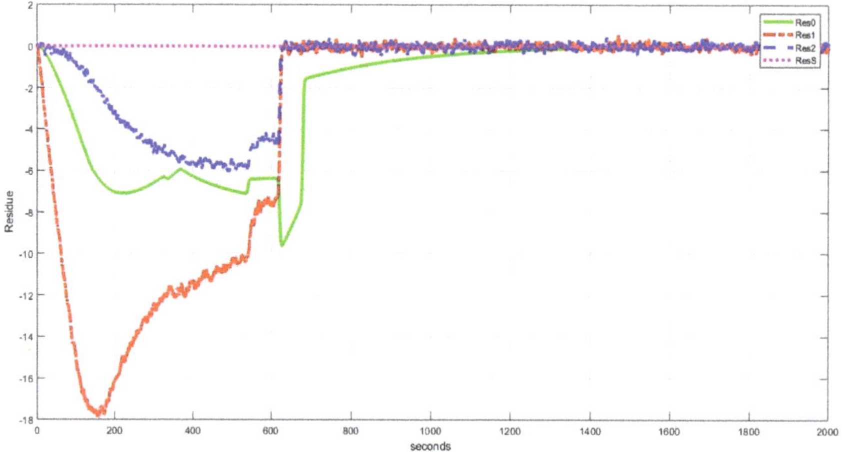

Figure 9. Pond residues for the saturation fault of sluice gate w_1.

As the magnitude and the scales of the power residue are very different from the other residues, the power residue is shown separately for all the faults discussed here. During the transient state when the pond levels are lower than the required levels, the power generated depends on the head pond level as the turbine wicket gate may be fully open. As the head pond level is lower than the normal transient state, these effects appear in the power residue in the transient state and disappear as soon as the system reaches steady state, as shown in Figure 10. In the case of the saturation fault of w_2, the power residue will remain the same.

6.2.2. Leakage Fault of the Sluice Gate

Equation (98) is used to compute the residue in case of the leakage fault f^2 of the sluice gate w_1 or w_2. As discussed before, the effects of that fault are only visible when the system is in steady-state mode or when the pond levels are greater than the required levels. When the ponds are in the steady-state mode, i.e., the current levels are equal to or greater than the required level, the sluice gate goes into leakage mode. Due to this leakage, the amount of water entering the pond is greater than the water leaving it, causing a slow rise in the water level. The effect is more apparent in the primarily affected pond, then transmitting to head pond 0 and forward to the power and surge tank and back to the other pond. The residue graphs for the leakage fault f^2 of the sluice gate w_1 are shown in Figure 11. In the case of fault of w_2, residues 1 and 2 are swapped.

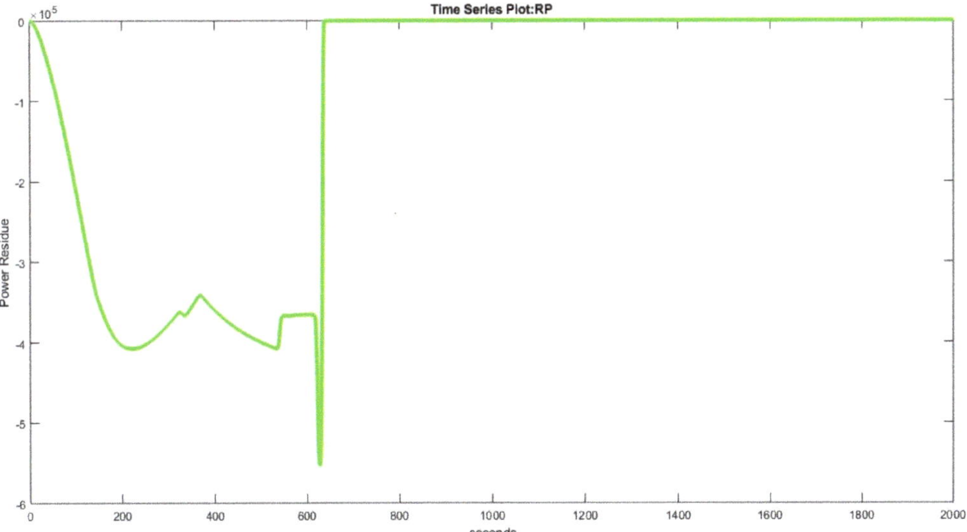

Figure 10. Power residue for the saturation fault of sluice gate w_1.

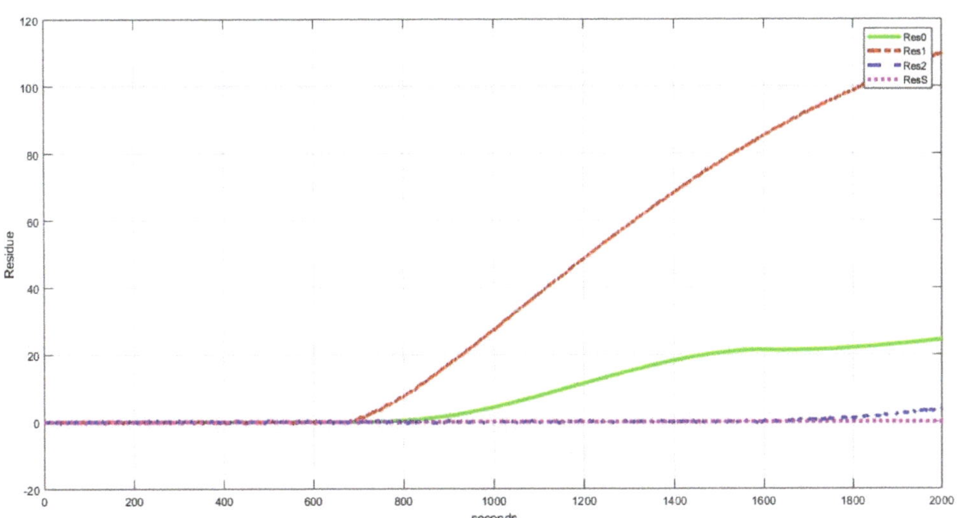

Figure 11. Pond residues for the leakage fault of sluice gate w_1.

The effects of the leakage fault are also transmitted to the output power of the turbine, and the output power will also start creeping up after the system has reached steady-state mode; therefore, the power will also creep up, as shown by the residue graph in Figure 12.

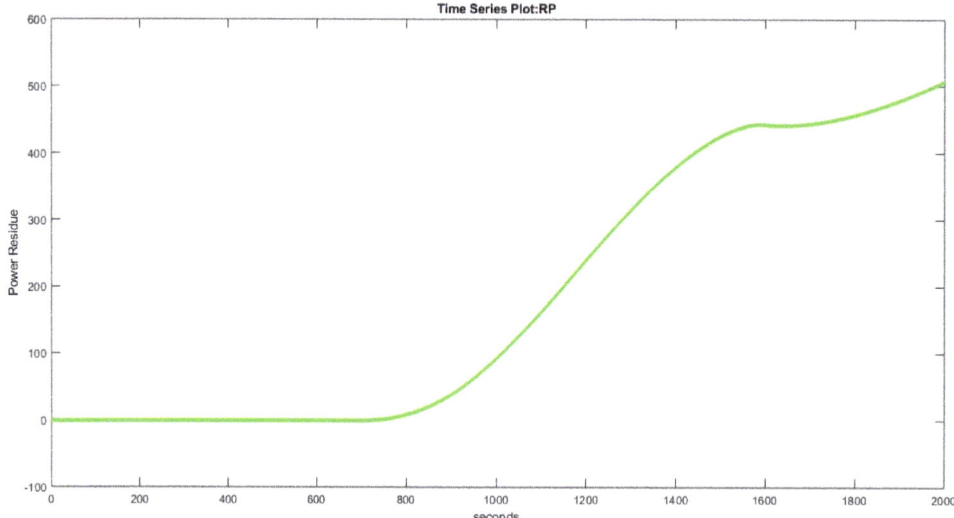

Figure 12. Power residue for the leakage fault of sluice gate w_1.

6.2.3. Saturation and Leakage Fault for the Sluice Gate

As discussed earlier, the f^3 fault is a combination of saturation and leakage faults, and its effect on the residue is during both the transient and steady-state regions. The residue graph for fault f^3 of the sluice gate w_1 is given in Figure 13. As stated before, for the same fault of w_2, residues 1 and 2 are swapped.

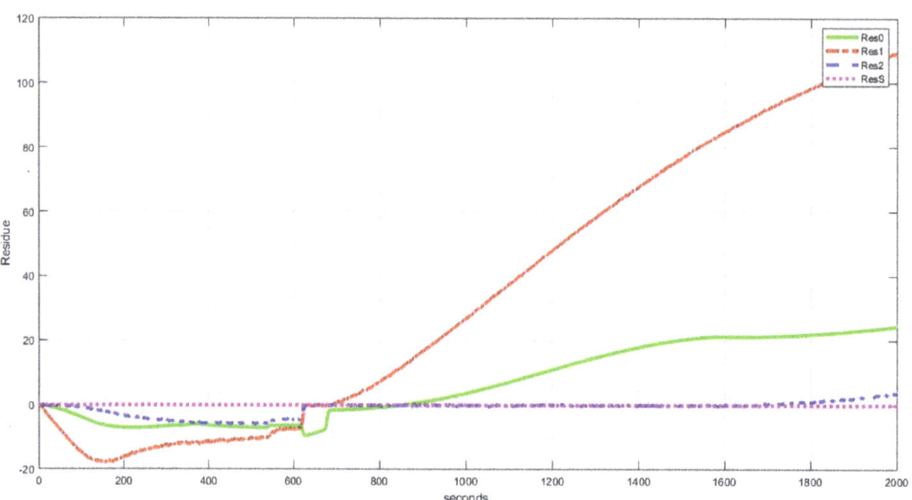

Figure 13. Pond residues for the saturation and leakage fault of sluice gate w_1.

Similarly, the power residue combines the previous two faults and is given in Figure 14. Although the power residue graph looks the same as in Figure 10, it is a combination of Figures 10 and 12, this is because the power residue in Figure 10 has a very large magnitude as compared with Figure 12, and its effect looks suppressed.

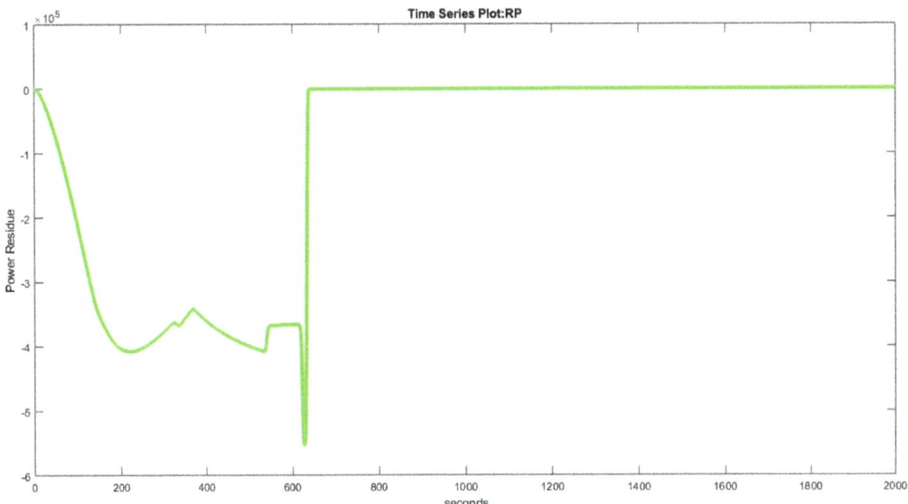

Figure 14. Power residue for the saturation and leakage fault of sluice gate w_1.

6.2.4. Saturation Fault of the Pond Valve

Equation (101) is used to compute the residue in case of the fault f^1 of the pond valves s_1 or s_2. As the effect of the saturation fault of the pond valve only appears when the system is in the transient region when the current pond level is lower than the required level, the valve cannot be fully opened and goes into saturation mode. Therefore, the effects of the saturation fault for the valve s_1 appear in the residue as shown below. As the effects of the leakage fault of the pond valve only appear when the system's required levels are lower than the current level, the leakage fault of the pond valve is also undetectable in the current residue generation mode, as the residue is unchanged. The effects of the saturation fault for valve s_1 appear in the residue as shown in Figure 15; also, both the fault type f^1 and f^3 have exactly the same residue for valve s_1.

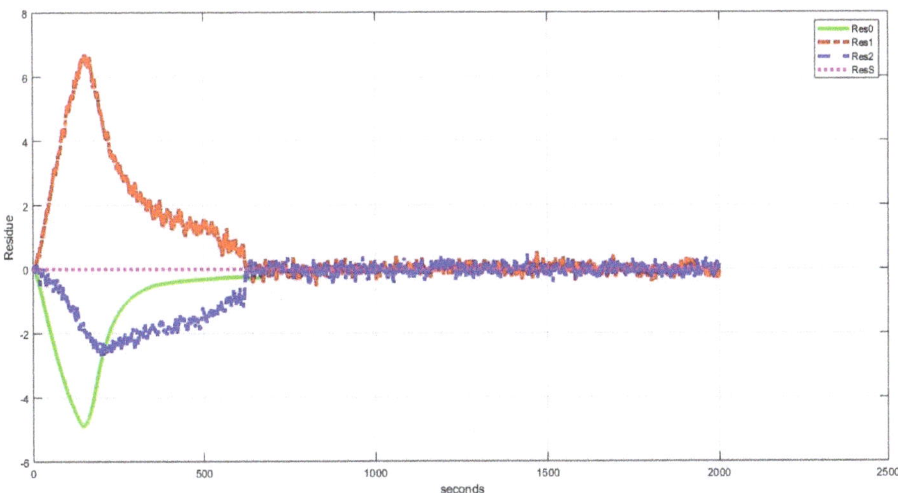

Figure 15. Pond residues for the saturation fault of the pond valve s_1.

Similarly, the effects of the fault f^1 of the pond valves appear only during the transient mode. The power residue graph for the f^1 fault of the pond valve s_1 is given in Figure 16; also, it is exactly the same as the f^3 fault.

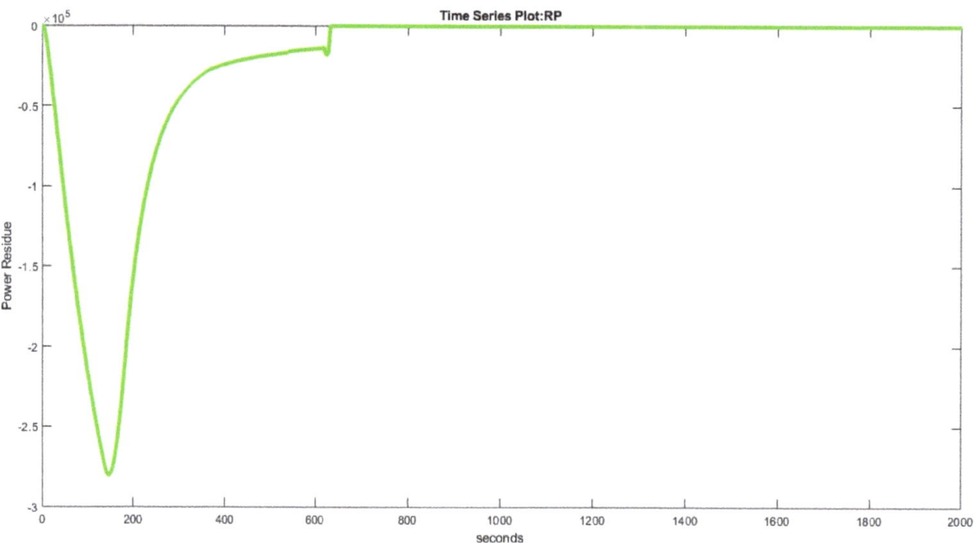

Figure 16. Power residue for the saturation fault of the pond valve s_1.

6.2.5. Saturation Fault of the Turbine Wicket Gate

Equation (106) gives the residue in case of the fault f^1 of the turbine wicket gate G. As stated before, the turbine operates at near or more than 90% gate opening for most of the time during operation. When the system is in transient state, and the current pond and power levels are below the required levels, at this time, the turbine gate is at nearly full throttle to try to reduce the power error as soon as possible. If a saturation fault exists in the turbine gate, its effects will appear in the power residue. When the system reaches its steady state, the turbine is still in its saturation mode, and the power residue will continue to be nonzero. Although the effects of this fault only appear in the power residue, they are not back propagated towards the ponds and hydraulic system. As all the saturation and leakage faults tested have a 10% magnitude, the effects of the fault are limited to the turbine. However, if the saturation fault increased significantly, then the effects of the fault will also back propagate towards the hydraulic system, starting with the surge tank. In the current situation, the hydraulic system's residue for the turbine gate's saturation fault is the same as a normal residue (near zero). However, the power residue is always nonzero. Furthermore, as the effects of the leakage fault do not appear in the normal residue generation mode, the power residue for fault f^3 of the turbine wicket gate G is the same as the saturation fault and is given in Figure 17.

6.2.6. Undetectable Faults

Due to the nature of the saturation and leakage faults, the system behavior is indistinguishable from a faultless system. As long as the faulty component of the system is not operating in a faulty (saturation or leakage) region, these faults will remain undetected. Examples of these are the leakage faults of the pond valves and the turbine wicket gate, whose effects are only visible if the pond level is needed to be lowered significantly or the turbine is shutdown. Therefore, another special residue generation mode needs to be implemented, forcing the pond valves and the turbine gate to go into leakage mode, thus making the leakage faults detectable. By implementing the special residue generation mode, all the discussed faults can be detected by running it in conjunction with the normal residue generation mode. Implementation of a special residue generation mode is left for future work as these faults do not affect the system in steady state and thus do not require the intervention of fault-tolerant control.

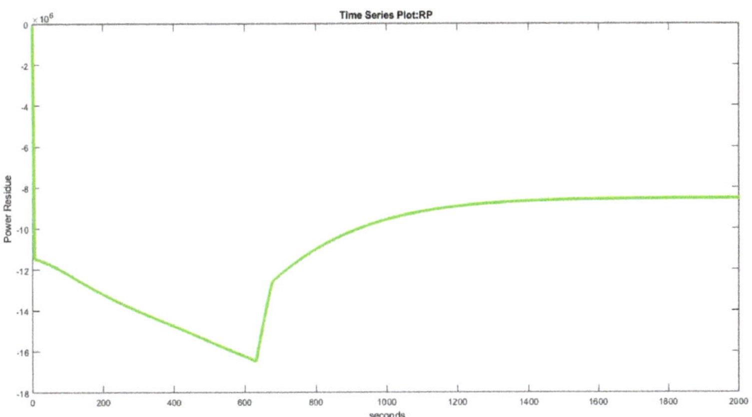

Figure 17. Power residue for the saturation fault of turbine gate *G*.

6.3. Effect of Fault-Tolerant Control

As stated in the fault description section, the faults affect the system in a transient or steady state. The requirement for fault tolerance arises if the system starts behaving undesirably in the steady state or its behavior is harmful to the system in transient state. In these cases, while any fault must be detected and taken care of during scheduled maintenance, it is often not feasible for the system to be stopped during the operation. For this reason, it is more efficient for the system to take countermeasure and operate the system in lower efficiency mode rather than shutting down the system or to keep running it in faulty mode. Although fault compensation or tolerant action is not required to occur in the case of detection of each and every fault, some faults have long-term effects on the system outputs. In case of the faults discussed above, fault tolerance is required only for two cases of faults. The first case is the leakage fault of the sluice gate, and the other is the saturation fault of the turbine. In the current system, the output power is most important for the system's proper operation. Therefore, for fault tolerance control, the output power is targeted for correction in case of the occurrence of a fault, while the pond levels are disregarded in favor of the output power. In the following section, the two cases that require fault tolerance are discussed below.

6.3.1. Sluice Gate Leakage Fault Tolerance

For the residue graphs for the sluice gate leakage faults, the power residue is nonzero in the steady state, and it keeps increasing. This will result in a slow but steady increase in the power output if left unchecked. The power residue graph shows that the residue is around 500 mark at the end of the simulation. This is due to the residue's very high residue gain (around 100). The undesirable increase in the level of the head pond is due to the uncontrolled increase in pond 1 or 2. Although the power output increases by 5–6 watts compared to the normal operation, it will tend to snowball later and cause the plant to behave in unpredictable ways. So, a fault-tolerant action is needed to avoid or at least delay that unpredictable behavior. One of the easiest ways to delay that undesirable behavior is to have the head pond at a lower level than normal and have the turbine compensate for it. The fault tolerant controller sets the desired level of the head pond to a lower level and has the turbine compensate for the lower pond level. As shown in the power graph (Figure 18), although the power level of the fault tolerant system reaches the desired level a little while after the faulty system, it will remain at the desired level a lot longer than the faulty system. The pond graphs for this case is shown in Figures 19 and 20, while the power graph is shown in Figure 18.

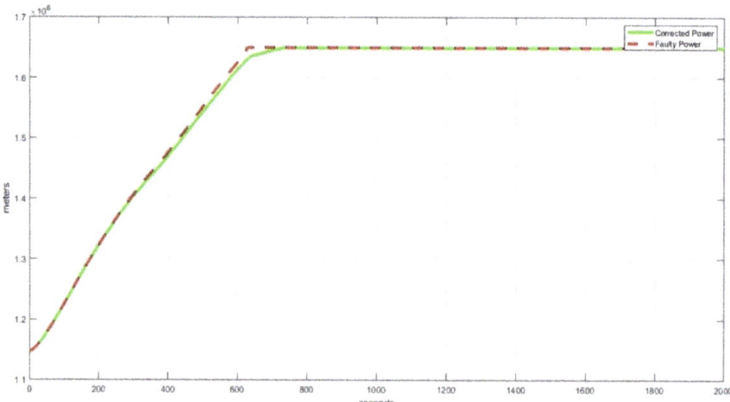

Figure 18. Power with faulty and fault-corrected operation given a sluice gate w_1 leakage fault.

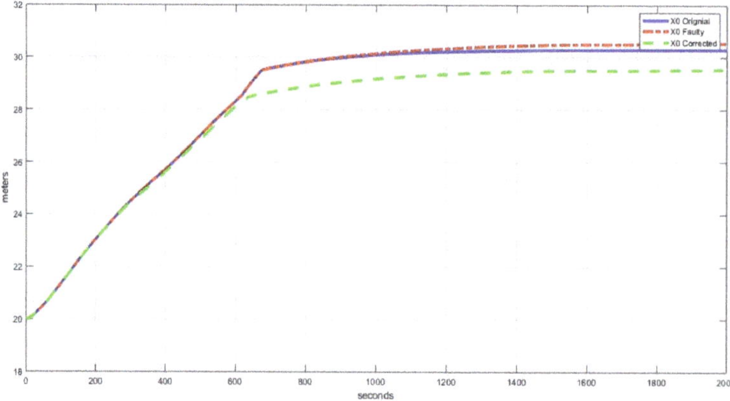

Figure 19. Pond 0 with normal, faulty, and fault-corrected operation given a sluice gate w_1 leakage fault.

6.3.2. Turbine Saturation Fault Tolerance

As discussed in the earlier section, the effects of turbine saturation faults are present 99% of the time when it occurs. The saturation fault of the turbine also directly affects the power of the turbine, which is the most important output of the system. Therefore, it is vital to include fault compensation or fixing the fault as soon as possible. The output power is reduced due to the reduction in the flow-rate through the turbine, as shown in Equation (34). Furthermore, it cannot be increased, so the water level needs to be increased to increase the output power of the turbine. Therefore, after identifying the turbine saturation fault, the fault-tolerant controller increases the required level of the pond as shown in Figure 21, and the system functions as usual. Looking at the power graphs in comparison to the normal and the faulty operation (Figure 22), we see that the fault tolerant mode takes a somewhat long time to reach the desired power level as compared to the faultless system, while the faulty system fails to reach the desired level.

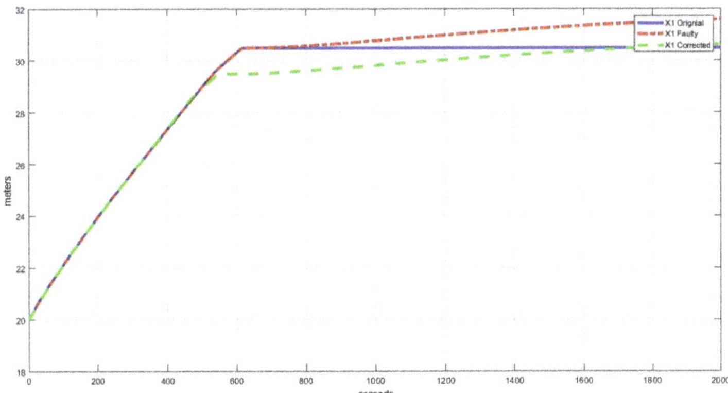

Figure 20. Pond 1 with normal, faulty, and fault-corrected operation given a sluice gate w_1 leakage fault.

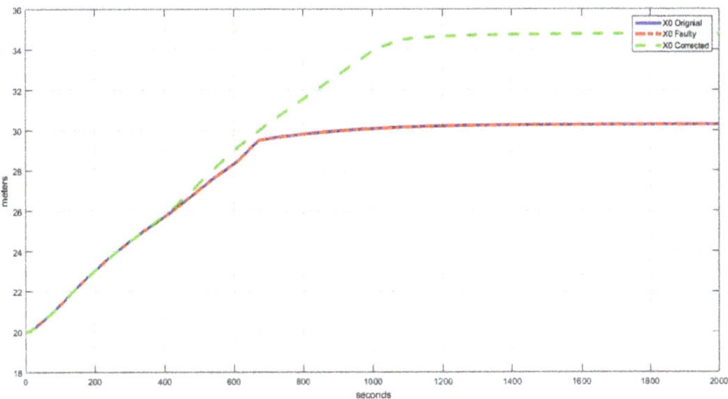

Figure 21. Pond 0 with normal, faulty, and fault-corrected operation given a turbine gate G saturation fault.

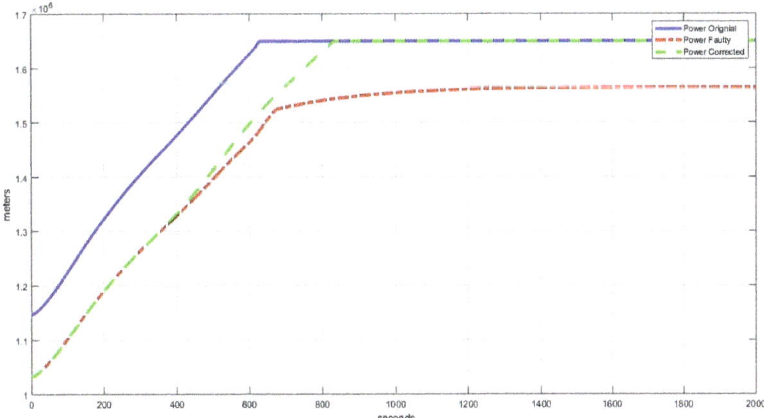

Figure 22. Power with normal, faulty, and fault-corrected operation given a turbine gate G saturation fault.

7. Conclusions and Future Directions

This paper analyzes the integration of a three-pond hydraulic system with a Francis turbine and adds two levels of control to the power generation. Next, some faults are added to the system, and the second part of the work deals with fault identification and then fault tolerance. The power control presented in this paper is adequate as it reduces the power error to nearly 100 watts of error. This error could be reduced further by introducing a different control scheme in the future. The main future direction of this work is to identify more types of faults in the system and a combination of faults. Another future direction is to discuss the detailed fault models in this paper, an example of which is discussed in the Appendix A in a separate paper. Furthermore, the fault tolerance is done by manipulating the required level of the head ponds to compensate for the faults; also, by having a fast technique to identify faults on the fly rather than running the system in fault identification mode periodically to identify the faults in the system. Another future direction is to have a rather large database of faults and fault combinations that could be identified by the fault identification system and the fault-tolerant control having more autonomy in manipulating the parameters of the system variables to run the system more effectively in fault-tolerant mode.

Author Contributions: Conceptualization, data curation, methodology, resources, validation, writing—original draft, writing—review and editing, A.S.; conceptualization, data curation, methodology, resources, validation, writing—original draft, writing—review and editing, formal analysis, funding acquisition, investigation, project administration, supervision, and visualization, A.U.K., M.I., G.H. and S.M.; writing—review and editing, formal analysis, funding acquisition, investigation, project administration, and visualization, E.S., A.W. and F.R.A. All authors have read and agreed to the published version of the manuscript.

Funding: The APC was funded by Taif University Researchers Supporting Project Number (TURSP-2020/331), Taif University, Taif, Saudi Arabia.

Institutional Review Board Statement: Not applicable.

Informed Consent Statement: Not applicable.

Data Availability Statement: Not applicable.

Acknowledgments: The authors would like to acknowledge the support from Taif University Researchers Supporting Project Number (TURSP-2020/331), Taif University, Taif, Saudi Arabia.

Conflicts of Interest: The authors declare no conflict of interest.

Appendix A

In the main paper, we discussed the explicit effects of the faults on the directly affected state variables. As an extension to the above fault models, we shall discuss the implicit effects of the faults on the state variables that are not directly affected. As an example, we shall be taking the effects of the saturation fault f^1 for the pond valve s_1. Rewriting the effected Equations (70) and (71):

The total effect of fault on pond 1 is given by rewriting Equation (70):

$$x_1 = -\frac{n_b a}{A} \int_0^t s_1 \ \sqrt{2g|x_1 - x_0|} sgn(x_1 - x_0) dt \\ -\frac{n_b a}{A} \int_0^t f_{s_1}^1 \sqrt{2g|x_1 - x_0|} sgn(x_1 - x_0) dt + \int_0^{At} \frac{w_1 l_1}{A} dt \tag{A1}$$

Similarly, the total effect of fault on pond 0 is given by rewriting Equation (71):

$$x_0 = \frac{n_b a}{A} \int_0^t \lceil s_1 \sqrt{2g|x_1 - x_0|} sgn(x_1 - x_0) + f_{s_1}^1 \sqrt{2g|x_1 - x_0|} sgn(x_1 - x_0) \\ + s_2 \sqrt{2g|x_2 - x_0|} sgn(x_2 - x_0) \rceil dt - \int_0^t \frac{Q_t}{A} dt \tag{A2}$$

Now for ease of notation, let us donate the x_1 and x_0 from the Equations (A1) and (A2) as $x_1^{f_{s_1}^1}$ and $x_0^{f_{s_1}^1}$, respectively. For the duration of the Appendix, let us assume the normal state variables, x_1 and x_0, as faultless, modifying Equations (A1) and (A2) in new notations as

$$x_1^{f_{s_1}^1} = -\frac{n_a}{A}\int_0^t s_1 \sqrt{2g|x_1-x_0|}\,sgn(x_1-x_0)dt \\ -\frac{n_a}{A}\int_0^t f_{s_1}^1 \sqrt{2g|x_1-x_0|}\,sgn(x_1-x_0)dt + \int_0^{At}\frac{w_1 l_1}{A}dt \tag{A3}$$

$$x_0^{f_{s_1}^1} = \frac{n_a}{A}\int_0^t \lceil s_1 \sqrt{2g|x_1-x_0|}\,sgn(x_1-x_0) + f_{s_1}^1 \sqrt{2g|x_1-x_0|}\,sgn(x_1-x_0) \\ + s_2 \sqrt{2g|x_2-x_0|}\,sgn(x_2-x_0)\rceil dt - \int_0^t \frac{Q_t}{A}dt \tag{A4}$$

Now let us rewrite Equations (A3) and (A4) in a simpler form as a combination of a normal value and faulty component:

$$x_0^{f_{s_1}^1} = x_0 + \frac{n_a}{A}\int_0^t f_{s_1}^1 \sqrt{2g|x_1-x_0|}\,sgn(x_1-x_0)dt \tag{A5}$$

Similarly,

$$x_1^{f_{s_1}^1} = x_1 - \frac{n_a}{A}\int_0^t f_{s_1}^1 \sqrt{2g|x_1-x_0|}\,sgn(x_1-x_0)dt \tag{A6}$$

Now, with the effects of the faults present in simplified terms in Equations (A5) and (A6), let us discuss the effects of the saturation fault of the pond valve on the other state variables. Now, rewriting Equation (31) for pond 2 with fault effects as

$$x_2^{f_{s_1}^1} = -\frac{n_a}{A}\int_0^t s_2 \sqrt{2g|x_2-x_0|}\,sgn\left(x_2 - x_0^{f_{s_1}^1}\right)dt + \int_0^t \frac{w_2 l_2}{A}dt \tag{A7}$$

Now, putting the value of $x_0^{f_{s_1}^1}$ in Equation (A7):

$$x_2^{f_{s_1}^1} = -\frac{n_a}{A}\int_0^t s_2 \sqrt{2g|x_2-x_0|}\,sgn \\ \left(x_2 - \left\{x_0 + \frac{n_a}{A}\int_0^t f_{s_1}^1 \sqrt{2g|x_1-x_0|}\,sgn(x_1-x_0)dt\right\}\right)dt + \int_0^t \frac{w_2 l_2}{A}dt \tag{A8}$$

Now simplifying the equation,

$$x_2^{f_{s_1}^1} = -\frac{n_a}{A}\int_0^t s_2 \sqrt{2g|x_2-x_0|}\,sgn\left(x_2 - x_0 - \frac{n_a}{A}\int_0^t f_{s_1}^1 \sqrt{2g|x_1-x_0|}\,sgn(x_1-x_0)dt\right)dt + \int_0^t \frac{w_2 l_2}{A}dt \tag{A9}$$

and

$$x_2^{f_{s_1}^1} = -\frac{n_a}{A}\int_0^t s_2 \sqrt{2g|x_2-x_0|}\,sgn(x_2-x_0)dt + \int_0^t \frac{w_2 l_2}{A}dt \\ + \frac{n_a}{A}\int_0^t s_2 \sqrt{2g|x_2-x_0|}\,sgn\left(\frac{n_a}{A}\int_0^t f_{s_1}^1 \sqrt{2g|x_1-x_0|}\,sgn(x_1-x_0)dt\right) \tag{A10}$$

Therefore, the expression for the level of pond 2 becomes

$$x_2^{f_{s_1}^1} = x_2 + \frac{n_a}{A}\int_0^t s_2 \sqrt{2g|x_2-x_0|}\,sgn\left(\frac{n_a}{A}\int_0^t f_{s_1}^1 \sqrt{2g|x_1-x_0|}\,sgn(x_1-x_0)dt\right)dt \tag{A11}$$

For finding the effect of the fault on the surge tank, we consider Equation (32), but it depends upon the flow rate through the headrace, which in turn depends on the level of pond 0. Therefore, we consider the equation of head race (14):

$$\frac{d}{dt}Q_t = \frac{gA_t}{L_t}(x_0 - x_s) - C_t Q_t|Q_t| \tag{A12}$$

Integrating Equation (A12) to find the value of Q_t as

$$Q_t = \frac{gA_t}{L_t}\left(\int_0^t x_0 dt - \int_0^t x_s dt\right) - \int_0^t C_t Q_t |Q_t| dt \tag{A13}$$

Now, with the explicit value of the flow-rate through the headrace, let us consider Equation (32) of the surge tank with the occurrence of the fault:

$$x_s^{f_{s1}^1} = \int_0^t \frac{Q_t^{f_{s1}^1}}{A_s} dt - \int_0^t \frac{Q}{A_s} dt \tag{A14}$$

Now, using the value of $Q_t^{f_{s1}^1}$ from Equation (A13) in the above Equation (A14):

$$x_s^{f_{s1}^1} = \int_0^t \frac{\frac{gA_t}{L_t}\left(\int_0^t x_0^{f_{s1}^1} dt - \int_0^t x_s dt\right) - \int_0^t C_t Q_t |Q_t| dt}{A_s} dt - \int_0^t \frac{Q}{A_s} dt \tag{A15}$$

Simplifying:

$$x_s^{f_{s1}^1} = \frac{1}{A_s}\int_0^t\left[\frac{gA_t}{L_t}\left(\int_0^t x_0^{f_{s1}^1} dt - \int_0^t x_s dt\right) - \int_0^t C_t Q_t |Q_t| dt\right] dt - \int_0^t \frac{Q}{A_s} dt \tag{A16}$$

Now putting the value of $x_0^{f_{s1}^1}$ in Equation (A16):

$$x_s^{f_{s1}^1} = \frac{1}{A_s}\int_0^t\left[\frac{gA_t}{L_t}\left(\int_0^t\left\{x_0 + \frac{n_a}{A}\int_0^t f_{s1}^1 \sqrt{2g|x_1 - x_0|}\,sgn(x_1 - x_0)dt\right\}dt - \int_0^t x_s dt\right) - \int_0^t C_t Q_t |Q_t| dt\right] dt$$
$$- \int_0^t \frac{Q}{A_s} dt \tag{A17}$$

After separating the faulty term from Equation (A17):

$$x_s^{f_{s1}^1} = \frac{1}{A_s}\int_0^t\left[\frac{gA_t}{L_t}\left(\int_0^t x_0 dt - \int_0^t x_s dt\right) - \int_0^t C_t Q_t |Q_t| dt\right] dt - \int_0^t \frac{Q}{A_s} dt$$
$$+ \int_0^t \frac{1}{A_s}\left[\int_0^t \frac{n_a}{A}\left\{\int_0^t f_{s1}^1 \sqrt{2g|x_1 - x_0|}\,sgn(x_1 - x_0)dt\right\}dt\right] dt \tag{A18}$$

Now Equation (A18) can be rewritten as a combination of faultless and faulty state using Equation (A16):

$$x_s^{f_{s1}^1} = x_s + \int_0^t \frac{1}{A_s}\left[\int_0^t \frac{n_a}{A}\left\{\int_0^t f_{s1}^1 \sqrt{2g|x_1 - x_0|}\,sgn(x_1 - x_0)dt\right\}dt\right] dt \tag{A19}$$

Now to find the effect of saturation fault of the pond valve on the output power, we consider the power Equation (34) as

$$P^{f_{s1}^1} = \frac{M_r x_0^{f_{s1}^1} n_r}{H_0} G\left[a_1 \frac{n^2}{n_r^2} + b_1 \frac{n}{n_r} + c_1\right] \tag{A20}$$

Now putting the value of $x_0^{f_{s1}^1}$ in the power equation:

$$P^{f_{s1}^1} = \frac{M_r\left\{x_0 + \frac{n_a}{A}\int_0^t f_{s1}^1 \sqrt{2g|x_1 - x_0|}\,sgn(x_1 - x_0)dt\right\}n_r}{H_0} G\left[a_1 \frac{n^2}{n_r^2} + b_1 \frac{n}{n_r} + c_1\right] \tag{A21}$$

Simplifying;

$$P^{f_{s1}^1} = \frac{M_r x_0 n_r}{H_0} G\left[a_1 \frac{n^2}{n_r^2} + b_1 \frac{n}{n_r} + c_1\right]$$
$$+ \left\{\frac{n_a}{A}\int_0^t f_{s1}^1 \sqrt{2g|x_1 - x_0|}\,sgn(x_1 - x_0)dt\right\}\frac{M_r n_r}{H_0} G\left[a_1 \frac{n^2}{n_r^2} + b_1 \frac{n}{n_r} + c_1\right] \tag{A22}$$

After further simplification the power becomes

$$P^{f_{s_1}^1} = P + \left\{ \frac{n_v a}{A} \int_0^t f_{s_1}^1 \sqrt{2g|x_1 - x_0|} sgn(x_1 - x_0) dt \right\} \frac{M_r n_r}{H_0} G \left[a_1 \frac{n^2}{n_r^2} + b_1 \frac{n}{n_r} + c_1 \right] \quad \text{(A23)}$$

Therefore, we can find the explicit and implicit effects of the saturation fault f^1 for the pond valve s_1 as Equations (A3), (A4), (A11), (A19), and (A23) as a linear combination of faultless state variables and fault effects. Similar calculations can be done for the other faults discussed in the paper.

References

1. Saeed, A.; Shahzad, E.; Aslam, L.; Qureshi, I.M.; Khan, A.U.; Iqbal, M. New Paradigm for Water Level Regulation using Three Pond Model with Fuzzy Inference System for Run of River Hydropower Plant. *arXiv* **2020**, arXiv:201113131.
2. Simani, S.; Alvisi, S.; Venturini, M. Fault tolerant control of a simulated hydroelectric system. *Control Eng. Pr.* **2016**, *51*, 13–25. [CrossRef]
3. He, J.; Yang, Q.; Wang, Z. On-line fault diagnosis and fault-tolerant operation of modular multilevel converters—A comprehensive review. *CES Trans. Electr. Mach. Syst.* **2020**, *4*, 360–372. [CrossRef]
4. Moloi, K.; Jordaan, J.A.; Abe, B.T. Development of a hybrid fault diagnostic method for power distribution network. In Proceedings of the 2019 IEEE AFRICON, Accra, Ghana, 25–27 September 2019; pp. 1–4.
5. Moloi, K.; Akumu, A.O. Power distribution fault diagnostic method based on machine learning technique. In Proceedings of the 2019 IEEE PES/IAS PowerAfrica, Abuja, Nigeria, 20–23 August 2019; pp. 238–242.
6. Trnka, P.; Hofreiter, M.; Sova, J. Combination of techniques for the fault diagnostics. In Proceedings of the 2017 18th International Carpathian Control Conference (ICCC), Sinaia, Romania, 28–31 May 2017; pp. 499–502.
7. Mattera, C.G.; Quevedo, J.; Escobet, T.; Shaker, H.R.; Jradi, M. A Method for Fault Detection and Diagnostics in Ventilation Units Using Virtual Sensors. *Sensors* **2018**, *18*, 3931. [CrossRef] [PubMed]
8. Rombach, K.; Michau, G.; Fink, O. Contrastive Learning for Fault Detection and Diagnostics in the Context of Changing Operating Conditions and Novel Fault Types. *Sensors* **2021**, *21*, 3550. [CrossRef] [PubMed]
9. De, V.; Lima, T.L.; Filho, A.C.L.; Belo, F.A.; Souto, F.V.; Silva, T.C.B.; Mishina, K.V.; Rodrigues, M.C. Noninvasive Methods for Fault Detection and Isolation in Internal Combustion Engines Based on Chaos Analysis. *Sensors* **2021**, *21*, 6925. [CrossRef] [PubMed]
10. Komorska, I.; Puchalski, A. Rotating Machinery Diagnosing in Non-Stationary Conditions with Empirical Mode Decomposition-Based Wavelet Leaders Multifractal Spectra. *Sensors* **2021**, *21*, 7677. [CrossRef] [PubMed]
11. Dong, L.; Guo, H.; Guo, Z.; Zheng, T. Fault Diagnosis Technique for Hydroelectric Generators using Variational Mode Decomposition and Power Line Communications. *J. Phys.* **2019**, *1176*, 062058.
12. Xin, X.; Ni, W.; Sang, Y. A novel analysis method for fault diagnosis of hydro-turbine governing system. *Proc. Inst. Mech. Eng. Part O J. Risk Reliab.* **2017**, *231*, 164–171.
13. Chen, W. Fault Detection and Isolation in Nonlinear Systems: Observer and Energy-Balance Based Approaches. Ph.D. Thesis, University of Duisburg-Essen, Duisburg, Germany, 2012.
14. Khan, A.Q. Observer-Based Fault Detection in Nonlinear Systems. Ph.D. Thesis, University of Duisburg-Essen, Duisburg, Germany, 2010.
15. Abid, M. Fault Detection in Nonlinear Systems: An Observer-Based Approach. Ph.D. Thesis, University of Duisburg-Essen, Duisburg, Germany, 2010.
16. Yadav, O.; Kishor, N.; Fraile-Ardanuy, J.; Mohanty, S.R.; Pérez, J.I.; Sarasúa, J.I. Pond head level control in a run-of-river hydro power plant using fuzzy controller. In Proceedings of the 2011 16th International Conference on Intelligent System Applications to Power Systems, Hersonissos, Greece, 25–28 September 2011; pp. 1–5.
17. Priyadharson, A.S.M.; Saravanan, M.S.; Gomathi, N.; Joshua, S.V.; Mutharasan, A. Energy efficient flow and level control in a hydro power plant using fuzzy logic. *J. Comput. Sci.* **2014**, *10*, 1703–1711. [CrossRef]

processes

MDPI

Article

Development of a Hydrokinetic Turbines Backwater Prediction Model for Inland Flow through Validated CFD Models

Chantel Monica Niebuhr [1,*], Craig Hill [2], Marco Van Dijk [1] and Lelanie Smith [3]

[1] Department of Civil Engineering, University of Pretoria, Pretoria 0001, South Africa; marco.vandijk@up.ac.za
[2] Mechanical and Industrial Engineering Department, University of Minnesota-Duluth, Duluth, MN 55812, USA; cshill@d.umn.edu
[3] Department of Mechanical Engineering, University of Pretoria, Pretoria 0001, South Africa; lelanie.smith@up.ac.za
* Correspondence: chantel.niebuhr@up.ac.za; Tel.: +27-79-427-5190

Abstract: Hydrokinetic turbine deployment in inland water reticulation systems such as irrigation canals has potential for future renewable energy development. Although research and development analysing the hydrodynamic effects of these turbines in tidal applications has been carried out, inland canal system applications with spatial constraints leading to possible blockage and backwater effects resulting from turbine deployment have not been considered. Some attempts have been made to develop backwater models, but these were site-specific and performed under constant operational conditions. Therefore, the aim of this work was to develop a generic and simplified method for calculating the backwater effect of HK turbines in inland systems. An analytical backwater approximation based on assumptions of performance metrics and inflow conditions was tested using validated computational fluid dynamics (CFD) models. For detailed prediction of the turbine effect on the flow field, CFD models based on Reynolds-averaged Navier–Stokes equations with Reynolds stress closure models were employed. Additionally, a multiphase model was validated through experimental results to capture the water surface profile and backwater effect with reasonable accuracy. The developed analytical backwater model showed good correlation with the experimental results. The model's energy-based approach provides a simplified tool that is easily incorporated into simple backwater approximations, while also allowing the inclusion of retaining structures as additional blockages. The model utilizes only the flow velocity and the thrust coefficient, providing a useful tool for first-order analysis of the backwater from the deployment of inland turbine systems.

Keywords: hydrokinetic; computational fluid dynamics; backwater; inland hydrokinetic; axial flow turbines

Citation: Niebuhr, C.M.; Hill, C.; Van Dijk, M.; Smith, L. Development of a Hydrokinetic Turbines Backwater Prediction Model for Inland Flow through Validated CFD Models. *Processes* **2022**, *10*, 1310. https://doi.org/10.3390/pr10071310

Academic Editors: Santiago Lain and Omar Dario Lopez Mejia

Received: 23 May 2022
Accepted: 20 June 2022
Published: 4 July 2022

Publisher's Note: MDPI stays neutral with regard to jurisdictional claims in published maps and institutional affiliations.

1. Introduction

Research and development of hydrokinetic (HK) devices in canal systems is increasing in popularity due to increasing electricity costs and the drive towards finding renewable energy sources with unconventional applications [1–3]. Although most development has focussed on tidal applications, multiple opportunities exist for the deployment of HK systems within inland water infrastructure (e.g., canal systems) [1]. However, the placement of such a device can have significant water level and hydrodynamic energy loss effects [4].

Prediction of the hydrodynamic effects of hydrokinetic turbines in canal systems remains an important pre-development objective. Due to the nature of canal design, these systems usually have flat slopes and subcritical flow regimes. Therefore, the analysis of backwater effects from blockages is critical for the prevention of flooding and water loss. This is especially important in array schemes where the cumulative effect of multiple devices can exceed the top of the channel and cause it to overtop.

Generally, backwater calculations utilize a blockage size (e.g., typical backwards-facing step, weir or pier shape) or energy loss function (quantified energy losses) to predict the backwater effect. Due to the novelty of HK energy, a streamlined procedure for determining the backwater effect has not yet been determined. This may be attributed to the variability of turbine types, operational conditions, and efficiencies, all of which result in a different effective blockage.

Previous studies have investigated the hydrodynamic effects of horizontal-axis hydrokinetic turbines (HAHTs) both experimentally (e.g., in canals [5], investigating boundary layers [6] and varying Reynolds numbers [7]) and computationally (e.g., both CFD applications in [4,8]). However, most of the studies are performed under constant operational conditions, and are site-specific (e.g., three-bladed [9] and two-bladed [9,10] turbines under optimal conditions). Some attempts have been made in the past to develop backwater models using roughness values [4] or analytical relationships [11], but the lack of experimental results over a range of turbine designs and operational conditions has resulted in site-specific models. Additionally, use of the models without in-depth knowledge of input variables limits their utility.

This study aimed to develop a simplified method for calculating the backwater effect of HK turbines in canal systems. An analytical approximation based on assumptions of performance metrics and inflow conditions was tested using validated computational fluid dynamics (CFD) models. These models allow a larger dataset and, thus, validation of the analytical approximation recommended.

2. Background

Placement of an HK device extracting energy in confined flow may affect water surfaces and water surface profiles. This is especially true in array schemes, and must be considered to ensure that the clearance between the rotor blades and the surface is sufficient. Myers and Bahaj [12] investigated a 1:30 scale HAHT model (rotor diameter, $d_T = 0.4$ m), and observed a clear difference in the water surface once energy was extracted. Water depths increased immediately upstream of the rotor and decreased downstream for about 2 d_T. Details of the water surface profiles can be seen in Figure 1. The results observed a standing wave 7–8 d_T downstream (it should be noted that this was for the high-freestream-velocity case).

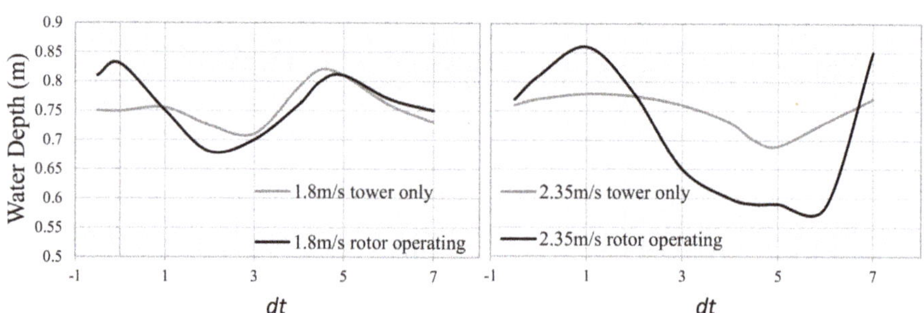

Figure 1. Water surface profile through a scaled turbine operating at 2 different velocities, compared to the no-energy-extraction stage [12].

Prediction of such occurrences requires understanding of the specific energy and flow regime (Froude number) to predict flow behaviour after HK deployment; here, a Froude number based on turbine diameter ($Fr_D = \frac{U}{\sqrt{gd_t}}$, where U is the mean velocity and d_t is the turbine diameter) may be more useful, which has also been found to govern the free-surface effects [13].

The flow effects observed can be explained using the specific energy of the flow section, which is a function of water depth and velocity. When ignoring friction losses (e.g., from the channel sides and bed), the specific energy may be defined as follows:

$$E = y + \frac{\alpha Q^2}{2g A^2}$$
(1)

where y is the water depth, α is the energy coefficient, Q is the volumetric flow rate, A is the cross-sectional area, and g is the gravitational acceleration. The energy coefficient can then be defined as follows:

$$\alpha = \frac{\sum u^3 \Delta A}{U^3 A}$$
(2)

where A is the total flow area and u is the velocity measured within an elemental area, ΔA. The flow regime (sub- or supercritical) and, thus, the Froude number of the flow govern the behaviour of the flow [14].

A parameter of specific interest is the critical depth of the channel in which the turbine is placed. When the water surface decreases to critical or subcritical depth, flow phenomena such as hydraulic jumps may form downstream to allow recovery to normal flow depth.

Simplification of crucial free-surface parameters in inland HK installations can be summarized by two fundamental effects:

- Free-surface effects in the form of a possible standing wave formed, or decreased water surface above the turbine (due to decreasing pressure).
- Potential backwater effects caused (e.g., damming upstream).

2.1. Free-Surface Effects of HK Turbines

Free-surface effects are a critical aspect in riverine and tidal turbine array design, as the standing wave formed downstream of the turbine affects additional downstream turbines. Previous studies have concluded that the depth of the downstream water surface is strongly dependant on the Fr_D. Additionally, the blockage ratio also affects this free-surface change, albeit not as strongly as Fr_D [13].

Myers and Bahaj [12] found that when imposing the typical wake expansion on the water surface, this coincided with the increased elevation observed 4–5 d_t downstream of the turbine rotor (as shown in Figure 2). In addition, due to the wake expansion coincident with the free surface, cumulative turbine placement at intervals smaller than the recovery length may cause the flow to approach critical depth, causing severe undulations in the water surface profiles (WSPs). Turbine operation and efficiency may also vary due to decreasing fluid velocity over the blades during operation. Accurate quantification of the WSPs around an array may be a challenge due to the multiple effects of turbulence, wake mixing, and superposition of WSP effects [12].

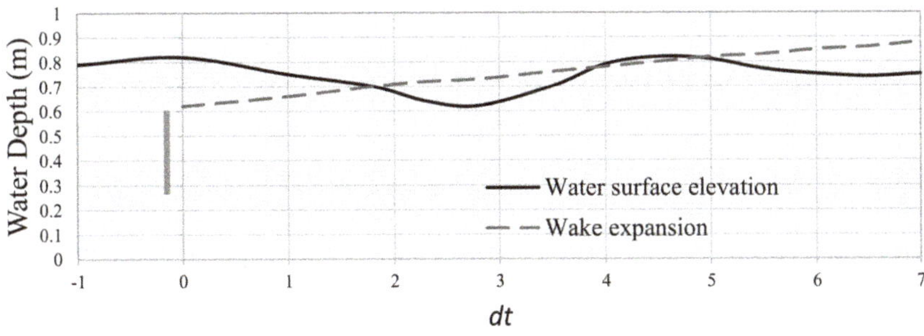

Figure 2. Wake expansion effect with free surface [12].

The presence of a support structure (tower/stanchion) also strongly affects the free-surface effect [12] (Figure 1). The same experiment also indicated the strong possibility of the formation of a hydraulic jump downstream of the turbine when flow is forced to a supercritical level due to the presence of the turbine and support structure.

Free-surface effects may be more pronounced for shallow turbines compared to turbines installed well below the free surface. It is also important to consider possible cross-sectional changes in the infrastructure where the turbine is placed, as this may alter/dampen/exaggerate these effects. Specified clearance coefficients have been investigated to limit the severity of decreased depths downstream of the turbine, or possible exposure of the turbine. Birjandi [15] proposed a clearance coefficient, C_h, defined as follows:

$$C_h = \frac{H}{L} \tag{3}$$

where H is the turbine submergence depth (i.e., the height between the top of the turbine and the water surface) and L is the rotor diameter. The recommended clearance coefficients for commercially available turbines can be seen in Table 1.

Table 1. Clearance coefficients for commercial HAHTs.

Turbine		Clearance Coefficient
Seaflow	2-Bladed, 300 kW	0.18–0.64
SeaGen	2-Bladed, 1.2 MW (2× 600 kW)	0.25–0.38
HS300	3-Bladed, 300 kW	0.75
AK-1000	3-Bladed, 1 MW	1.02

2.2. Backwater Effect

A turbine acts as a blockage in the channel and results in energy loss in inland flow infrastructure (where flow is constrained). When Fr << 1 and subcritical flow is prevalent, the backwater effect (damming upstream) may extend a large distance upstream as well as causing significant damming. This depends greatly on the blockage ratio, which is a function of the turbine swept area (A_T), channel flow area (A_o) ($BR(\%) = \frac{A_T}{A_o}$), and additional constrictions [16], as well as the theoretical to actual efficiency [17].

In a study on a pilot HK installation in an irrigation canal, the backwater effect from the presence of the turbine extended up to 2.7 km upstream, due to the flat slope and subcritical flow present in the channel [1] (Figure 3). The clearance coefficient for this installation had not yet been defined. The specific turbine studied in that project contained grids upstream of the turbine, which trapped debris and caused a further increase in the backwater effect, to the point of channel overtopping. The blockage ratio for the installation was around 12.5%.

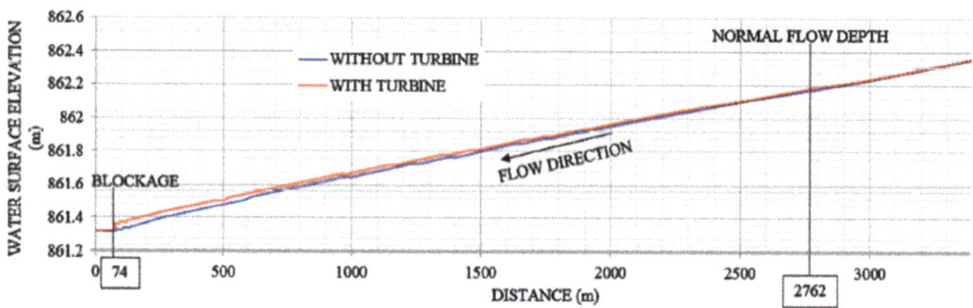

Figure 3. Backwater effect due to turbine blockages [1].

Additional to the blockage ratio, the Froude number (Fr) or Froude number based on turbine diameter (Fr_D) of the flow can influence the backwater effect. A previous study analysing this effect drew the following conclusions [13]:

1. The upstream free-surface deformation increased with Fr_D.
2. The location of maximum damming (i.e., the highest water level) moved closer to the turbine as Fr_D increased.

Previous studies have attempted to quantity the effective blockage of an HK device. Some have addressed this through the relationship of power extracted to total power dissipated by the devices [18], analytical relationships [11], and even enhanced Manning n-values quantifying the energy loss as a friction loss [4]. However, a simple formula quantifying the effective blockage for different turbine types and operational conditions is yet to be determined, and provides the motivation for this paper.

2.3. Backwater Calculations

The extraction of energy resulting from the HK device may also be analytically incorporated through the use of the momentum equation [19], where the power extraction term is added as a shear stress component (added to the effective shear stress caused by bed friction). Assuming gradually varied steady-state flow, the conservation of mass and momentum can be used to adjust the standard open-channel flow equation [20] with the addition of a term for artificial energy extraction [19]:

$$\left(1 - \frac{Q^2}{h^3 b^2 g}\right)\frac{\partial h}{\partial x} = \frac{\partial h}{\partial x}\frac{Q^2}{gh^2 b^3} - \frac{1}{\rho g b h} P\, \tau_{eff} \tag{4}$$

The effective shear (τ_{eff}) is defined as a combination of the bed shear (τ_o) and power extraction added as a shear term (τ_{add}):

$$\tau_{eff} = \tau_o + \tau_{add} = \rho \frac{g}{C^2} U^2 + \frac{P_x R}{U} \tag{5}$$

where P is the wetted perimeter, C is the Chezy friction coefficient, and P_x is the term added for power extraction, which may be more useful to express in terms of P_A being the power extracted per unit area, as the flow passes through a plane where $P_x = \frac{P_A}{\Delta x}$ (Δx being the change in distance). Such an effect can be seen graphically in Figure 4, where a 10% energy extraction term has been added.

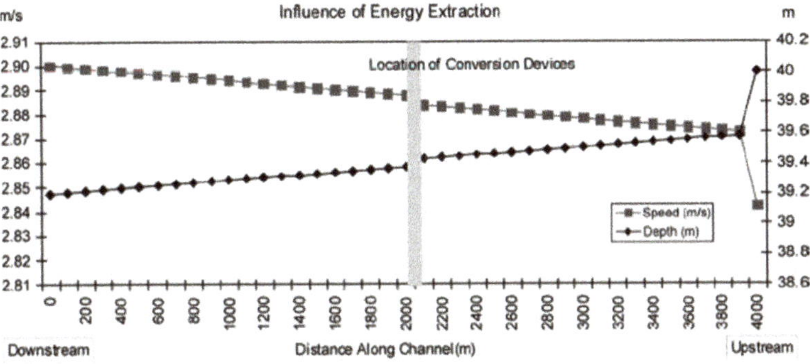

Figure 4. Influence of artificial energy extraction on speed and depth of flow [19].

A previous study [21] attempted to determine the backwater curve for instances when either the cross-section varied, the channel slope changed, or there was an obstacle in the channel and gradually varied flow was present. The model was based on the Bernoulli equation between two cross-sections. The energy loss between the cross-sections (related to distance) was termed the hydraulic loss, I, which can be calculated as follows:

$$I = \frac{n^2 U^2}{R_H^{\frac{4}{3}}} \tag{6}$$

where n is the Manning roughness ($s/m^{1/3}$), R_h is the hydraulic radius of the channel (m), and U is the velocity of the water (m/s). The change in water levels (Δz) between two sections can then be determined between two significant cross-sections (e.g., 0 and 1) and calculated as shown in Equation (7), where α is the Coriolis coefficient and U_0 and U_1 are the average velocities over distance ΔL:

$$\Delta z = \Delta L \left(\frac{I_0 + I_1}{2} \right) + \frac{\alpha}{2g} \left(U_0^2 - U_1^2 \right) \tag{7}$$

Very few models have been developed to attempt to predict the backwater effect— mostly 1D analytical models [4,11]. Most studies have focussed on tidal arrays and using free-surface effects to determine the optimal number and placement of turbines [22,23]. Within tidal applications, free-surface effects are only of concern for tip clearance; therefore, these models have limitations within the application in steady inland channels where spatial constraints are of primary concern.

In a study by Kartezhnikova and Ravens [4], an increased Manning roughness coefficient was used on the channel section representing the hydrokinetic device. The n-value used was a function of the actual channel n-value, slope, water depth, device efficiency, blockage ratio, and device deployment density. This method can then be used to determine the hydraulic impact, as well as the impact of various device configurations.

The head loss associated with the channel friction (h_{Lt}) (used in the energy conservation equation) can be written as shown in Equation (8), as a function of cross-sectional area (A_0), channel hydraulic radius (R_h), discharge (Q_n), and the length over which the loss is applied (L).

$$h_{Lt} = \left(\frac{Q_n}{A_0 R_h^{2/3}} \right)^2 L \tag{8}$$

Based on the assumption that the upstream and downstream velocity and pressure heads are equal, and assuming that the drag loss is negligible, the following equation for an enhanced bottom roughness (n_t) can be derived as a function of the total power dissipated (h_p), change in elevation (Δz), and channel Manning roughness coefficient (n).

$$n_t = n \left(1 - \frac{h_p}{\Delta z} \right)^{-1/2} \tag{9}$$

Lalander and Leijon [11] investigated the use of numerical and analytical models to determine the effects on upstream water levels in a river. The analytical models are dependent on the channel blockage (of the HK device) determining energy loss from the energy capture, as well as the energy losses in the wake. The numerical models are based on the same theory—that energy is removed, causing a power loss; thus, energy capture is a component of the total friction in the channel. The total head loss can be determined as the sum of the friction loss (Δh_f) and head loss caused by the turbine (Δh_t), and P_t is the total power of the turbine (W). The formulation of the stress term τ_f is shown in Equation (11), where f is equal to the Darcy–Weisbach coefficient (unitless) and U is the velocity of water (m/s).

$$\Delta h_{tot} = \Delta h_f + \Delta h_t = \frac{L}{\rho g R_h} * \tau_f + \frac{P_t}{\rho g Q} \tag{10}$$

$$\tau_f = \frac{f\rho}{8} * U^2 \tag{11}$$

It is also important to consider the blockage effect, which can increase the turbine power output [22,24].

2.4. Summary of Literature

When HK devices are placed in array schemes in inland channels/rivers, the cumulative effect and inter-effect of these devices should be well understood to avoid unfavourable free-surface effects. As shown in the organogram (Figure 5), the blockage resulting from the HK device (in typical subcritical, flat-sloped channels) may have multiple subsequent effects influencing downstream installations (within the array) as well as upstream flow conditions. Neglecting the free-surface effects in high-blockage cases may result in exposed downstream turbines (i.e., freeboard reduced), hydraulic jump formation (enforced critical flow), and upstream damming effects, and may also lead to potential blade-tip cavitation problems. Accurately quantifying the water surface effects is a challenge, and guidelines to avoid unfavourable conditions may be extremely useful.

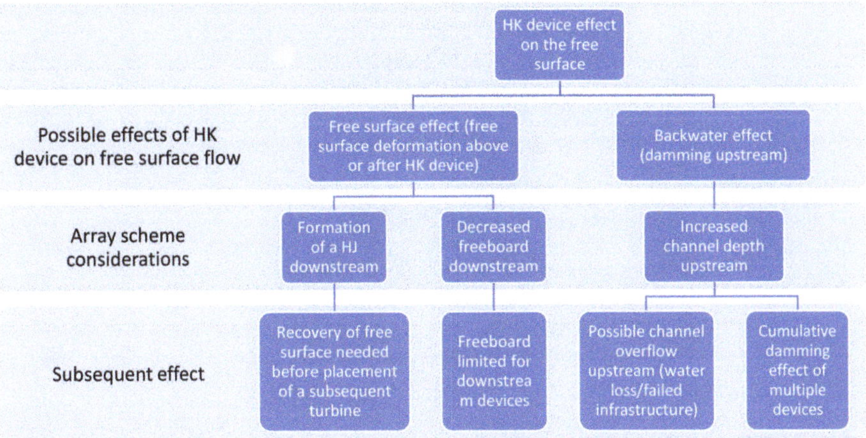

Figure 5. Water surface deformation from HK devices in inland flow infrastructure.

For inland systems, the backwater effect and the calculation thereof are of primary concern. Existing models include multiple unknowns and assumptions of values required as inputs that may not be available to users. Development of a clear, simple, effective blockage approximation of an HK device that can be applied to typical backwater calculations is necessary.

3. Validation of CFD Models

Comprehensive validation of a developed analytical model requires a dataset from a range of turbine styles under various operational conditions. Computational models offer an alternative to physical scale models to simulate the complex flow physics around HK devices. However, the importance of model validation through a physical model dataset should not be neglected. A number of studies have tested and validated modelling approaches for HK turbines [25,26].

Developments in tidal energy have led to multiple large-scale analyses of tidal turbine arrays, where simplified numerical models are used for array schemes [27–30]. With the ever-increasing availability and reliability of computational power, computational fluid dynamics (CFD) simulations are being used to model complex external effects and more accurately resolve fluid dynamic and wall effects [31]. Additionally, blade element momentum (BEM) theory is often used as a rotor modelling technique that also significantly reduces computational load [32].

CFD models may be used to resolve the effects of turbulence at the sub-grid-scale level, and are being used more often for first-order analysis or design. They have the potential to offer more comprehensive solutions and insights when their limitations are understood. Accurately representing turbulent flow in CFD is imperative due to its strong dependence on initial conditions, as well as the wide range of scales (eddies) present in the flow. Most often, statistical approaches based on the Reynolds-averaged Navier–Stokes (RANS) equations, with eddy viscosity models for turbulence closure, are used [33].

CFD models are valuable tools in flow-field analysis, especially at the sub-grid-scale level and for resolving turbulent length scales. When they are correctly applied and their limitations are understood, they offer useful insights and an alternative to laboratory testing. Correctly representing the turbulent flow, along with prescription of initial conditions, is important—especially due the wide range of scaled eddies present in the flow. Approaches based on the Reynolds-averaged Navier–Stokes (RANS) equations are most often used with various turbulence closure models incorporated, such as eddy viscosity models (k-Epsilon and SST k-ω).

A number of recent studies have used the BEM embedded in CFD method [8,34,35], which is also widely used in wind turbine array models, with good representation in terms of experimental results [36,37]. Various authors [38,39] have used this method to analyse the flow field of turbines arranged in arrays—specifically for tidal turbine optimization. A study analysing the accuracy of RANS approaches revealed good correlation with the experimental data found when using a Reynolds stress turbulence model (RSM) coupled with a BEM blade modelling technique (RSM-BEM) [40]. The study also highlights the importance of using an RSM rather than standard eddy viscosity models, due to the strong anisotropic flow in the wake.

3.1. CFD Models

Three different turbines (shown in Table 2) were modelled using CFD and validated with experimental results (a variation of free-surface, wake, and performance measurements). The primary validation case used was the U.S. Department of Energy's Reference Model 1 (RM1) dual-rotor axial flow turbine, which was modelled in the St. Anthony Falls Laboratory (SAFL) at the University of Minnesota [10]. Free-surface measurements allowed validation of the multiphase CFD model and free-surface deformation.

Additionally, two three-bladed turbine models with experimental results found in the literature [41] and tested on site [1] were modelled as specified in [40], for further measurements and validation of the backwater calculation methods. Where possible, BEM-VD (virtual disk) models were used to reduce the computational expense required. However, for the turbines which customized NACA profiles (Smart Hydropower turbine used in [1]) the full rotor geometry was modelled using a sliding mesh.

Table 2. Turbines modelled in CFD.

Turbine	Name	Blades	Diameter (m)	CFD Model
T1	RM1 [10]	2-Bladed NACA4415	0.5	Multiphase RSM-BEM model
T2	IFREMER [9]	3-Bladed NACA63418	0.7	Single-phase RSM-BEM model
T3	SHP [1]	3-Bladed custom blade	1	Single-phase RSM-FRG model

3.2. RM1 Model Validation

A 0.5 m diameter dual-rotor axial flow tidal turbine was investigated in a laboratory setup. The relevant details of the experiment can be seen in Table 3, with additional details available in [42]. The experimental results [42] provide high-resolution wake measurements of the near- and far-wake flow field surrounding the turbine from -5 to $10\ d_t$. These were used previously to validate the CFD procedure for a single-phase analysis [40].

Table 3. RM1 laboratory setup details [42].

Description	Variable
Rotor diameter	0.5 m
Blade profile	NACA 4415
Flow depth	1 m
Flow rate	2.425 m^3/s
Tip speed ratios measured	1 to 9
Flow velocity (U_{hub})	1.05 m/s
Turbulence intensity	5%
Froude number	0.28
Reynolds number (chord)	~3.0 $\times$ 10^5

Although the velocity profiles and performance metrics were adequately modelled with both the single-phase and multiphase models, correct approximation of the backwater effect through the multiphase modelling required validation. For this, free-surface measurements were collected for the optimal operational case ($TSR = 5.1$, $U = 1.05$ m/s) of the RM1 experimental setup, at a resolution of $1\ d_t$ (diameters) in the streamwise direction, and $0.4\ d_t$ in the cross-stream direction. The measurement zone was $-5\ dt$ to $10\ dt$ downstream. Elevation data were sampled at 50 Hz for 120 seconds at each location using a Massa ultrasonic range sensor, allowing for both time-averaged and fluctuating water surface elevation analysis and CFD validation.

Siemens STAR-CCM+ Commercial modelling software was used to simulate the turbines. The computational domain representing the RM1 laboratory model is shown in Figure 6. A wall-bounded model was used, extending from $-14\ d_t$ upstream to $16\ d_t$ downstream of the axis of rotation. The specified inlet length allowed full flow development prior to reaching the turbine axis of rotation. The outlet length ensured that no effects from the downstream boundary condition affected the near-wake behaviour. Previous studies have found that around $15\ d_t$ is usually adequate for the outlet boundary length [43,44].

A velocity inlet and downstream pressure outlet were specified as boundary conditions. The laboratory test turbulence and velocity values that were measured experimentally were specified at the inlet. Full development of the boundary layer on all surfaces (i.e., boundary walls, blades, and stanchion) was ensured through the specific turbulence model wall treatment and mesh resolution for each test case (details seen in [40]).

A virtual disk (VD) rotor modelling technique was used, and a BEM model was employed over the VD. This VD-BEM modelling approach has demonstrated good accuracy in the past, also significantly reducing computational costs [8,34,35,45,46]. A BEM tip-loss correction was incorporated using the Prandtl tip-loss correction method [47].

A Reynolds stress model (RSM) was used to allow more accuracy in prediction of possible flow anisotropy. The RS linear pressure strain two-layer (RS-LPS2) model [48,49] was found to model the near wake accurately, with a low y^+ wall treatment on the turbine and turbine structure ($y^+ < 1$).

Only transient simulations were performed, where unsteady terms were discretized using a 2nd-order implicit scheme. A time step ensuring Courant numbers of less than 1 over the domain was ensured as far as possible (some cells exceeded 1 at the blade tips, where smaller time steps did not change results and, therefore, larger Courant numbers were allowed to reduce computational time). The time steps were around 0.003 s for the BEM-CFD models. However, the time steps varied over each approach, depending on the results of the grid convergence index (GCI).

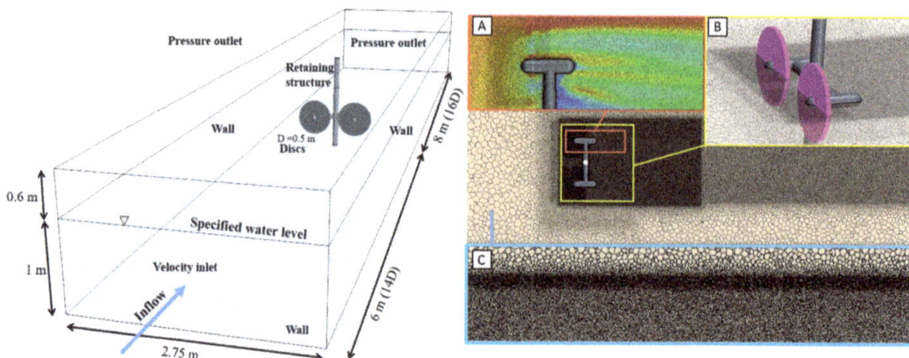

Figure 6. Computational domain with grid refinements: (**A**) near wake, (**B**) blades, and (**C**) free surface.

The computational domain (Figure 6) consisted of a polyhedral mesh with grid refinements in the free surface, near wake, far wake, and surrounding the turbine structure. In the past, results indicated the importance of fine grids to track the tip vortices when using simplified RANS models [31]. For grid refinement, an adapted GCI method [50] was used, due to the variance in grid sensitivity in the different regions. Separate regions' mesh sizes were incrementally decreased, until no changes in turbine performance, free surface, or wake behaviour were observed. As depicted in Figure 6, the final mesh sizes were around 14 million cells.

The smallest cell sizes were specified at the near wake and free surface, where a 10 mm minimum size proved to be adequate. Refinement in the near-wake region is imperative for accurate development of the complex near-wake behaviour, where both separated and attached flow exist [51]. Gibson and Launder [52] investigated the pressure fluctuation effects of capturing the boundary layer surrounding the turbine, and noted the importance of an accurate capture of boundary layer formation. For the VD rotor modelling technique, a minimum of four cells over the blade thickness was ensured for all meshes, which is the recommended minimum when using the VD method. A mesh base size of 14 mm (2.8% d_t) proved adequate in the far-wake region.

The channel walls around the turbine were modelled as non-slip walls using a high-y+ wall treatment to ensure that the effects of the wall boundary layer were included in the simulation. A two-layer formulation may be applied to the linear pressure strain model (LPS2 model). All analyses used the two-layer formulation.

Simulation of the air–water interface (multiphase flow) may be approached in various ways. A multiphase analysis ensures a robust approach, but demands a higher computational load; therefore, a symmetry boundary condition is often used in a single-phase model [40].

Due to the necessity of free-surface measurements in this analysis, the free surface was modelled using a volume-of-fluid approach. This approach uses a 2nd-order discretization to compute a clear interface between the air and water. A volume-fraction variable is used to specify the spatial distribution of each phase. Cells with multiple phases are treated as mixtures, and the method is highly dependent on adequate mesh resolution. A high-resolution mesh (Figure 6C) was ensured on the free surface, with the final cell size determined through a GCI test focussed on a free-surface profile analysis.

Validation of the modelled wake and performance can be seen in [40]. Due to the importance of the backwater approximation, and correct modelling of the backwater effect caused by the turbine blockage, the free-surface measurements from a multiphase model were compared in this paper. Comparison of the experimental and CFD results can be seen in Figure 7. The results correlated well, with the computed free-surface behaviour fitting well with the experimental tests. The CFD results were recorded over one rotation and plotted as an envelope. The backwater effect was predicted well, with a maximum of 12 mm damming occurring upstream.

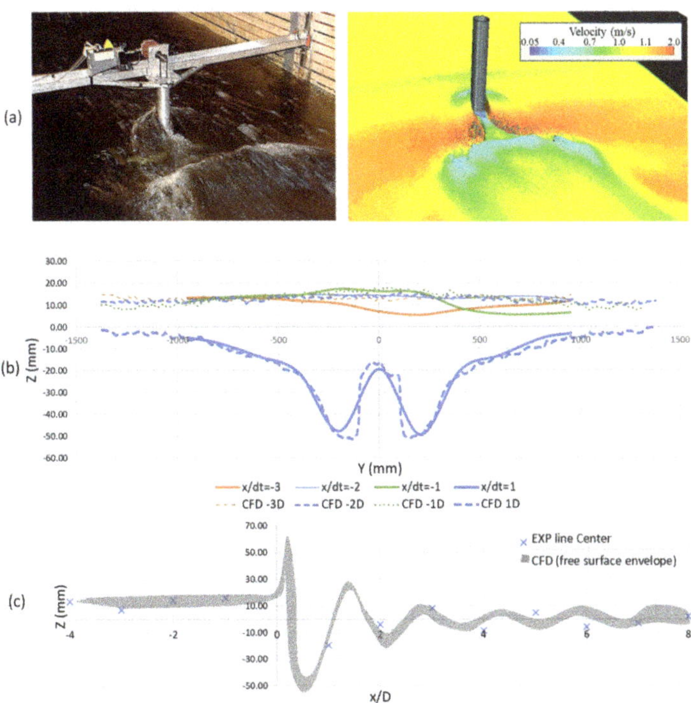

Figure 7. Comparison of experimental and computational water surface profiles for the RM1 tests: (**a**) experiment and CFD water surface graphics; (**b**) lateral WSE comparison; (**c**) longitudinal centre-line WSE comparison.

4. Methods

The primary objective of the development of a backwater model is to allow a usable model with only basic inputs required. Thus, a mathematical formulation based on existing models, basic hydrodynamic principles, and a set of recommended assumptions (where/if information is not yet available) was carried out. The workflow for the backwater model's development can be seen in Figure 8, and is explained in the following subsections.

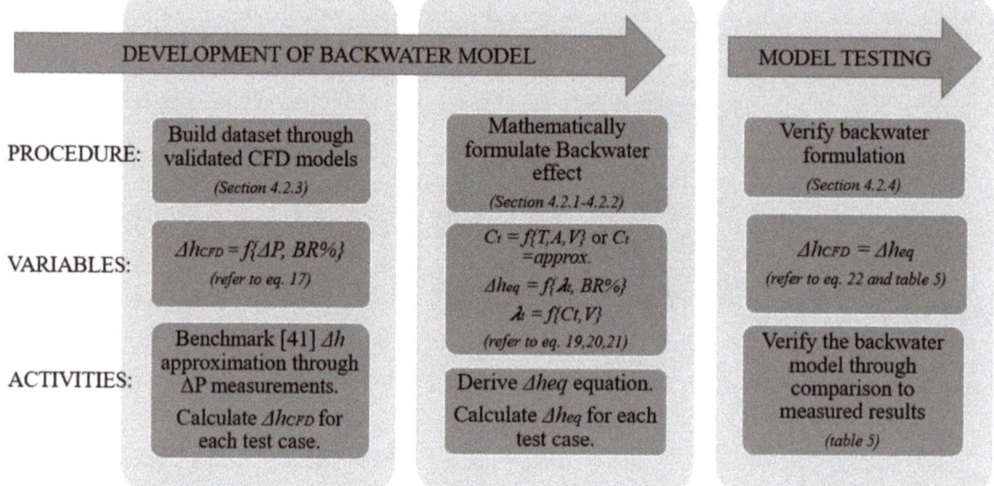

Figure 8. Workflow of backwater model development.

4.1. Assumptions and Exclusions

It is important to state the limitations of the model. Operational condition boundaries were set to within the typical canal operating conditions, where deployment of HK devices would be considered. Scenarios outside of these boundaries were not considered, as it was assumed that the use of such a model would not be necessary outside of these conditions (e.g., at supercritical flow conditions) [1,14,53].

The following limitations on flow conditions were set:

- Subcritical flow regime (Fr < 1);
- 5000 < Re < 1,500,000;
- Typical operational velocities of channels (0.8–2.8 m/s);
- Manning n-value around 0.016–0.023 s/m$^{1/3}$ (lined channel).

4.2. Mathematical Formulation

Energy losses in a channel are categorized and included with various approximations. For channel roughness, the friction losses can be accounted for by the Manning equation. Additionally, sudden losses due to channel features such as piers, bends, and drop structures have also been defined/estimated empirically, and can be included as form losses [14]. These are typically defined as eddy losses (h_e), included as an energy loss:

$$h_e = C_L \frac{V^2}{2g}$$

(12)

where C_L is the loss coefficient predefined for typical losses in a channel. The drop in water level due to a particular loss can be quantified/included by applying either the momentum or energy equation over a channel section, and the upstream and downstream sections (in which the energy loss exists). Additionally, an empirical approach may be used, where experimental results are used to determine an empirical relationship, such as that done by Yarnell in 1934 for bridge piers [54].

4.2.1. Approach 1: Momentum Approach

A possible approach often used to determine the effect of an object/structure on the free surface (backwater effect) is the momentum approach. Energy losses occur due to flow separation, vortex generation, friction, and turbulence—all associated with the changes in velocity due to the presence of the turbine. In some cases, the presence of the turbine may also result in the formation of a hydraulic jump on the water surface, resulting in additional energy losses. Applying the momentum approach avoids inclusion of these individual energy losses by considering the change in momentum between an upstream and downstream section. Additionally, if drag can be quantified, the momentum approach may be favourable.

The simple momentum formulation between two flow sections can be used with the resistance of the turbine represented as a drag coefficient, as shown in Figure 9 and Equation (13), where the change in momentum is quantified by the hydrostatic forces upstream (F_1) and downstream (F_2) of the device (water level change), as well as the friction from the channel bed and walls (F_f) and the force due to the turbine (F_D). This can then be rewritten to Equation (14), in terms of the drag force (F_D) due to the presence of the turbine.

$$F_1 - F_2 - F_D - F_f = \rho Q(\beta_2 V_2 - \beta_1 V_1) \tag{13}$$

$$F_D = \frac{1}{2}\rho g B h_1{}^2 - \frac{1}{2}\rho g B h_2{}^2 - \rho Q(\beta_2 V_2 - \beta_1 V_1) \tag{14}$$

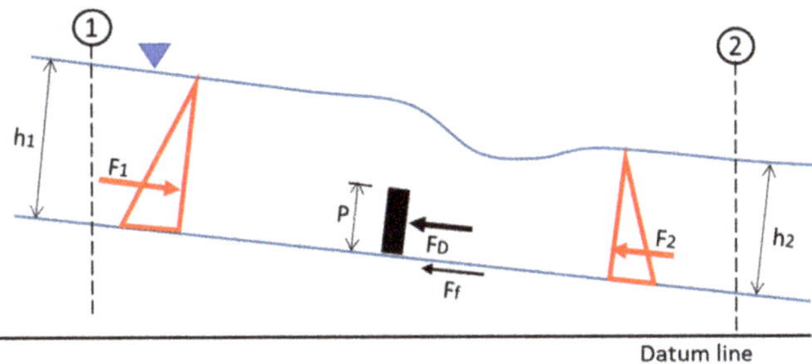

Figure 9. Momentum approach schematic (adapted form [55]).

The drag coefficient for the HK device can be rewritten, which allows the determination of h_1 (the upstream water level) through knowledge of the drag coefficient and downstream flow conditions. However, this requires knowledge of the downstream conditions, which is not always possible in feasibility studies in the design phase. An alternative conservative analysis would be to use the normal flow depth as the downstream value.

4.2.2. Approach 2: Energy Approach

The energy approach is more often used to determine backwater effects in typical open-channel flow scenarios. This is also a common approach in bridge pier modelling [14] when modelling the backwater effect of arch bridges [56], bridge piers, and even irregular structures such as wood jams [57].

The energy approach can be seen in Equation (15) and Figure 10. All energy losses between a point upstream and downstream of an HK device or blockage are quantified as terms that contribute to either the friction losses (h_f) or local losses (h_l).

$$\frac{V_1{}^2}{2g} + h_1 + Z_1 = \frac{V_4{}^2}{2g} + h_4 + Z_4 + \sum h_{f1-4} + \sum h_{l1-4} \tag{15}$$

All terms are defined in terms of velocity U), water depth (h), and distance from a datum (Z). The total energy (*TEL*), energy grade line (*EGL*), and water level/hydraulic grade line (*HGL*) are shown in Figure 10.

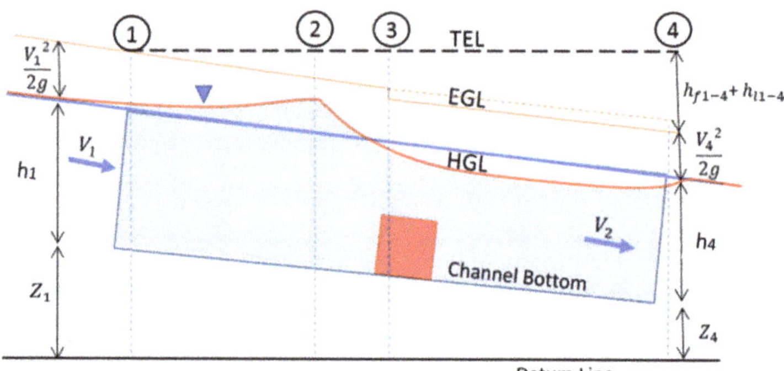

Figure 10. Energy approach.

Approximations of the loss due to the presence of the turbine have been attempted, such as those mentioned in Sections 2 and 3. However, preliminary tests of these methods indicated inaccuracies over a range of varying turbines and operational scenarios.

A method of including the turbine loss in the energy equation includes quantifying it in terms of a loss coefficient that has been calibrated to the turbine type and operating conditions. The energy loss due to the presence of the turbine (h_t) can be written as a function of a loss coefficient (α), the freestream velocity (U), and the blockage ratio of the turbine, as shown in Equation (16).

$$h_t = \alpha \frac{U^2}{2g} \times \frac{A_t}{A} \tag{16}$$

Although this approach is most often used in the literature, the loss (h_t) may also be quantified as a pressure drop, which is then directly converted to an energy loss as follows:

$$h_t = \frac{\Delta P_t}{\rho g} \times \frac{A_t}{A} \tag{17}$$

The pressure change (ΔP_t) is measured in the computational models as the pressure drop across the turbine, and applied (with consideration to the blockage ratio) to the channel area. This approach was followed to allow the use of CFD models to approximate energy loss due to the presence of the turbine. Other than the losses due to the direct pressure drop over the turbine, there are additional losses in the near wake, due to the turbulent flow. Using the ΔP_t approximation (Equation (17)) allows measurement of the pressure drop over the turbine and near-wake area, thus including additional losses.

4.2.3. Validation of Pressure Drop Measurement in CFD Results

CFD has previously been used to measure the backwater effects from blockages such as bridge piers [58], with computed and measured levels showing almost identical results. Multiple methods have also been analysed and validated for determining the backwater effects of common structures found in river channels [59–61]. To validate whether the approximation for h_t shown in Equation (17) holds true, the ΔP_t was measured in the CFD model for the RM1 validation case. The subsequently calculated loss (h_t) was then compared to the measured backwater effects in the laboratory tests (as well as multiphase CFD analysis). Inclusion of the support structure blockage was incorporated using the Yarnell approximation. The Yarnell approximation for a single circular bridge pier (similar to the support stanchion) was implemented:

$$\left[\frac{\Delta y}{y}\right]_{empirical} = K\left(K + 5F_r^2 - 0.6\right)\left(15\alpha^4\right)Fr^2 \tag{18}$$

where Δy is the backwater generated by the pier, y is the undisturbed flow depth, F_r is the downstream Froude number, and α is the ratio of the flow area obstructed by the pier to the total flow area downstream of the pier (also referred to as the blockage ratio). K is used as a coefficient reflecting the pier's shape. To ensure that the Yarnell approximation and pressure loss (ΔP_t) calculation work independently, the RM1 model free-surface deformation was measured with and without the stanchion structure (Figure 11), and the results were compared to the backwater calculation using only the pressure drop, as well as including the stanchion through the Yarnell approximation.

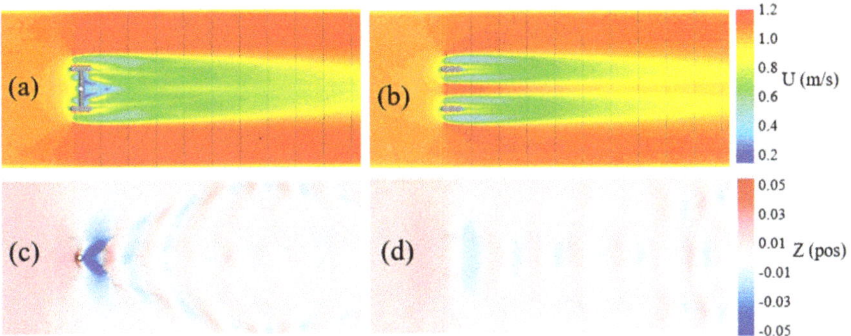

Figure 11. (**a,b**) Velocity and (**c,d**) surface water measurements graphics for the RM1 full model vs. the RM1 rotor and nacelle only.

The pressure drop due to the presence of the HK device was found to be maximal when measured over the size of the turbine-swept area from 1 D upstream to 1.5 D downstream (Figure 12). The calculated backwater (Equation (17)) was then compared to the measured backwater (Table 4), which was calculated from the average disk pressure drop, as shown in Figure 13.

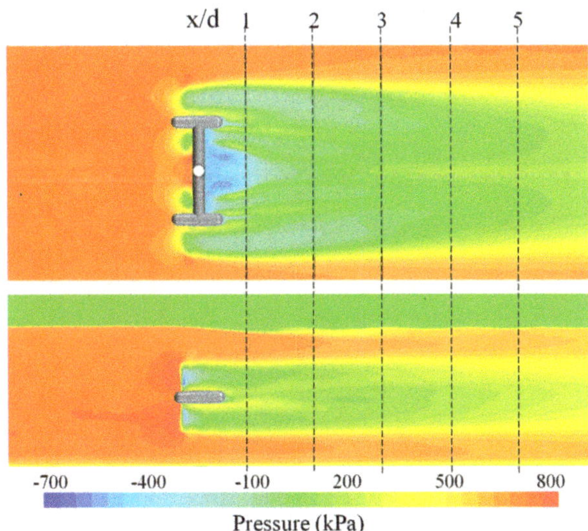

Figure 12. Pressure measurements over the horizontal and vertical planes (at the turbine hub height centerline).

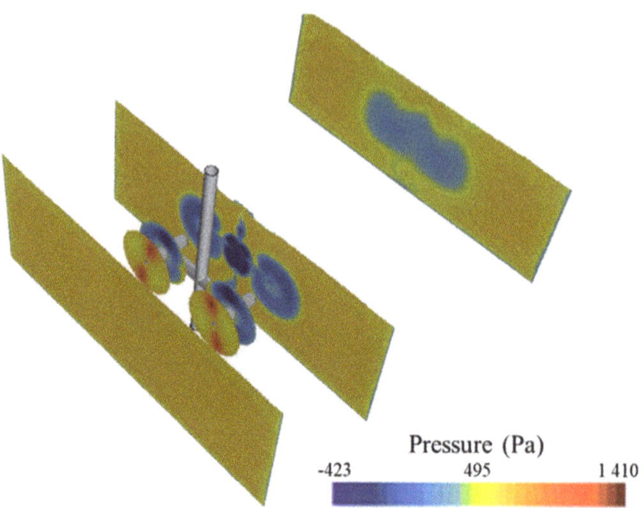

Figure 13. Pressure measurements over the disk and planes upstream and downstream of the RM1 turbine and retaining structure.

Table 4. Comparison of measured and predicted backwater levels.

	ΔP_t Disk (Pa)	ΔP_t Plane (Pa)	Calculated h_t (mm)	Yarnell Approx. (mm)	Measured h_t (mm)	$\frac{ht\,meas - ht\,calc}{y}$ (%)
RM1 (no stanchion)	570	57.73	8.30	-	9.60	0.13%
RM1 (with stanchion)	530	74.09	7.72 / 12.66	4.36	12.00	0.01%

As shown in Table 4, the calculated analytical backwater values from Equation (17) and the computational reading for disk ΔP result in a very similar h_t, as would be the result of the RM1 device in channel flow.

4.2.4. Lambda Approximation

To allow a simple empirical model for the determination of the energy loss due to the turbine, λ_T was selected as the energy loss coefficient used in the energy equation:

$$h_t = \lambda_T \times \frac{A_t}{A} \frac{U^2}{2g} \tag{19}$$

where h_t is included as a loss in the energy equation (Equation (15)), and λ_T is calculated as a function of the thrust coefficient (C_t):

$$\lambda_T = \frac{C_t \times U^2}{2g} \tag{20}$$

where C_t is a value that can be obtained from the manufacturer, calculated, or assumed in the pre-feasibility stage. For HAHTs, these thrust coefficients (C_t) usually range from 0.52 to 0.89 [9,62–64]. According to the actuator disk theory, C_t may be written in terms of the induction factor a [65]. It is also known that ideally, according to the Betz limit, $a = \frac{1}{3}$; therefore, the ideal and highest attainable C_t would be 0.88. Theoretically, according to the BEM theory, this should result in the highest velocity deficit in the near wake and, therefore, the "worst case" scenario for the operational conditions. Realistically, the values lie at an upper limit of $C_t = 0.8$. The thrust coefficient can be calculated directly if the thrust force (T), inlet velocity (U), and swept area (A) are known:

$$C_t = \frac{T}{\left(\frac{1}{2}\right)\rho U^2 A}$$ (21)

To justify the use of the λ_T approximation, the validated CFD models were analysed, the pressure drop/total thrust was measured, and the subsequent backwater effect was determined. The calculated h_t (through Equation (19)) was then compared to the h_t determined through the ΔP_t (Equation (17)) results, as validated in Section 4.2.3.

Two approximations for λ_T were included (calculated and assumed C_t). The model should be usable with only basic knowledge of the turbine installation and operating parameters; therefore, simple available metrics could be used to obtain a conservative result. Acceptable correlation between the experimental and calculated values created confidence to proceed with the model and build a larger dataset to analyse the model's accuracy at a larger operational variance from optimal conditions.

A regression to the mean approach was used to accumulate the necessary dataset for the analysis of the aforementioned calculation procedure and assumptions. This was required to reduce computational costs, and due to the lack of available data on various input parameters. The dataset was created through results from three models of turbines typically used for inland installations (Table 2). These CFD models were validated through benchmark validation using experimental results obtained at optimal performance points. The models were then varied in five primary operational states, namely:

1. Inlet velocity changes (0.4 < U < 2.8);
2. Blockage ratio changes (Swept area to flow area) (4% < BR < 23%);
3. Tip speed ratio changes (lower or higher load applied) (3 < TSR < 6);
4. Froude number (0.18 < Fr < 0.34) (within the subcritical flow regime);
5. Froude number based on turbine diameter (0.15 < Fr_D < 0.9).

These primary variables have been previously investigated and shown to influence the turbine thrust imposed on the flow area, and may therefore influence the backwater effect. Constraints were set to the variation of these variables to ensure that the computational models remained within realistic scenarios, whilst allowing insight into the effects of changes. The primary objective of the model validation was to ensure that a relative level of accuracy was obtained and, more importantly, a conservative approach to predicting the possible backwater effects caused by such a device.

A measure of accuracy of the methods is indicated by the mean square absolute error. The absolute error (MAE), rather than relative error (RMSE), was used to place greater emphasis of the larger backwater values (at higher blockage ratios) rather than uniform predictions over the range of backwater predictions (h_t). Additionally, as the sample size changed in the analysis, the strength of the sample size effect was minimized when comparing MAE. The variance was also included to give an indication of the test conditions with greater variability, and under which test conditions the model (and assumptions) performed best.

$$\sigma = \sqrt{\frac{1}{N}\sum\left(\frac{h_t}{y}_{exp} - \frac{h_t}{y}_{calc}\right)^2}$$ (22)

The MAE (σ) and variance values calculated for various scenarios are shown in Table 5. The variations between experimental and approximated h_t for variations in blockage ratio (BR) velocity (U) and Fr_D are shown in Figure 14.

Table 5. Mean absolute error and variance for test cases.

Test Condition	C_t	N	MAE	Variance
	Equation (21)	14	1.45	2.26
All tests conducted	0.8	14	1.42	2.17
	0.89	14	1.99	4.27
Optimal operational point	Equation (21)	3	1.26	1.85
	0.8	3	1.25	1.83
Variation of blockage ratios (BR = 4–22%) at optimal operational point	Equation (21)	4	0.35	0.16
	0.8	4	0.27	0.09
Variation of inlet velocities at optimal tip speed ratios	Equation (21)	4	2.4	7.65
	0.8	4	1.36	2.47

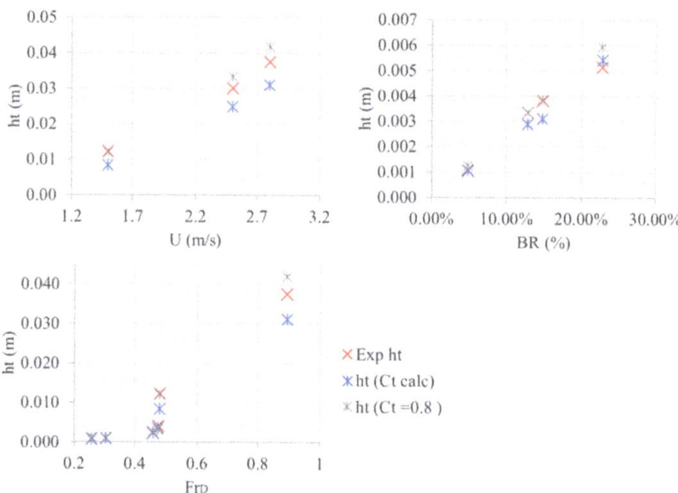

Figure 14. Effects of U, BR, and F_{rD} on the determined h_t values.

From the results in Table 5 and Figure 14, the following observations can be drawn:

1. At turbine optimal operational points, a maximum deviation of 13% from the predicted backwater was obtained when using the correct C_t value. This deviation increased to 19% for the $C_t = 0.8$ approximation.
2. When utilizing the C_t assumption of 0.8, a conservative result was obtained, with the backwater estimation generally overestimating the measured blockage.
3. Calculating C_t based on the turbine thrust (measured thrust) lowered the h_t approximation. However, for test cases operating close to the optimal performance and highest C_t value, the backwater was underestimated by up to 20%.
4. Test cases at low operational velocities (low Froude numbers) resulted in larger errors in approximating h_t; however, it is important to note that these are unfavourable installation conditions and far from typical installations. The turbines may have low performance at these low operational velocities and, therefore, pose an unrealistic scenario. Here, the C_t calculation resulted in a more realistic value, due to the reduced performance.
5. The C_t approximation resulted in large overestimations of the h_t at lower TSRs. However, the C_t equation (Equation (22)) performed well in these scenarios, as the turbine thrust was significantly lower, and the C_t assumption did not hold.
6. The C_t approximation gave significantly better results for the three-bladed turbines. The two-bladed (T1) case predicted better results with the C_t calculation, which

was also higher than the 0.8 approximation, indicating that the turbine operates closer to the Betz limit and ideal induction factor (a), which could be further tested and calibrated. The C_t calculation performed better in this case, predicting $C_t = 0.89$. Therefore, utilizing this assumption may be favourable for avoiding errors—especially when turbines with higher operational tip speed ratios are used.

Based on the small dataset obtained from the three turbine models, a C_t value for each turbine can be determined empirically, which could be improved with a larger dataset.

The recommended model also performed significantly better than models found in the literature, as well as needing less input data and background knowledge on specific turbine performance. This indicates the usefulness of this approximation.

5. Conclusions

Quantifying the backwater resulting from the effective blockage caused by the operating turbine remains a challenge for the deployment of HK turbines in inland infrastructure, such as canal systems. This paper shows a simple analytical model to estimate the effective blockage and backwater effect from HK devices. The development of the model followed a similar approach to what has been previously used for bridge pier modelling and quantifying blockages from such structures.

Three variations of typical examples of commercially available HK devices were modelled using a recommended CFD approach, and conditions were varied to allow testing over a range of operational conditions. The water surface profile measurements from the Reference Model 1 (RM1) scaled model experiments allowed validation of the water surface deformation obtained when using a VOF approach coupled with an RS-LPS2 Reynolds stress closure model and BEM-VD blade modelling approach. This highlights the usefulness of CFD models in HK energy development.

The developed backwater model allows a conservative approach with various levels of certainty attainable, depending on the input parameters installed. Although the recommended procedure is an extremely simplified approximation, the results obtained were significantly closer to available dataset of backwater effects than the methods found in the literature. The ease of use also makes this method useful for engineers and developers when detailed numerical models are not feasible.

This model also allows cumulative estimation of backwater, with simple inclusion of blockage structures or multiple turbines using available approximations such as the Yarnell equation. Additionally, due to the nature of flow in canals (flat slopes), subcritical conditions govern, and backwater effects extend a large distance upstream. This allows simple cumulative inclusion of blockages without complex computational modelling, making this approach simple and relatively accurate (or at least favourably conservative).

This approach allows room for further development and determination of calibrated thrust coefficients for typical turbines or typical operational conditions (similar to what was previously done for pier shapes, etc.). Additionally, a similar approach may be investigated for cross-flow turbines where experimental results are available.

Although this paper focussed on backwater determination, other free-surface effects— such as the water level drop over the turbine, or possible hydraulic jumps downstream—are important considerations for array designs. Recommendations for clearance coefficients and turbine spacing (due to wake recovery) should be carefully considered prior to deployment.

Author Contributions: Conceptualization, C.M.N. and M.V.D.; methodology, C.M.N., M.V.D. and L.S.; software, C.M.N. and L.S.; validation, C.M.N. and C.H.; formal analysis, C.M.N., C.H., M.V.D. and L.S.; investigation, C.M.N. and C.H.; resources, C.H.; data curation, C.M.N. and C.H.; writing—original draft preparation, C.M.N.; writing—review and editing, C.H., M.V.D. and L.S.; visualization, C.M.N. and C.H.; supervision, M.V.D. and L.S.; project administration, C.M.N. All authors have read and agreed to the published version of the manuscript.

Funding: This research received no external funding. Experimental results used are open-access and can be found in [65].

Data Availability Statement: Not applicable.

Acknowledgments: This work was supported by the University of Pretoria. The computational capabilities were made possible due to academic hours allocated by the Center for High-Performance Computing (CHPC) South Africa. Siemens STAR-CCM+ Simcenter software and support were provided by Aerotherm Computational Dynamics (Pty) Ltd. Experimental results were obtained from the Sandia Laboratories repository. All contributors are thanked for their kind assistance and support.

Conflicts of Interest: The authors declare no conflict of interest.

References

1. Niebuhr, C.M.; van Dijk, M.; Bhagwan, J.N. Development of a design and implementation process for the integration of hy-drokinetic devices into existing infrastructure in South Africa. *Water SA* **2019**, *45*, 434–446. [CrossRef]
2. Riglin, J.D. Design, Manufacture and Prototyping of a Hydrokinetic Turbine Unit for River Application. Master's Thesis, Lehigh University, Bethlehem, PA, USA, 2016.
3. Runge, S. Performance and Technology Readiness of a Freestream Turbine in a Canal Environment. Ph.D. Thesis, Cardiff University, Cardiff, UK, 2018.
4. Kartezhnikova, M.; Ravens, T.M. Hydraulic impacts of hydrokinetic devices. *Renew. Energy* **2014**, *66*, 425–432. [CrossRef]
5. Gunawan, B.; Roberts, J.; Neary, V. Hydrodynamic Effects of Hydrokinetic Turbine Deployment in an Irrigation Canal. In Proceedings of the 3rd Marine Energy Technology Symposium, Washington, DC, USA, 27–29 April 2015; pp. 1–6.
6. Bahaj, A.S.; Myers, L.E.; Rawlinson-Smith, R.I.; Thomson, M. The effect of boundary proximity upon the wake structure of horizontal axis marine current turbines. *J. Offshore Mech. Arct. Eng.* **2011**, *134*, 021104. [CrossRef]
7. Bachant, P.; Wosnik, M. Effects of Reynolds Number on the Energy Conversion and Near-Wake Dynamics of a High Solidity Vertical-Axis Cross-Flow Turbine. *Energies* **2016**, *9*, 73. [CrossRef]
8. Turnock, S.R.; Phillips, A.B.; Banks, J.; Nicholls-Lee, R. Modelling tidal current turbine wakes using a coupled RANS-BEMT approach as a tool for analysing power capture of arrays of turbines. *Ocean Eng.* **2011**, *38*, 1300–1307. [CrossRef]
9. Mycek, P.; Gaurier, B.; Germain, G.; Pinon, G.; Rivoalen, E. Experimental study of the turbulence intensity effects on marine current turbines behaviour. Part II: Two interacting turbines. *Renew. Energy* **2014**, *68*, 876–892. [CrossRef]
10. Hill, C.; Neary, V.S.; Guala, M.; Sotiropoulos, F. Performance and Wake Characterization of a Model Hydrokinetic Turbine: The Reference Model 1 (RM1) Dual Rotor Tidal Energy Converter. *Energies* **2020**, *13*, 5145. [CrossRef]
11. Lalander, E.; Leijon, M. In-stream energy converters in a river—Effects on upstream hydropower station. *Renew. Energy* **2011**, *36*, 399–404. [CrossRef]
12. Myers, L.; Bahaj, A.S. Wake studies of a 1/30th scale horizontal axis marine current turbine. *Ocean Eng.* **2007**, *34*, 758–762. [CrossRef]
13. Adamski, S.J. Numerical Modeling of the Effects of a Free Surface on the Operating Characteristics of Marine Hydrokinetic Turbines. Ph.D. Thesis, University of Washington, Washington, DC, USA, 2013.
14. Henderson, F.M. *Open Channel Flow*; The Mcmillan Company: New York, NY, USA, 1966.
15. Birjandi, A.H.; Bibeau, E.L.; Chatoorgoon, V.; Kumar, A. Power measurement of hydrokinetic turbines with free-surface and blockage effect. *Ocean Eng.* **2013**, *69*, 9–17. [CrossRef]
16. Niebuhr, C.; van Dijk, M.; Neary, V.; Bhagwan, J. A review of hydrokinetic turbines and enhancement techniques for canal installations: Technology, applicability and potential. *Renew. Sustain. Energy Rev.* **2019**, *113*, 109240. [CrossRef]
17. Whelan, J.I.; Graham, J.M.R.; Peiró, J. A free-surface and blockage correction for tidal turbines. *J. Fluid Mech.* **2009**, *624*, 281–291. [CrossRef]
18. Polagye, B.L. *Hydrodynamic Effects of Kinetic Power Extraction by In-Stream Tidal Turbines*; University of Washington: Washington, DC, USA, 2009.
19. Bryden, I.; Grinsted, T.; Melville, G. Assessing the potential of a simple tidal channel to deliver useful energy. *Appl. Ocean Res.* **2004**, *26*, 198–204. [CrossRef]
20. Chanson, H. *Hydraulics of Open Channel Flow*, 2nd ed.; Elsevier Science & Technology: Amsterdam, The Netherlands, 2004.
21. Mańko, R. Ranges of Backwater Curves in Lower Odra. *Civ. Environ. Eng. Rep.* **2018**, *28*, 25–35. [CrossRef]
22. Garrett, C.; Cummins, P. The efficiency of a turbine in a tidal channel. *J. Fluid Mech.* **2007**, *588*, 243–251. [CrossRef]
23. Garrett, C.; Cummins, P. The power potential of tidal currents in channels. *Proc. R. Soc. A Math. Phys. Eng. Sci.* **2005**, *461*, 2563–2572. [CrossRef]
24. Ross, H.; Polagye, B. An experimental assessment of analytical blockage corrections for turbines. *Renew. Energy* **2020**, *152*, 1328–1341. [CrossRef]
25. López, Y.; Contreras, L.; Laín, S. CFD Simulation of a Horizontal Axis Hydrokinetic Turbine. *Renew. Energy Power Qual. J.* **2017**, *1*, 512–517. [CrossRef]
26. Laín, S.; Contreras, L.T.; López, O. A review on computational fluid dynamics modeling and simulation of horizontal axis hydrokinetic turbines. *J. Braz. Soc. Mech. Sci. Eng.* **2019**, *41*, 375. [CrossRef]

27. Adcock, T.A.; Draper, S.; Nishino, T. Tidal power generation—A review of hydrodynamic modelling. *J. Power Energy* **2015**, *229*, 755–771. [CrossRef]

28. Nishino, T.; Willden, R.H. Effects of 3-D channel blockage and turbulent wake mixing on the limit of power extraction by tidal turbines. *Int. J. Heat Fluid Flow* **2012**, *37*, 123–135. [CrossRef]

29. Nishino, T.; Willden, R.H.J. Two-scale dynamics of flow past a partial cross-stream array of tidal turbines. *J. Fluid Mech.* **2013**, *730*, 220–244. [CrossRef]

30. Gotelli, C.; Musa, M.; Guala, M.; Escauriaza, C. Experimental and Numerical Investigation of Wake Interactions of Marine Hydrokinetic Turbines. *Energies* **2019**, *12*, 3188. [CrossRef]

31. Sanderse, B.; van der Pijl, S.P.; Koren, B. Review of computational fluid dynamics for wind turbine wake aerodynamics. *Wind Energy* **2011**, *14*, 799–819. [CrossRef]

32. Whale, J.; Anderson, C.; Bareiss, R.; Wagner, S. An experimental and numerical study of the vortex structure in the wake of a wind turbine. *J. Wind Eng. Ind. Aerodyn.* **2000**, *84*, 1–21. [CrossRef]

33. Pyakurel, P.; Tian, W.; VanZwieten, J.H.; Dhanak, M. Characterization of the mean flow field in the far wake region behind ocean current turbines. *J. Ocean Eng. Mar. Energy* **2017**, *3*, 113–123. [CrossRef]

34. Masters, I.; Chapman, J.C.; Willis, M.R.; Orme, J.A.C. A robust blade element momentum theory model for tidal stream tur-bines including tip and hub loss corrections. *J. Mar. Eng. Technol.* **2014**, *10*, 25–35. [CrossRef]

35. Guo, Q.; Zhou, L.; Wang, Z. Comparison of BEM-CFD and full rotor geometry simulations for the performance and flow field of a marine current turbine. *Renew. Energy* **2015**, *75*, 640–648. [CrossRef]

36. Malki, R.; Masters, I.; Williams, A.J.; Croft, N. The variation in wake structure of a tidal stream turbine with flow velocity. In Proceedings of the MARINE 2011, IV International Conference on Computational Methods in Marine Engineering, Lisbon, Portugal, 28–30 September 2011. [CrossRef]

37. Edmunds, M.; Williams, A.; Masters, I.; Croft, N. An enhanced disk averaged CFD model for the simulation of horizontal axis tidal turbines. *Renew. Energy* **2017**, *101*, 67–81. [CrossRef]

38. Masters, I.; Williams, A.; Croft, T.N.; Togneri, M.; Edmunds, M.; Zangiabadi, E.; Fairley, I.; Karunarathna, H. A Comparison of Numerical Modelling Techniques for Tidal Stream Turbine Analysis. *Energies* **2015**, *8*, 7833–7853. [CrossRef]

39. Masters, I.; Malki, R.; Williams, A.J.; Croft, T.N. The influence of flow acceleration on tidal stream turbine wake dynamics: A numerical study using a coupled BEM–CFD model. *Appl. Math. Model.* **2013**, *37*, 7905–7918. [CrossRef]

40. Niebuhr, C.; Schmidt, S.; van Dijk, M.; Smith, L.; Neary, V. A review of commercial numerical modelling approaches for axial hydrokinetic turbine wake analysis in channel flow. *Renew. Sustain. Energy Rev.* **2022**, *158*, 112151. [CrossRef]

41. Mycek, P.; Gaurier, B.; Germain, G.; Pinon, G.; Rivoalen, E. Experimental study of the turbulence intensity effects on marine current turbines behaviour. Part I: One single turbine. *Renew. Energy* **2014**, *66*, 729–746. [CrossRef]

42. Hill, C.; Neary, V.S.; Gunawan, B.; Guala, M.; Sotiropoulos, F.U.S. *Department of Energy Reference Model Program RM1: Experimental Results*; University of Minnesota: Minneapolis, MN, USA, 2014.

43. Nasef, M.H.; El-Askary, W.A.; AbdEL-hamid, A.A.; Gad, H.E. Evaluation of Savonius rotor performance: Static and dynamic studies. *J. Wind Eng. Ind. Aerodyn.* **2013**, *123*, 1–11. [CrossRef]

44. Franke, J.; Hirsch, C.; Jensen, A.G.; Krus, H.W.; Schatzmann, P.S.; Miles, S.D.; Wisse, J.A.; Wright, N.G. Recommendations on the use of CFD in wind engineering. In Proceedings of the CWE2006 Fourth International Symposium Computational Wind Engineering, Yokohama, Japan, 16–19 July 2006.

45. Malki, R.; Williams, A.; Croft, T.; Togneri, M.; Masters, I. A coupled blade element momentum—Computational fluid dynamics model for evaluating tidal stream turbine performance. *Appl. Math. Model.* **2013**, *37*, 3006–3020. [CrossRef]

46. Bekker, A.; Van Dijk, M.; Niebuhr, C.M. A review of low head hydropower at wastewater treatment works and development of an evaluation framework for South Africa. *Renew. Sustain. Energy Rev.* **2022**, *159*, 112216. [CrossRef]

47. Shen, W.Z.; Mikkelsen, R.; Sørensen, J.N.; Bak, C. Tip loss corrections for wind turbine computations. *Wind Energy* **2005**, *8*, 457–475. [CrossRef]

48. Speziale, C.G.; Sarkar, S.; Gatski, T.B. Modelling the pressure-strain correlation of turbulence: An invariant dynamical systems approach. *J. Fluid. Mech.* **1991**, *227*, 245–272. [CrossRef]

49. Sarkar, S.; Lakshmanan, B. Application of a Reynolds stress turbulence model to the compressible shear layer. *AIAA J.* **1991**, *29*, 743–749. [CrossRef]

50. Roache, P.J. Perspectvie: A method for Uniform Reporting of Grid Refinement Studies. *J. Fluids Eng. Trans. ASME* **1994**, *116*, 405–413. [CrossRef]

51. Silva, P.A.S.F.; De Oliveira, T.F.; Brasil Junior, A.C.P.; Vaz, J.R.P.P.; Oliveira, T.F.D.E.; Junior, A.C.P.B.; Vaz, J.R.P.P. Numerical Study of Wake Characteristics in a Horizontal-Axis Hydrokinetic Turbine. *Ann. Braziian Acad. Sci.* **2016**, *88*, 2441–2456. [CrossRef]

52. Gibson, M.M.; Launder, B.E. Ground effects on pressure fluctuations in the atmospheric boundary layer. *J. Fluid Mech.* **1978**, *86*, 491–511. [CrossRef]

53. Neary, V.S.; Gunawan, B.; Hill, C.; Chamorro, L.P. Near and far field flow disturbances induced by model hydrokinetic tur-bine: ADV and ADP comparison. *Renew. Energy* **2013**, *60*, 1–6. [CrossRef]

54. Yarnell, D. *Bridge Piers as Channel Obstructions*; United States Department of Agriculture: Washington, DC, USA, 1934.

55. Azinfar, H.; Kells, J.A. Backwater Prediction due to the Blockage Caused by a Single, Submerged Spur Dike in an Open Channel. *J. Hydraul. Eng.* **2008**, *134*, 1153–1157. [CrossRef]

56. Martin-Vide, J.; Prio, J. Backwater of arch bridges under free and submerged conditions. *J. Hydraul. Res.* **2005**, *43*, 515–521. [CrossRef]
57. Follett, E.; Schalko, I.; Nepf, H. Momentum and Energy Predict the Backwater Rise Generated by a Large Wood Jam. *Geophys. Res. Lett.* **2020**, *47*, e2020GL089346. [CrossRef]
58. Kocaman, S. Prediction of Backwater Profiles due to Bridges in a Compound Channel Using CFD. *Adv. Mech. Eng.* **2014**, *6*, 905217. [CrossRef]
59. Azinfar, H.; Kells, J.A. Drag force and associated backwater effect due to an open channel spur dike field. *J. Hydraul. Res.* **2011**, *49*, 248–256. [CrossRef]
60. Raju, K.R.; Rana, O.; Asawa, G.; Pillai, A. Rational assessment of blockage effect in channel flow past smooth circular cylinders. *J. Hydraul. Res.* **1983**, *21*, 289–302. [CrossRef]
61. Morandi, B.; Di Felice, F.; Costanzo, M.; Romano, G.; Dhomé, D.; Allo, J. Experimental investigation of the near wake of a horizontal axis tidal current turbine. *Int. J. Mar. Energy* **2016**, *14*, 229–247. [CrossRef]
62. Jeffcoate, P.; Whittaker, T.; Boake, C.; Elsaesser, B. Field tests of multiple 1/10 scale tidal turbines in steady flows. *Renew. Energy* **2016**, *87*, 240–252. [CrossRef]
63. Stallard, T.; Collings, R.; Feng, T.; Whelan, J. Interactions between tidal turbine wakes: Experimental study of a group of three-bladed rotors. *Philos. Trans. R. Soc. A Math. Phys. Eng. Sci.* **2013**, *371*, 20120159. [CrossRef] [PubMed]
64. Lam, W.-H.; Chen, L. Equations used to predict the velocity distribution within a wake from a horizontal-axis tidal-current turbine. *Ocean Eng.* **2014**, *79*, 35–42. [CrossRef]
65. Sandia National Laboritories: Refernce Model Porject (RMP). Available online: https://energy.sandia.gov/programs/renewable-energy/water-power/projects/reference-model-project-rmp/ (accessed on 21 January 2022).

Article

A Method for the Integrated Optimal Design of Multiphase Pump Based on the Sparse Grid Model

Cancan Peng [1,*], Xiaodong Zhang [1], Yongqiang Chen [2], Yan Gong [1], Hedong Li [1] and Shaoxiong Huang [1]

[1] Institution of Mechanical and Electrical Engineering, Southwest Petroleum University, Chengdu 610500, China; zxd123420@126.com (X.Z.); gongyan0101@163.com (Y.G.); 202031030010@stu.swpu.edu.cn (H.L.); xkarz52116@outlook.com (S.H.)
[2] Jiyuan Huaxin Petroleum Machinery Co., Ltd., Jiyuan 454650, China; xiaoxuan425@sina.com
* Correspondence: 201811000059@stu.swpu.edu.cn

Abstract: Multiphase pumps are used as an important tool for natural gas hydrate extraction owing to their excellent gas–liquid mixing and transport properties. This paper proposes an adaptive response surface-based integrated optimization design method. A model pump is designed based on the axial flow pump design theory. The model pump is numerically simulated and analyzed to obtain its performance parameters. Then the structural and performance parameters of the pump are parameterized to establish a closed-loop input–output system. Based on this closed-loop system, a sensitivity analysis is performed on the structural parameters of the impeller and guide vane, and the parameters that affect the performance of the gas–liquid hybrid pump the most are derived. The Sparse Grid method was introduced to design the experiment and construct the approximate model. The structural parameters of the impeller and guide vane are used as design variables to optimize the pressure increment and efficiency of the pump. After optimization, the pressure increment of the multiphase pump was increased by 10.78 KPa and the efficiency was increased by 0.89% compared to the original model. Finally, we validate the accuracy of the optimized model with tests.

Keywords: multiphase pump; integrated design; Sparse Grid method; numerical analysis; flow field characteristics

Citation: Peng, C.; Zhang, X.; Chen, Y.; Gong, Y.; Li, H.; Huang, S. A Method for the Integrated Optimal Design of Multiphase Pump Based on the Sparse Grid Model. *Processes* **2022**, *10*, 1317. https://doi.org/10.3390/pr10071317

Academic Editor: Blaž Likozar

Received: 22 June 2022
Accepted: 1 July 2022
Published: 5 July 2022

Publisher's Note: MDPI stays neutral with regard to jurisdictional claims in published maps and institutional affiliations.

1. Introduction

With the development of the world and the advancement of technology, traditional resources such as coal and oil can no longer meet people's pursuit of a green, eco-friendly, and low-carbon lifestyle. Natural gas hydrates are a clean energy resource and have huge global reserves of approximately 1.5×10^{16} m^3 [1]. More than 230 hydrate deposits have been discovered globally on the seafloor and in permafrost zones, where low-temperature and high-pressure conditions are suitable for the formation of natural gas hydrates [2]. Currently, natural gas hydrate extraction relies heavily on a hydraulic lift and requires a substantial reduction in downhole pressure to prevent the collapse of the seabed [3]. Traditional axial flow pumps are effective in reducing downhole pressure and perform well with liquid media but are less effective in lifting mixed media such as natural gas hydrates. Multiphase pumps are multiphase fluid transfer devices that play a critical role in the hydraulic lift of gas hydrates. Therefore, research on downhole multiphase pumps is very important.

In recent years, several scholars have investigated numerical simulation methods and internal flow characteristics of multiphase pumps. Suh et al. [4] proposed a developed Euler–Euler method to investigate the effect of different GVF, interphase forces, and bubble diameter on the internal flow characteristics of the multiphase pump. Yu et al. [5,6], Liu et al. [7], and Zhang et al. [8] have studied the interphase forces and obtained a rule of influence between the variation of the interphase forces and the flow field characteristics of the multiphase pump. Zhang et al. [9] applied a non-uniform bubble model to the numerical simulation process of a multiphase pump. The obtained results of the non-uniform bubble model did not differ much from those of the conventional model at low GVF. However, when the GVF is high, the numerical result of the non-uniform bubble model is closer to the experimental data. Li et al. [10] investigated the distribution of bubbles in a multiphase pump. The size of bubbles was calculated according to the bubble number density equation, and the distribution characteristics of bubbles at different GVF were analyzed to obtain the movement pattern of bubbles in the impeller and guide vane. Zhang et al. [11] analyzed the formation process of the gas pocket in the multiphase pump through non-constant numerical simulations and visualized experiments, and the results of the study provide some guidance for the future design of multiphase pump impellers. Wang et al., Huang et al., Jiang et al., and Shi et al. [12–15] studied the flow characteristics of the multiphase pump from the perspective of energy loss and obtained the influence law of structure parameters and working condition parameters on the performance of the multiphase pump. Xu et al. [16] studied the internal flow pattern of a multiphase pump and developed a multiphase pump with splitting vanes. The efficiency and head of the pump were improved, which provided guidance for the further design and optimization of the multiphase pump. Yu et al. [17] analyzed the transport process of a non-stationary multiphase pump using a two-fluid model and found the main causes of pressure and head fluctuations in a multiphase pump. Xu et al. [18] and Zhang et al. [19] investigated the pressure pulsation characteristics of the multiphase pump and obtained the distribution and variation pattern of pressure fluctuations in the multiphase pump. Zhang et al. [20] and Shi et al. [21] studied the tip clearance of the multiphase pump and obtained the effect of tip clearance on energy characteristics, flow characteristics, and pressure fluctuations, respectively. Shi et al. [22] and Li et al. [23] investigated the effect of GVF on the energy conversion characteristics, pressure fluctuations, blade surface load, gas distribution, and external characteristics of a multiphase pump. Others have investigated the performance of the multiphase pump under high GVF conditions. Zhang et al. [24] proposed improvements to the performance of the multiphase pump under high GVF conditions. Yang et al. [25] developed a comprehensive SRMP model to predict the performance of pumps at high GVF, which included steady-state behavior and transient performance during pump operation. Shi et al. [26,27] and Liu et al. [28] investigated the cavitation effects of the multiphase pump, and the effect of different GVFs on the energy loss characteristics for each cavitation phase was obtained.

There have also been some studies on the design and optimization of multiphase pumps. Zhang et al. [29] proposed a 3D Blade Hydraulic Design Method to design the impeller of a multiphase pump. The new impeller is more conducive for the transport of gas–liquid mixtures and is more suitable for large volume rate conditions. Cao et al. [30] proposed a method that combines Inverse Design and CFD Analysis. The mixture of the two species is considered a homogeneous fluid, and the geometry of the vanes is designed according to the velocity–torque distribution. This method designed a pump that can operate over a wide range of flow rates. Xiao et al. [31] proposed a controlled velocity moment design method for multiphase pumps combined with the singularity method, and the optimized design has a large performance improvement over the model pump. Liu et al. [32] proposed a hydraulic design method with a controllable blade angle to design and optimize multiphase pumps, and the optimized GVF and pressure distribution are more uniform. The transmission performance of the multiphase pump was improved. Shi et al. [33], Kim et al. [34], and Suh et al. [35] all used the design method of orthogonal experiments to optimize the structural parameters of the multiphase pump, and the results show that the head and efficiency of the optimized multiphase were improved, respectively. Zhang et al. [36] proposed an optimization method combining an artificial neural network and non-dominated sorting genetic algorithm-ii to optimize the structural parameters of a multiphase pump, and the head and efficiency of the optimized pump were greatly improved compared with the original model pump. Liu et al. [37] presented a method for optimizing the design of multi-stage multiphase pumps based on the Oseen vortex. The method can predict the smooth velocity moment downstream of the impeller and apply it to the optimization of the inlet vane angle of the impeller in the next stage. They carried out a case study of a three-stage multiphase pump, in which the head and efficiency of the optimized pump were slightly improved.

In general, scholars have conducted a lot of research on the internal flow characteristics of the multiphase pump, but less research has been conducted on the design of the multiphase pump blades. In this paper, an integrated optimal design based on the adaptive response surface method is proposed, which provides a new approach to thinking about the design method of multiphase pumps. The chapters of this paper are organized as follows. In Chapter 2, an original model is designed based on the design theory of the axial flow pump. In Chapter 3, we start with a numerical simulation of the original model to obtain its pressure increment and efficiency. The shape parameters of the original model and the resulting parameters of the numerical simulation are parametrized to create a closed-loop calculation system. Based on the closed-loop calculation system, parameter correlation analysis is carried out to obtain several parameters that have the greatest influence on the performance of the multiphase pump. Next, we carried out the experimental design of the parameters and the construction of the approximate model based on the closed-loop calculation system. The optimal solution was obtained by optimizing the approximate model using the multi-objective genetic algorithm (MOGA) method. In Chapter 4, we compare the flow field characteristics of the model before and after the optimization. In Chapter 5, an experimental test is carried out to verify the accuracy of the numerical simulations of the optimized model. In Chapter 6, we summarize the work in the full text.

2. Design and Calculation of the Original Model

The working parameters for the blade design of a multiphase pump are shown in Table 1.

Table 1. Multiphase pump design working parameters.

Design Parameter	Value	Unit
Rotation speed	4500	rpm/min
Flow rate	35	L/s
Head	20	m
Specific speed	325	

The calculation formula of specific speed is shown as follows.

$$n_0 = \frac{3.65n\sqrt{Q}}{H^{3/4}}, \tag{1}$$

where n is the rotation speed, Q is the flow rate, and H is the head.

(1) Head coefficient Ψ

$$\Psi = gH/U_t^2, \tag{2}$$

where h is the head, U_t is the circumferential velocity of the shroud, and g is the gravitational acceleration.

(2) Inlet flow coefficient Φ

$$\Phi = V_m/U_t, \tag{3}$$

where V_m is the axial component of absolute velocity. The flow coefficient has a great influence on the efficiency of the pump. Each airfoil corresponds to an optimal flow coefficient range.

(3) Shroud diameter a_1

$$a_1 = \frac{60U_t}{n\pi}, \tag{4}$$

The shroud diameter is determined by the head coefficient and should meet the corresponding flow coefficient requirements. For the multiphase pump, due to the compressibility of the gas, the volume flow of the gas–liquid mixture decreases from the inlet to the outlet of the pump impeller, so the impeller diameter at the rear end should be smaller than that in the first half.

(4) Inlet hub ratio h_t and Hub inlet diameter a_2

Inlet hub ratio h_t is the ratio of hub inlet diameter to shroud diameter.

$$a_2 = a_1 h_t, \tag{5}$$

(5) Shroud length diameter ratio h_v and Hub length of impeller a_4

$$a_4 = h_v a_1, \tag{6}$$

(6) Hub half cone angle γ and Hub outlet diameter a_3

$$a_3 = a_2 + 2a_4 tan\gamma, \tag{7}$$

where a_4 is the axial length of the impeller and γ is the half cone angle of the hub.

A conical hub is generally used for the multiphase pump. This structure can avoid the reduction of the velocity component on the shaft surface caused by gas compression.

(7) Hub inlet flow angle a_6

$$a_6 = a_6' + \Delta a_6, \tag{8}$$

where a_6' is the inlet angle and $\triangle a_6$ is the attack angle.

In the design process, the inlet flow angle is calculated according to the inlet speed and circumferential speed.

$$tana_6 = V/U_t, \tag{9}$$

where V is the velocity at which the fluid enters the impeller.

Then, according to the principle of equal radial guide path, the flow angle of the blade inlet at other radial positions is determined.

$$Const = rtana_6, \tag{10}$$

where r is the radius of the blade and a_6 means the hub inlet flow angle.

(8) Hub outlet flow angle a_7

According to the design theory of axial flow pump, the blade outlet installation angle (a_7) can be calculated, and then the blade outlet flow angle of other diameters can be determined according to the principle of equal diameter and guide path.

(9) Tip clearance a_9

The tip clearance is the radial clearance between the impeller shroud and the pump shell. Excessive clearance will lead to increased leakage. In this paper, the clearance is taken as 0.5 mm.

(10) Maximum blade thickness

$$\delta_{max} = (0.012 \sim 0.015)a_1 \sqrt{H}, \tag{11}$$

where δ_{max} is the maximum thickness of the hub and H is the head.

After the above design parameters are determined, the impeller of the multiphase pump is designed, and the parameter of the blade is shown in Table 2.

Table 2. Design and construction parameters of the multiphase pump.

	Design Parameter	Value	Unit	Expression
	Head coefficient	0.23		Ψ
	Circumferential velocity of the shroud	29.2	m/s	U_t
	Shroud diameter	0.138	m	a_1
	Inlet hub ratio	0.7		h_t
	Hub inlet diameter	0.096	m	a_2
	Hub half cone angle	7	°	γ
	Shroud length diameter ratio	0.39		h_v
Impeller	Hub length of impeller	0.054	m	a_4
	Hub outlet diameter	0.108	m	a_3/P7
	Hub inlet flow angle	16.1	°	a_6/P1
	Shroud inlet flow angle	13.4	°	P2
	Hub outlet flow angle	15.8	°	a_7/P3
	Shroud outlet flow angle	14.4	°	P4
	Number of blades	3		a_5
	Wrap angle	173	°	a_8/P8 (Hub), P9 (Shroud)
	Hub inlet diameter	0.108	m	P7
	Hub outlet diameter	0.096	m	-
	Hub inlet flow angle	41.66	°	P5
Guide vane	Shroud inlet flow angle	36.9	°	P6
	Axial length of guide vane	0.054	m	P10 (inlet position), P11 (outlet position)
	Number of blades	11		-
	Wrap angle	32.73	°	-

The blade is modeled according to the structural parameters of Table 1, and the geometric model is shown in Figure 1.

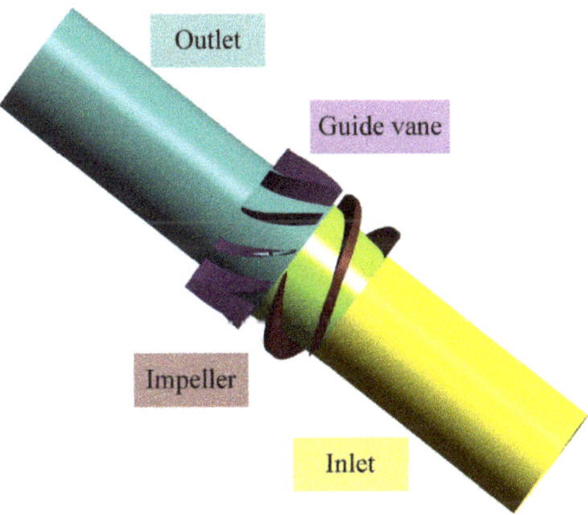

Figure 1. Model diagram of the impeller and guide vane of the multiphase pump.

3. Numerical Simulation and Optimization

3.1. Numerical Simulation of the Original Model

CFD methods are widely used in the fields of turbulence [38], multiphase flow [39], and heat transfer [40] owing to the advantages of low research costs and the ability to simulate complex or ideal processes. In this chapter, the CFD method is used to design, mesh, and numerically simulate an original model of a multiphase pump based on the ANSYS 2020R2 Workbench platform. We consider this process as the creation of an input–output channel. The structural parameters of the original model and the calculated performance parameters are parametrized. We construct a parameter set for the input and output separately and then connect them via the channels we created earlier. When we modify the input parameter set, and the system automatically applies this computational channel that we established earlier to calculate the output performance parameters. The solution of the data points involved in the parameter correlation analysis, design of experiments, construction of the agent model, and validation of the optimization results required in the following section is carried out on the closed-loop system of this calculation channel. The inner circle of Figure 2 represents our numerical simulation of the original model. The outer ring represents a closed-loop calculation channel created after parameterization of the structural and performance parameters of the multiphase pump.

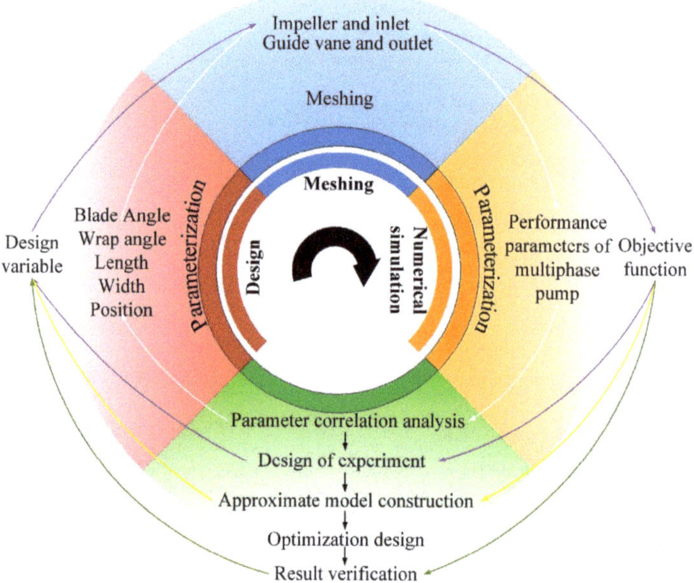

Figure 2. The flow chart of the integrated optimal design approach.

The numerical simulation in this paper is solved by the Euler–Euler method for the gas–liquid two-phase flow model. The gas is considered as a dispersive particle of equal particle size for the solution and the liquid is considered a continuous medium. The two-phase flow model is a non-homogeneous model, which considers the velocity slip and momentum transfer between two phases. The particle model is used for interphase transfer and the Schiller–Nauman model is used for the traction model. The velocity inlet condition and the opening outlet condition are used in the numerical simulation process. The no-slip wall condition is used for the hub and wall, and the counter-rotating wall condition is used for the impeller's shroud. The frozen rotor model is used for the dynamic–static interface. The SST model combines the advantages of the k-ε and k-ω models to provide a more accurate and reliable description of the flow of a fluid and to correspond to a wider range of flow models [41]. In this paper, the SST model is used for the solution of the liquid phase, and the zero-equation model is used for the solution of the gas phase. The convergence residuals are taken as 1×10^{-5}. During the calculation of the numerical simulation, the flow rate changes at the inlet and outlet are detected, and the convergence curve of the residuals is combined to determine whether the calculation process converges.

To ensure the accuracy of the numerical simulation, a mesh-independent analysis is performed. The grid in this paper is refined by controlling the y+ value of the first layer boundary to refine the global grid. The y+ value is taken as 10, and the grid diagram is shown in Figure 3.

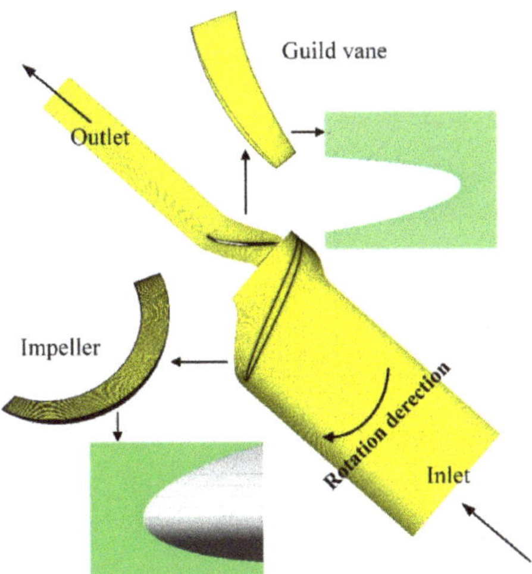

Figure 3. Grids for the numerical simulation the multiphase pump.

The grid-independent analysis was performed for the GVF = 0 working condition. Figure 4 shows that the efficiency and head curves tend to level off when the number of grids reaches 1.96 million, so we use a grid of 1.96 million for the subsequent numerical simulations.

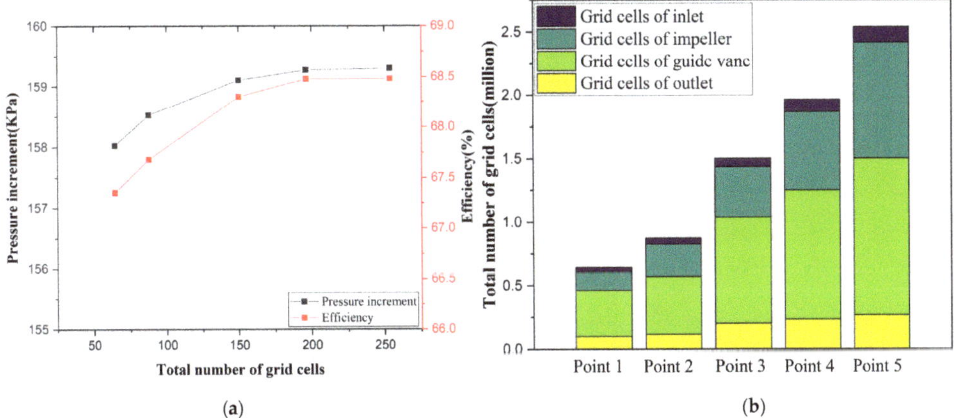

Figure 4. Verification of grid independence for numerical simulation of the multiphase pump. (**a**) Curves of grid number growth with performance parameters; (**b**) comparison of the number of grids in different areas of the multiphase pump.

After numerical calculations, we obtain a pressure increment of 159.3 KPa and an efficiency of 68.48% for the pump under the design conditions.

3.2. Parameter Correlation Analysis

We parameterize the input structural parameters and the output pump performance parameters separately. A calculation channel connecting the input and output parameters is established for the working condition of GVF = 0. Then, a parametric sensitivity analysis is performed on the structural parameters.

The Person correlation coefficient method [42] is a statistical method to measure quantitatively the correlation between variables. It has a wide range of applications in computational linguistics and is mainly used in research on text classification and information extraction. Using the vector space model of feature words, the distance between two pairs of texts is calculated to derive the similarity, which leads to the classification and further study of the text collection.

When calculated for a sample, the magnitude of the Pearson correlation coefficient is determined by the value of σ, which reflects the degree to which the two variables are linearly correlated. σ has a range of values from -1 to $+1$. A correlation coefficient of $+1$ indicates a perfect positive linear correlation between the variables. The Pearson correlation coefficient is calculated as follows [43].

$$\sigma = \frac{\sum XY - \frac{\sum X \sum Y}{N}}{\sqrt{\left(\sum X^2 - \frac{(\sum X)^2}{N}\right)\left(\sum Y^2 - \frac{(\sum Y)^2}{N}\right)}} \tag{12}$$

where variable X is the set of x-coordinates of all points, variable Y is the set of y-coordinates of all points, and N denotes the total number of points.

The Pearson correlation coefficient is the ratio of the product of the covariance and the standard deviation of the two variables and is a dimensionless, standardized covariance. A linear change does not affect the results of the Pearson correlation coefficient. Therefore, a unit change in the horizontal or vertical coordinates does not change the value of σ, so the σ values are comparable for data with different units.

Figure 5 gives the parameter sensitivity relationships between the design parameters and the optimization objectives, in which red and blue colors indicate high correlation and the white color indicates low correlation. The distribution of colors in the graph shows a high correlation between the variables P8, P9, P2, P4, P6, and the optimization objective. Among them, we consider P8 and P9 to be equal, so we finally determine P8, P2, P4, and P6 as the design variables.

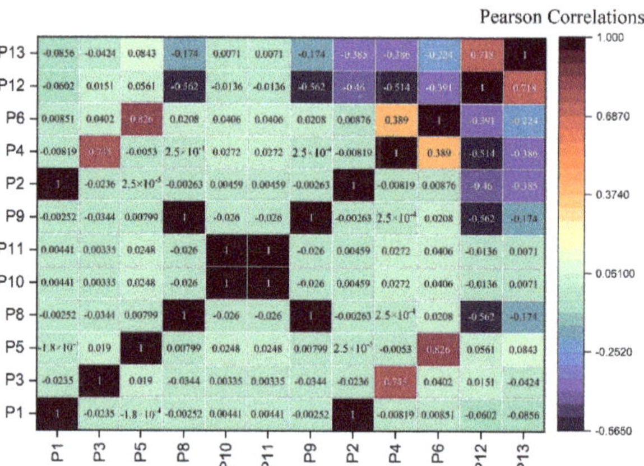

Figure 5. Parameter correlation chart.

3.3. DOE

The Sparse Grid model is a hierarchical Sparse Grid interpolation algorithm based on segmented multilinear basis functions. As its initial sampling points are related to the hierarchical basis functions, the number of sampling points will be significantly reduced compared to a full grid model. The Sparse Grid relies on a hierarchical decomposition of the underlying approximation space. The hierarchical basis is a function based on the following, which we express in Formula (13).

$$\varphi(x) = \begin{cases} 1 - |x| & x \in [-1,1] \\ 0 & x \notin [-1,1] \end{cases},\tag{13}$$

The basis function builds on this function with the following expression.

$$\varphi_{l,i}(x) = \varphi\left(\frac{x - ih_l}{h_l}\right) \quad x \in [x_{l,i} - h_l, x_{l,i} + h_l],\tag{14}$$

where l is the number of levels and i is the space location. Figure 6 shows the basis functions and response space for one-dimensional interpolation at the $l = 3$ condition [44].

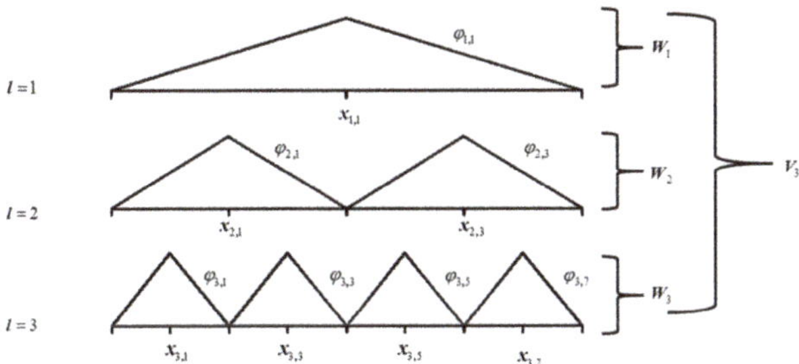

Figure 6. Basis functions and response space for one-dimensional interpolation at $l = 3$ condition.

The layered incremental space is shown in Equation (15)

$$W_l = span\{\varphi_{l,i} : i \in I_l\}$$
$$I_l = \left\{i \in N, 1 \le i \le 2^l - 1\right\}\tag{15}$$

The response function space for the full grid is shown below.

$$V_l = \bigoplus_{k \le l} W_k,\tag{16}$$

The Sparse Grid model, on the other hand, ignores subspaces from the full grid that have less influence on local differences, and its response function space is shown below [37].

$$V_{O,n}^S = \bigoplus_{|i| \le n+d-1} W_l,\tag{17}$$

The one-dimensional interpolation approximation model can be expressed as

$$f(x) \approx u(x) = \sum_{k=1}^{l} \sum_{i \in I_l} \alpha_{k,i} \varphi_{k,i}(x),\tag{18}$$

When dealing with multi-dimensional problems, the Sparse Grid model can significantly reduce the dependence on the sampling points in each dimension and improve computational efficiency. Sparse Grid experimental design enables the construction of experimental designs with fewer points while ensuring that the test points are distributed as evenly as possible in the design space. Compared with CCD and Latin hypercube algorithms, it is less computationally intensive and more efficient. Therefore, the Sparse Grid model is used for the experimental design method in this paper.

The range of values for each parameter of the multiphase pump is shown in Table 3. The experimental design table obtained is shown in Table 4.

Table 3. The range of values of structural parameters.

Structure Parameters	Lower Value	Upper Value	Unit	Expression
Shroud inlet flow angle of impeller	13	19		P2
Shroud outlet flow angle of impeller	15	21		P4
Wrap angle of impeller	155	195		P8
Shroud inlet angle of guide vane	39	45		P6

Table 4. Experimental design based on the Sparse Grid method.

Name	P2 (°)	P4 (°)	P6 (°)	P8 (°)	Pressure Increment (KPa)	Efficiency (%)
1	16	18	42	175	111.680	47.8541
2	13	18	42	175	117.246	48.8855
3	19	18	42	175	105.822	46.4865
4	16	15	42	175	107.892	47.1617
5	16	21	42	175	116.008	48.7249
6	16	18	39	175	111.877	47.8882
7	16	18	45	175	111.609	47.8679
8	16	18	42	155	148.415	44.3653
9	16	18	42	195	62.7643	43.5081

3.4. Construction of Approximate Models

The Sparse Grid method can provide refinement for continuous parameters, which means that it can automatically optimize itself. The dimensional adaptive algorithm allows it to determine which dimensions are most important to the objective function. When updating the response surface, the Sparse Grid uses an automatic local optimization process to determine the areas of the response surface that are most in need of further optimization. Then, it concentrates refinement points in these regions, allowing for the response surface to reach the specified accuracy faster with fewer design points. This results in a reduction in computational effort.

Figure 7 shows the residual convergence curve for the computation of the Sparse Grid. It can be seen that the head converges faster during the calculation, and the specified convergence requirement is reached at 40 data points. The efficiency comes the convergence requirement at 96 data points. Figure 8 shows the goodness-of-fit curve of the Sparse Grid, which refers to the degree of fit of the regression curve to the observed values. The statistical measure of goodness of fit is the coefficient of determination (also called coefficient of determination) R^2, the maximum value of which is 1. In summary, the goodness of fit is used to measure how well the fitted curve fits the original data, and the closer the data points are to the line y = x, the better the goodness of fit. From the data in Figure 8, we can see that almost all data points are very close to the line y = x, so we can judge that the goodness of fit is very good.

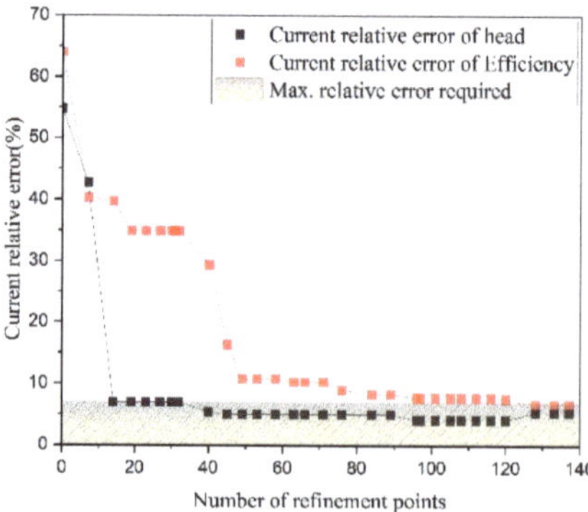

Figure 7. Convergence curves for the Sparse Grid approximation model.

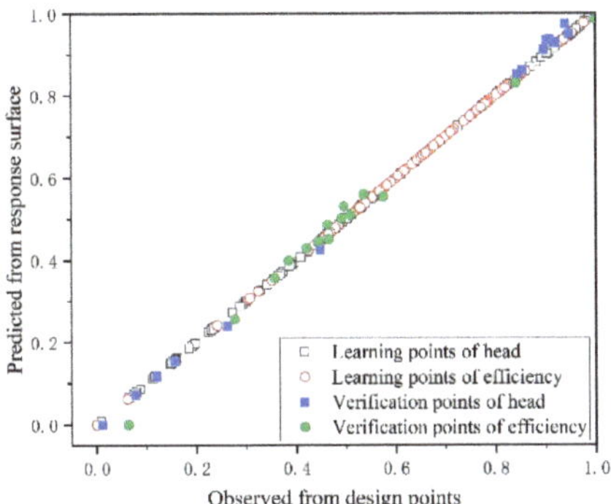

Figure 8. Goodness of fit curve for the Sparse Grid method.

3.5. Optimization

For the optimization of impeller parameters in this paper, a multi-objective genetic algorithm (MOGA) is used. The estimated number of evaluations is 2000. The number of initial samples is 100. The number of samples per iteration is 100. The maximum allowable Pareto percentage is 70%. The convergence stability percentage is 2%. The maximum number of iterations is 100. When the calculation reaches the 9th iteration step, the calculation converges, and the number of evaluations reaches 823.

The optimal set of Pareto solutions obtained from the optimization is shown in Figure 9. From the data in the figure, it can be seen that when the head increases, the efficiency is decreasing. When the efficiency increases, the head also decreases. The relationship between the two is mutually constrained. The goal of optimization is to maximize both head and efficiency, and we can find the optimal solution that satisfies both goals from these 13 optimization results. The final results of the optimal solutions are shown in Table 5.

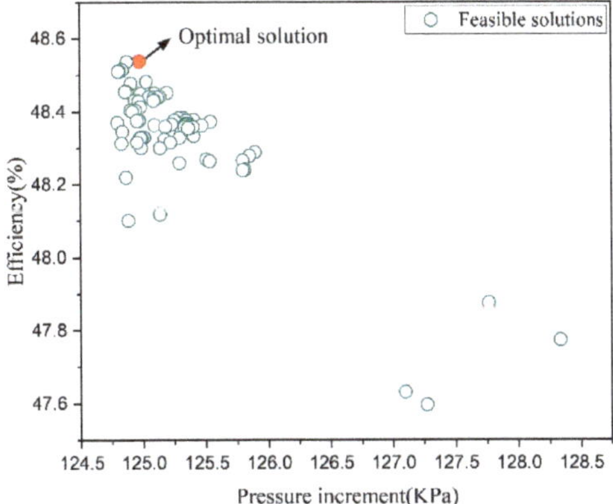

Figure 9. Pareto fronts of the optimization process.

Table 5. Optimization result parameters.

Name	Candidate Point	Calibrated Point of Optimized Model	Unit
P2	13.008	13	°
P4	20.15	20	°
P6	39.494	39.5	°
P8	170.05	170	°
Head	124.87	124.72	KPa
Efficiency	48.455	48.501	%

4. Discussion

We constructed an integrated optimized design model for the multiphase pump and completed the optimized design of the multiphase pump to obtain the optimized model of the multiphase pump structure based on the closed-loop calculation system we built. In this chapter, we compare the models before and after optimization, and analyze the changes in the internal flow field of the two different models.

Figure 10 gives the static pressure distribution of the blade surface under the water condition. Figure 10a,b shows that the area of the high-pressure region of the pressure surface of the optimized model is higher than that of the original model, especially in the position near the leading and trailing edges of the impeller is more obvious. The higher static pressure on the impeller surface indicates that the fluid medium is converted into more energy, so the performance of the impeller is better. Figure 10c,d shows that the pressure change of the suction surface before and after optimization is not obvious. Figure 11 gives the static pressure distribution on the blade surface with GVF = 0.2. From Figure 11a,b, it can be concluded that the optimized blade surface pressure is larger than the original model, and the area of pressure increase is mainly shown in the outer edge of the impeller pressure surface, which shows that the optimized blade has better energy conversion characteristics. From Figure 11c,d, it can be concluded that the pressure change of the optimized suction surface is small, and there is a slight increase at the inlet of the impeller.

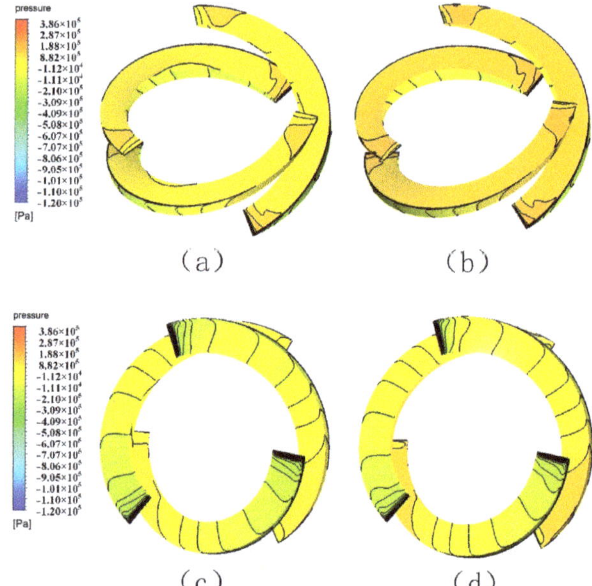

Figure 10. Surface static pressure distribution of the impeller in water condition. (**a**) Pressure side of original model; (**b**) pressure side of optimized model; (**c**) suction side of original model; (**d**) suction side of optimized model.

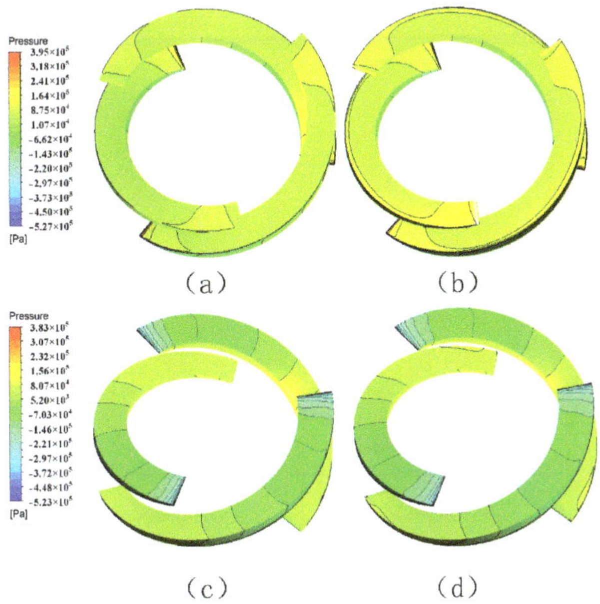

Figure 11. Distribution of surface static pressure on impeller at GVF = 20% working condition. (**a**) Pressure side of original model; (**b**) pressure side of optimized model; (**c**) suction side of original model; (**d**) suction side of optimized model.

Figure 12 gives the distribution of blade load under the water working condition. We can conclude that the surface load of the blade increases with the increase of span-wise and then decreases, and the pressure in the outer area of the blade is obviously higher than that in the inner area. It can be seen from the figure that the pressure of the blade suction surface is lower and the pressure difference is larger after optimization. Therefore, it can produce higher head. The area enclosed by the optimized blade load curve is larger than that of the original model, so the energy conversion characteristics of the blade are better. As can be seen from Figure 13, the overall surface load of the blade at GVF = 0.2 is lower than the water working condition, and the changing trend of the curve is not much different from the water condition. The energy conversion characteristics of the optimized impeller are better.

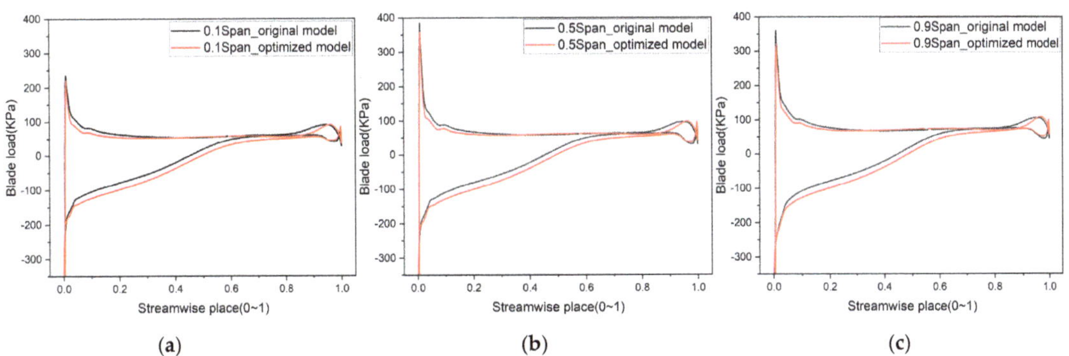

Figure 12. Distribution of blade loads along the span-wise place under the water condition. (**a**) 0.1 span, (**b**) 0.5 span, and (**c**) 0.9 span.

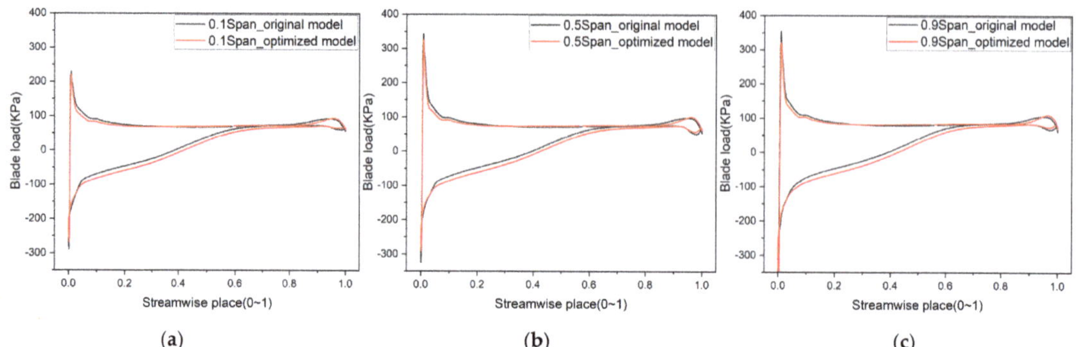

Figure 13. Distribution of blade loads along the span-wise place at 20%GVF condition. (**a**) 0.1 span, (**b**) 0.5 span, and (**c**) 0.9 span.

A swirling strength criterion was used to construct a vortex intensity cloud in the impeller region. Figure 14 gives the distribution of the vortex structure under the different working conditions. It can be concluded that the internal vortex of the multiphase pump is mainly caused by the impeller rotation, which is mainly distributed at the impeller inlet, outlet, and between the gap of the impeller and the shroud. From Figure 14a,e, it can be concluded that the vortex length d_2 at the blade exit of the optimized model is smaller than d_1 of the original model. The vortex intensities in Figure 14f–h are smaller than those in Figure 14b–d. This indicates that the optimization achieved better results.

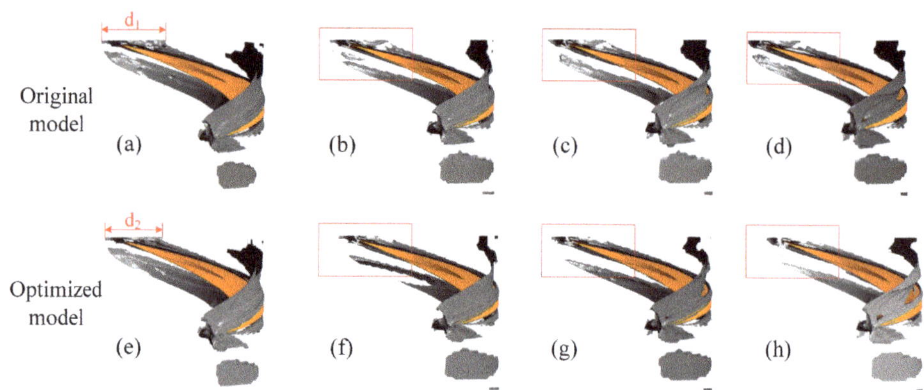

Figure 14. Vortex distribution diagram of the optimized model and the original model. (**a**) GVF = 0 of original model; (**b**) GVF = 10% of original model; (**c**) GVF = 20% of original model; (**d**) GVF = 30% of original model; (**e**) GVF = 0 of optimized model; (**f**) GVF = 10% of optimized model; (**g**) GVF = 20% of optimized model; (**h**) GVF = 30% of optimized model.

The distributions of turbulent energy dissipation rates for the original and optimized models under the water condition are given in Figure 15. The overall turbulent energy dissipation of the optimized model is lower than that of the original model, and the dissipation in both the outlet region shown in Figure 15a and the inlet region shown in Figure 15b of the impeller is improved. The turbulent energy dissipation rate distribution of the impeller and guide vane for GVF = 0.2 is presented in Figure 16. The turbulent energy dissipation at the intersection of the impeller and guide vane is significantly improved by the optimized model. The dissipation in the flow channel region in the middle of the impeller also becomes smaller. It indicates that the optimized performance is better.

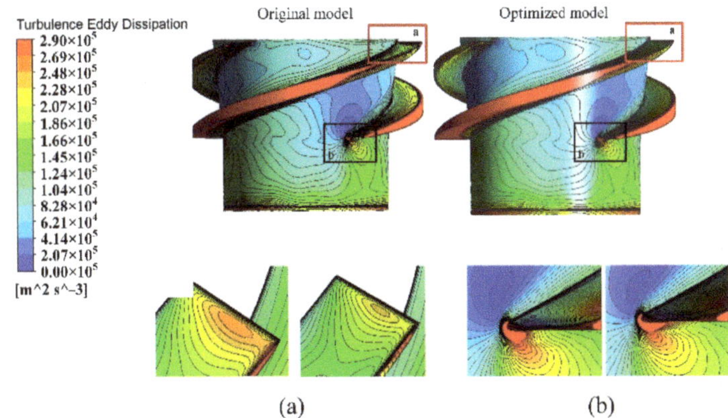

Figure 15. (**a**,**b**) Distribution of turbulent energy dissipation rate under the water condition.

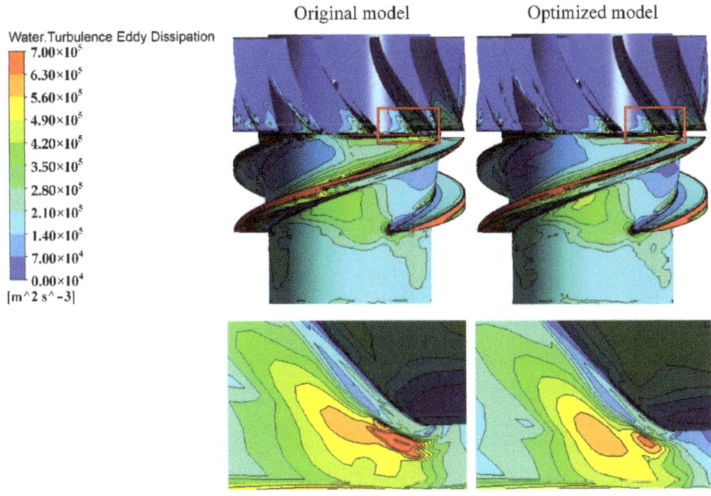

Figure 16. Distribution of turbulent energy dissipation rate under the 20%GVF condition.

Figure 17 provides an internal flow diagram of the impeller and guide vane of a multiphase pump under the water conditions. We can conclude that the flow velocity on the blade surface increases from the hub to the shroud. Compared to the original model, the flow velocity in the impeller region of the optimized model decreases, and the flow velocity in the guide vane region increases. The increased velocity in the guide vane region leads to more severe flow separation, especially near the hub surface. The increased flow velocity results in higher kinetic energy, which is the main reason for the increased head of the optimized multiphase pump. Figure 18 shows the flow lines inside the multiphase pump and guide vane for the condition of 20%GVF. Due to the increased GVF, the flow velocity of the liquid in both the impeller and the guide vane is lower than in the water condition, which leads to a reduction in the performance of the multiphase pump. The optimized model has a lower flow rate in the impeller and a higher flow rate in the guide vane, with increased flow separation near the hub side. This is consistent with the results for the water condition.

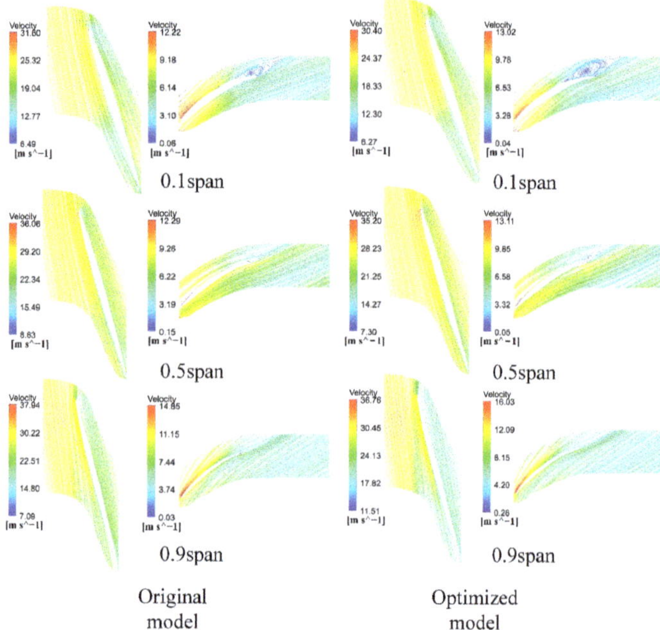

Figure 17. Streamline distribution of impeller and guide vane before and after optimization under the water condition.

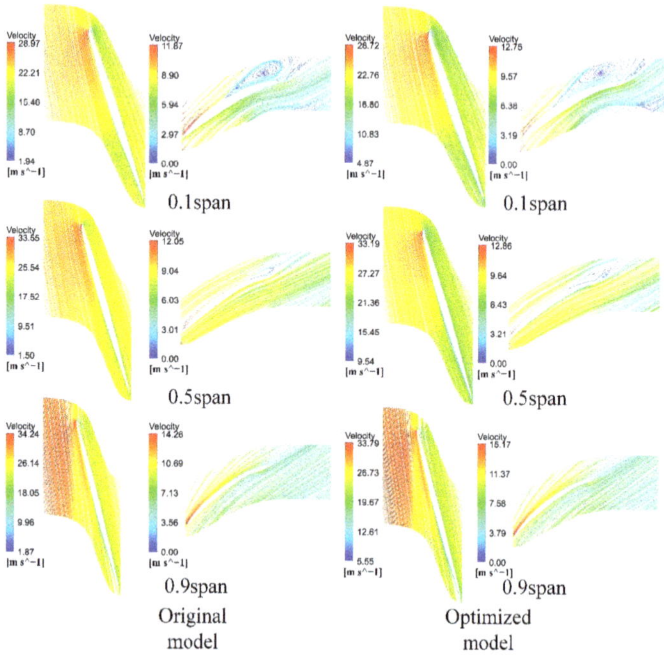

Figure 18. Streamline distribution of impeller and guide vane before and after optimization under the 20%GVF condition.

Figure 19 shows the performance curves of the multiphase pump for different GVF cases. When GVF = 0, the flow-head curve of the multiphase pump shows an obvious hump trend, and the efficiency curve shows a trend of first increasing and then decreasing, which is consistent with the trend of the performance curve of the general axial flow pump. When the gas content gradually increases from 0.0 to 0.3, the optimized model has a higher head than the original model. When the flow rate is small, the efficiency change is not obvious. When the flow rate is larger than the design condition, the optimized model has higher efficiency than the original model. The larger the flow rate is, the more obvious the efficiency improvement is.

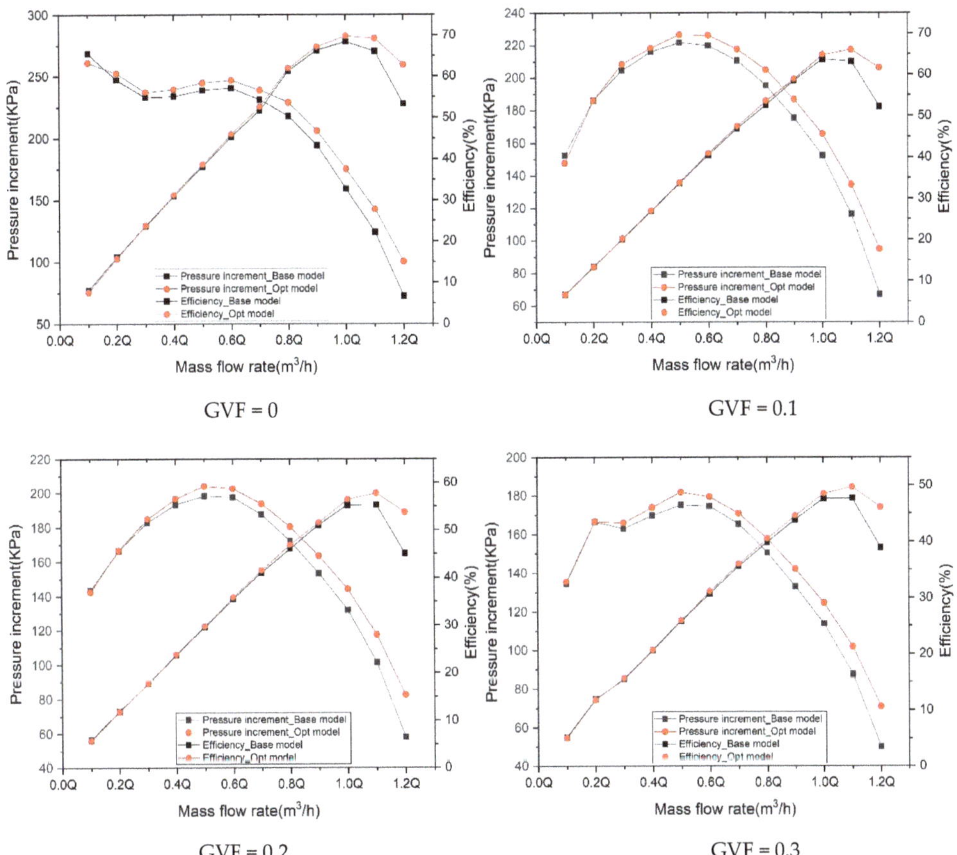

Figure 19. Characteristic curves of the multiphase pump under different working conditions.

A comparison of our work with other similar works is given in Table 6. The optimization results show an increase in pressure increment and efficiency of 9.4% and 1.8%, respectively. Our working conditions and specific size are different, so there will be some differences in the results. However, we used an integrated design–grid–numerical simulation approach for the optimization process. Only an initial set of data must be calculated to build up a computational framework. Then, the design of experiments, the construction of an approximate model, and the relevant parameters of the optimization method are set to produce the optimal solution. Such design methods are less computationally intensive and more intelligent than their work. Therefore, we believe that the integrated optimization design method proposed in this paper is effective.

Table 6. A comparison of our work and similar works.

		Ming Liu [32]	Zhang [36]	Kim [34]	Suh [35]	Our Work
Optimization process			Step by step			Integrated
Working Condition		GVF = 0 ~10%	GVF = 0 ~40%	GVF = 0	GVF = 0 ~15%	GVF = 30%
Blade size	Diameter	0.15 m	-	0.15 m	0.15 m	0.138 m
	Length	0.075 m	-	0.075 m	-	0.054 m
Pressure Increment (Growth Value/Percentage)		12.8 KPa/-	-/10%	30.9 KPa/-	48.89 KPa/-	10.78 KPa/9.4%
Efficiency (Growth Value/Percentage)		-/-	-/3%	1.9%/-	1.78%/-	0.892%/1.8%

5. Experimental Validation

In order to verify the accuracy of our optimization scheme, a multiphase pump performance test rig was set up to test the performance of the optimized blades. The flow chart of the experiment is shown in Figure 20, and the process of the experiment is shown in Figure 21.

We tested the performance of multiphase pumps under the water condition and GVF = 0.2 condition. The performance curves are shown in Figure 22. The process of numerical simulation is idealized for individual blades and does not take into account the errors caused by the friction between bearings, packing seals, and shafts that exist in the real experimental process. The performance parameters obtained from the tests are slightly lower compared to the results obtained from numerical simulation, which are within the acceptable range. Therefore, we consider that the integrated and optimized design approach used in this paper is effective.

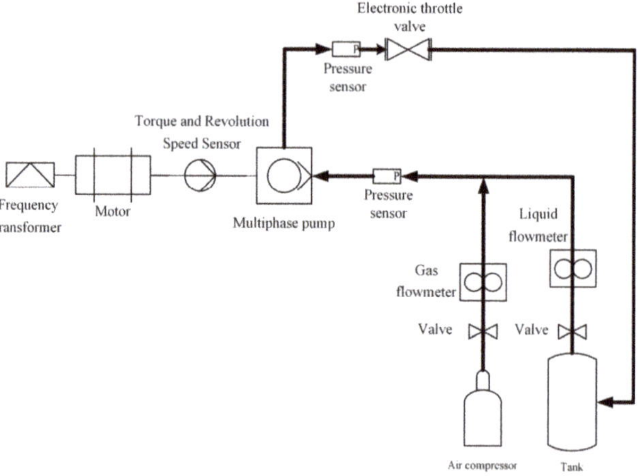

Figure 20. Experimental scheme for the multiphase pump.

Figure 21. Experimental process diagram for the multiphase pump. (1) Motor; (2) frequency transformer; (3) torque and revolution speed sensor; (4) multiphase pump; (5) pressure sensor; (6) liquid flowmeter; (7) gas flowmeter; (8) air compressor; (9) guide vane; (10) impeller; (11) electronic throttle valve; (12) tank; (13) outlet pipe; (14) valve; (15) inlet pipe.

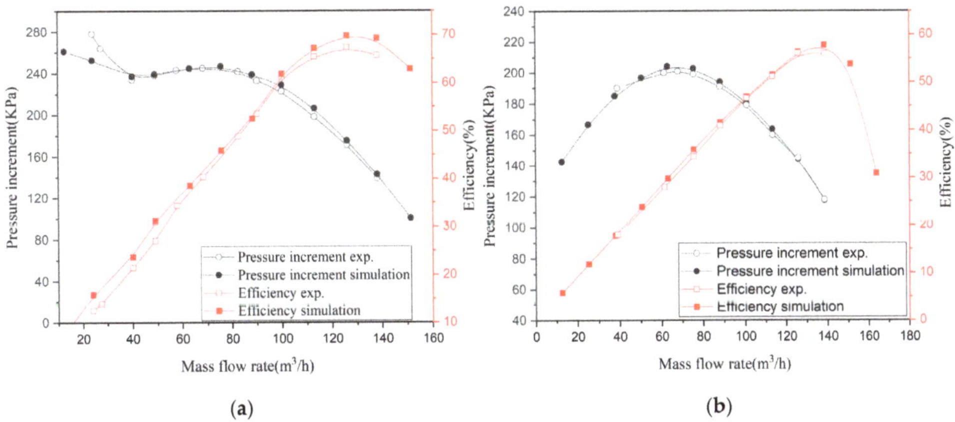

Figure 22. Experimental results of the multiphase pump performance parameters compared with simulation results. (**a**) The water condition; (**b**) GVF = 20%.

6. Conclusions

Firstly, this paper completes the design and numerical simulation of a multiphase pump and derives several pump performance parameters for the water condition. The structural parameters of the blade and the parameters of the results obtained from the numerical simulations are parametrized. A framework integrating the design–meshing–numerical simulation is built up. Within this framework, a parametric correlation analysis was carried out to obtain the parameters that have the greatest influence on the performance of the multiphase pump. They are the shroud inlet angle, the shroud outlet angle, the wrap angle of the impeller, and the shroud inlet angle of the guide vane. Then, the design of experiments and approximate model building was carried out using the Sparse Grid adaptive method in an integrated framework. The MOGA method was used for the optimization analysis in this paper. The final optimized model parameters were obtained, with 0.89% increase in efficiency and a 10.78 KPa increase in pressure increment compared to the original model.

The internal flow characteristics of the original model and the optimized model were compared and analyzed. We obtained a slight increase in the blade surface load of the optimized model and a slight increase in the static pressure at the pressure surface of the blade compared to the original model. In particular, the pressure increase near the impeller shroud is more obvious. The flow velocity in the impeller is reduced compared to the original model, and the flow velocity in the guide vane is slightly increased, which leads to an increase in the vortex in the exit area of the guide vane. The vortex within the impeller occurs mainly at the inlet, outlet, and rotating walls of the impeller. The vortex of the optimized model is slightly improved at the outlet. The turbulent energy dissipation in the optimized model is smaller compared to the original model, so the optimized model has higher efficiency.

We finally carried out an experimental study of the optimized model and obtained that the numerical simulation parameters of the optimized model matched the experimental parameters. Therefore, we conclude that the integrated design–meshing–numerical simulation proposed in this paper is effective.

Author Contributions: Conceptualization, methodology, software, writing—original draft, writing—review and editing, investigation, and software, C.P; Funding acquisition, X.Z. and Y.G., Methodology, C.P., X.Z., Y.C. and Y.G.; Visualization, Y.C., C.P., H.L. and S.H.; All authors have read and agreed to the published version of the manuscript.

Funding: This work was funded by National Key R&D Program of China (Grant No.2019YFC 0312305-02).

Conflicts of Interest: The authors declare no conflict of interest.

Nomenclature

n_s	Specific speed	[rpm/min:m^3/s, m]
Q	Flow rate	[L/s]
H	Head	[m]
U_t	The circumferential velocity of the impeller shroud	[m/s]
V_m	Axial component of absolute velocity	[m/s]
n	Rotation speed of impeller	[rpm/min]
a_1	Shroud diameter of impeller	[m]
a_2	Hub inlet diameter of impeller	[m]
h_t	Inlet hub ratio of impeller	[-]
a_3	Hub outlet diameter of impeller	[m]
h_v	Shroud length diameter ratio of impeller	[-]
a_4	Hub length of impeller	[m]
a_5	Number of impellers	[-]
a_6	Inlet angle of impeller	[°]
V	Velocity at which the fluid enters the impeller	[m]
r	Radius of the impeller	[m]
a_7	Hub outlet flow angle of impeller	[°]
a_8	Wrap angle of impeller	[°]
a_9	Tip clearance of impeller	[mm]
σ_{max}	Maximum blade thickness	[mm]
X	The set of x-coordinates of all points	[-]
Y	The set of y-coordinates of all points	[-]
N	The total number of points	[-]
W_l	The layered incremental space	[-]
V_l	The response function space for the full grid model	[-]
$V_{O,n}^S$	Response function space of the Sparse Grid model	[-]
$y+$	Dimensionless wall distance	[-]

Greek Letters

Ψ	Head coefficient	[-]
Φ	Inlet flow coefficient	[-]
γ	Hub half cone angle	[°]
σ	Pearson correlation coefficient	[-]

Abbreviations

GVF	Gas volume fraction	[%]
SRMP	Synchronal rotary multiphase pump	[-]
CFD	Computational fluid dynamics	[-]
MOGA	Multi-objective genetic algorithm	[-]
SST	Shear stress transport	[-]
CCD	Central composite design	[-]

References

1. Makogon, Y.F. Natural gas hydrates—A promising source of energy. *J. Nat. Gas Sci. Eng.* **2010**, *2*, 49–59. [CrossRef]
2. Yongchen, S.; Lei, Y.; Jiafei, Z.; Weiguo, L.; Mingjun, Y.; Yanghui, L.; Yu, L.; Qingping, L. The status of natural gas hydrate research in China: A review. *Renew. Sust Energ Rev.* **2014**, *31*, 778–791.
3. Shimizu, T.; Yamamoto, Y.; Tenma, N. Experimental Analysis of Two-Phase Flows and Turbine Pump Performance. *Int. J. Offshore Polar* **2016**, *26*, 371–377. [CrossRef]
4. Suh, J.; Kim, J.; Choi, Y.; Kim, J.; Joo, W.; Lee, K. Development of numerical Eulerian-Eulerian models for simulating multiphase pumps. *J. Pet. Sci. Eng.* **2018**, *162*, 588–601. [CrossRef]
5. Yu, Z.; Zhu, B.; Cao, S. Interphase force analysis for air-water bubbly flow in a multiphase rotodynamic pump. *Eng. Comput.* **2015**, *32*, 2166–2180. [CrossRef]
6. Yu, Z.; Zhu, B.; Cao, S.; Liu, Y. Effect of Virtual Mass Force on the Mixed Transport Process in a Multiphase Rotodynamic Pump. *Adv. Mech. Eng.* **2014**, *6*, 958352. [CrossRef]
7. Liu, M.; Cao, S.; Cao, S. Numerical analysis for interphase forces of gas-liquid flow in a multiphase pump. *Eng. Comput.* **2018**, *35*, 2386–2402. [CrossRef]
8. Zhang, W.; Yu, Z.; Zahid, M.; Li, Y. Study of the Gas Distribution in a Multiphase Rotodynamic Pump Based on Interphase Force Analysis. *Energies* **2018**, *11*, 1069. [CrossRef]
9. Zhang, W.; Yu, Z.; Li, Y. Application of a non-uniform bubble model in a multiphase rotodynamic pump. *J. Petrol. Sci. Eng.* **2019**, *173*, 1316–1322. [CrossRef]
10. Li, Y.; Yu, Z.; Zhang, W. Analysis of bubble distribution characteristics in a multiphase rotodynamic pump. *IOP Conf. Ser. Earth Environ. Sci.* **2019**, *240*, 62026. [CrossRef]
11. Zhang, J.Y.; Li, Y.J.; Cai, S.J.; Zhu, H.W.; Zhang, Y.X. Investigation on the gas pockets in a rotodynamic multiphase pump. *IOP Conf. Ser. Mater. Sci. Eng.* **2016**, *129*, 12007–12015. [CrossRef]
12. Wang, C.; Zhang, Y.; Zhang, J.; Zhu, J. Flow pattern recognition inside a rotodynamic multiphase pump via developed entropy production diagnostic model. *J. Pet. Sci. Eng.* **2020**, *194*, 107467. [CrossRef]
13. Huang, Z.; Shi, G.; Liu, X.; Wen, H. Effect of Flow Rate on Turbulence Dissipation Rate Distribution in a Multiphase Pump. *Processes* **2021**, *9*, 886. [CrossRef]
14. Jiang, Z.; Li, H.; Shi, G.; Liu, X. Flow Characteristics and Energy Loss within the Static Impeller of Multiphase Pump. *Processes* **2021**, *9*, 1025. [CrossRef]
15. Shi, G.; Liu, Z.; Xiao, Y.; Wang, Z.; Luo, Y.; Luo, K. Energy conversion characteristics of multiphase pump impeller analyzed based on blade load spectra. *Renew. Energy* **2020**, *157*, 9–23. [CrossRef]
16. Xu, Y.; Cao, S.L.; Reclari, M.; Wakai, T.; Sano, T. Multiphase performance and internal flow pattern of helico-axial pumps. *IOP Conf. Ser. Earth Environ. Sci.* **2019**, *240*, 32029. [CrossRef]
17. Yu, Z.Y.; Zhang, Q.Z.; Huang, R.; Cao, S.L. Numerical analysis of gas-liquid mixed transport process in a multiphase rotodynamic pump. *IOP Conf. Ser. Earth Environ. Sci.* **2012**, *15*, 32062. [CrossRef]
18. Xu, Y.; Cao, S.; Sano, T.; Wakai, T.; Reclari, M. Experimental Investigation on Transient Pressure Characteristics in a Helico-Axial Multiphase Pump. *Energies* **2019**, *12*, 461. [CrossRef]
19. Zhang, W.; Yu, Z.; Li, Y.; Yang, J.; Ye, Q. Numerical analysis of pressure fluctuation in a multiphase rotodynamic pump with air–water two-phase flow. *Oil Gas Sci. Technol.—Rev. Difp Energ. Nouv.* **2019**, *74*, 18. [CrossRef]
20. Zhang, J.; Fan, H.; Zhang, W.; Xie, Z. Energy performance and flow characteristics of a multiphase pump with different tip clearance sizes. *Adv. Mech. Eng.* **2019**, *11*, 2072051423. [CrossRef]
21. Shi, G.; Liu, Z.; Liu, X.; Xiao, Y.; Tang, X. Phase Distribution in the Tip Clearance of a Multiphase Pump at Multiple Operating Points and Its Effect on the Pressure Fluctuation Intensity. *Processes* **2021**, *9*, 556. [CrossRef]
22. Shi, G.; Yan, D.; Liu, X.; Xiao, Y.; Shu, Z. Effect of the Gas Volume Fraction on the Pressure Load of the Multiphase Pump Blade. *Processes* **2021**, *9*, 650. [CrossRef]

23. Li, C.; Luo, X.; Feng, J.; Zhu, G.; Xue, Y. Effects of Gas-Volume Fractions on the External Characteristics and Pressure Fluctuation of a Multistage Mixed-Transport Pump. *Appl. Sci.* **2020**, *10*, 582. [CrossRef]
24. Zhang, J.Y.; Zhu, H.W.; Ding, K.; Qiang, R. Study on measures to improve gas-liquid phase mixing in a multiphase pump impeller under high gas void fraction. *IOP Conf. Ser. Earth Environ. Sci.* **2012**, *15*, 62023. [CrossRef]
25. Yang, X.; Hu, C.; Hu, Y.; Qu, Z. Theoretical and experimental study of a synchronal rotary multiphase pump at very high inlet gas volume fractions. *Appl. Eng.* **2017**, *110*, 710–719. [CrossRef]
26. Shi, J.; Tao, S.; Shi, G.; Song, W. Effect of Gas Volume Fraction on the Energy Loss Characteristics of Multiphase Pumps at Each Cavitation Stage. *Water-Sui* **2021**, *13*, 2293. [CrossRef]
27. Shi, G.; Wang, S.; Xiao, Y.; Liu, Z.; Li, H.; Liu, X. Effect of cavitation on energy conversion characteristics of a multiphase pump. *Renew. Energ.* **2021**, *177*, 1308–1320. [CrossRef]
28. Liu, X.; Hu, Q.; Wang, H.; Jiang, Q.; Shi, G. Characteristics of unsteady excitation induced by cavitation in axial-flow oil–gas multiphase pumps. *Adv. Mech. Eng.* **2018**, *10*, 2072046214. [CrossRef]
29. Zhang, Y.; Zhang, J.; Zhu, H.; Cai, S. 3D Blade Hydraulic Design Method of the Rotodynamic Multiphase Pump Impeller and Performance Research. *Adv. Mech. Eng.* **2014**, *6*, 803972. [CrossRef]
30. Cao, S.; Peng, G.; Yu, Z. Hydrodynamic Design of Rotodynamic Pump Impeller for Multiphase Pumping by Combined Approach of Inverse Design and CFD Analysis. *J. Fluids Eng.* **2005**, *127*, 330–338. [CrossRef]
31. Xiao, W.; Tan, L. Design method of controllable velocity moment and optimization of pressure fluctuation suppression for a multiphase pump. *Ocean Eng.* **2021**, *220*, 108402. [CrossRef]
32. Liu, M.; Tan, L.; Cao, S. Design Method of Controllable Blade Angle and Orthogonal Optimization of Pressure Rise for a Multiphase Pump. *Energies* **2018**, *11*, 1048. [CrossRef]
33. Shi, G.; Li, H.; Liu, X.; Liu, Z.; Wang, B. Transport Performance Improvement of a Multiphase Pump for Gas–Liquid Mixture Based on the Orthogonal Test Method. *Processes* **2021**, *9*, 1402. [CrossRef]
34. Kim, J.; Lee, H.; Kim, J.; Choi, Y.; Yoon, J.; Yoo, I.; Choi, W. Improvement of Hydrodynamic Performance of a Multiphase Pump Using Design of Experiment Techniques. *J. Fluids Eng.* **2015**, *137*, 081301. [CrossRef]
35. Suh, J.; Kim, J.; Choi, Y.; Joo, W.; Lee, K. A study on numerical optimization and performance verification of multiphase pump for offshore plant. *Proc. Inst. Mech. Eng. Part A J. Power Energy* **2017**, *231*, 382–397. [CrossRef]
36. Zhang, J.; Zhu, H.; Yang, C.; Li, Y.; Wei, H. Multi-objective shape optimization of helico-axial multiphase pump impeller based on NSGA-II and ANN. *Energy Convers. Manag.* **2011**, *52*, 538–546. [CrossRef]
37. Liu, M.; Tan, L.; Xu, Y.; Cao, S. Optimization design method of multi-stage multiphase pump based on Oseen vortex. *J Pet. Sci. Eng.* **2020**, *184*, 106532. [CrossRef]
38. Mycek, P.S.S.; Gaurier, B.I.; Sup, T.S.; Germain, G.E.; Sup, G.S.; Pinon, G.E.; Sup, G.S.; Rivoalen, E.S.S. Experimental study of the turbulence intensity effects on marine current turbines behaviour. Part I: One single turbine. *Renew. Energy Int. J.* **2014**, *66*, 729–746. [CrossRef]
39. Ma, D.; Duan, H.; Liu, W.; Ma, X.; Tao, M. Water–Sediment Two-Phase Flow Inrush Hazard in Rock Fractures of Overburden Strata During Coal Mining. *Mine Water Env.* **2020**, *39*, 308–319. [CrossRef]
40. Ramezanizadeh, M.R.M.S.; Nazari, M.N.M.A.; Ahmadi, M.A.M.H.; Chau, K.C.K.S. Experimental and numerical analysis of a nanofluidic thermosyphon heat exchanger (Article). *Eng. Appl. Comp. Fluid Mech.* **2019**, *13*, 40–47.
41. Wu, D.W.D.S.; Yuan, S.Y.S.S.; Ren, Y.R.Y.S.; Mu, J.M.J.S.; Yang, Y.Y.Y.S.; Liu, J.L.J.S. CFD investigation of the influence of volute geometrical variations on hydrodynamic characteristics of circulator pump. *Chin. J. Mech. Eng.* **2016**, *29*, 315–324. [CrossRef]
42. Gravier, J.; Vignal, V.; Bissey-Breton, S.; Farre, J. The use of linear regression methods and Pearson's correlation matrix to identify mechanical–physical–chemical parameters controlling the micro-electrochemical behaviour of machined copper. *Corros. Sci.* **2008**, *50*, 2885–2894. [CrossRef]
43. Sup, M.T.S.; Aemail, D.Z.S.S.; Sup, D.W.S.; Sup, J.D.S.; Sup, D.K.S.; Sup, H.Z.S. Performance prediction of 2D vertically stacked MoS_2-WS_2 heterostructures base on first-principles theory and Pearson correlation coefficient. *Appl. Surf. Sci.* **2022**, *596*, 153498.
44. Butnaru, D.; Pflüger, D.; Bungartz, H. Towards High-Dimensional Computational Steering of Precomputed Simulation Data using Sparse Grids. *Procedia Comput. Sci.* **2011**, *4*, 56–65. [CrossRef]

Article

Influence of Guide Vane Profile Change on Draft Tube Flow Characteristics of Water Pump Turbine

Qifei Li [1,2,*], Lu Xin [1], Gengda Xie [1], Siqi Liu [1] and Qifan Wang [3]

1   School of Energy and Power Engineering, Lanzhou University of Technology, Lanzhou 730050, China; xinlu1557669@163.com (L.X.); xiegengda2000@163.com (G.X.); liusiqi0722@163.com (S.L.)
2   State Key Laboratory of Fluid Machinery and Systems, Lanzhou 730050, China
3   Huaneng Gansu Hydropower Development Co., Ltd., Lanzhou 730050, China; 212019080700005@stu.xhu.edu.cn
*   Correspondence: lqfy@lut.cn

Abstract: In order to study the influence of the change of the guide vane airfoil on the flow characteristics in the draft tube of a reversible hydraulic turbine, a reversible hydraulic turbine was used as the object of study, and the effect of the change on the flow pattern, energy loss, and pressure pulsation in the draft tube area was studied based on the SST k-ω turbulence model. The results show that under low flow conditions, the modified movable guide vane directly affects the direction and speed of water entering the draft tube, reduces the density of vortex in the draft tube area, reduces the impact on the near wall of the draft tube during the rotation of the vortex belt, and improves the stability of the unit operation. The turbulent energy comparison graph shows that the energy loss in the bent elbow section and the diffusion section of the draft tube is reduced, and the energy return coefficient of the draft tube is improved by calculating that the energy recovery level of the draft tube is improved under different operating conditions. A comparative analysis of the pressure pulsation in the draft tube area before and after the modification in combination with the development of the vortex belt shows that the modified movable guide vane effectively reduces the vibration intensity in the draft tube area and improves the stable operation threshold of the unit.

Keywords: reversible water turbines; guide vane profile change; draft tube vortex belt; pressure pulsation; energy recovery factor

Citation: Li, Q.; Xin, L.; Xie, G.; Liu, S.; Wang, Q. Influence of Guide Vane Profile Change on Draft Tube Flow Characteristics of Water Pump Turbine. *Processes* **2022**, *10*, 1494. https://doi.org/10.3390/pr10081494

Academic Editors: Santiago Lain and Omar Dario Lopez Mejia

Received: 7 July 2022
Accepted: 27 July 2022
Published: 29 July 2022

Publisher's Note: MDPI stays neutral with regard to jurisdictional claims in published maps and institutional affiliations.

1. Introduction

Under the "dual carbon" target (China clearly proposes the goal of "carbon peaking" in 2030 and "carbon neutrality" in 2060), the hydropower industry is highly valued by the country [1]. The pumped storage power station is a special power source in the power system with multiple functions, such as peak regulation, valley filling, frequency regulation, phase regulation, and accident backup, which has become an important part of China's power system [2]. As the performance indexes of reversible turbine with high and ultra-high head become the need of industrial development, the stability of turbine operation also draws attention. For many years, domestic and foreign research scholars have been attaching great importance to the study of the draft tube area and have achieved many work results. In the early years, scholars used experimental and numerical methods to carry out experimental analysis and theoretical research on the draft tube area of hydraulic turbine units, trying to investigate the relationship between the water flow pattern and pressure pulsation in the draft tube area of hydraulic turbines in order to conduct a preliminary discussion on the pulsation mechanism [3–5]. In the 1970s, Kubota et al. conducted a systematic study on the relationship between the amplitude of pressure pulsation in the draft tube and the cavitation coefficient and flow rate using model tests, and proposed that the amplitude of pressure pulsation in the draft tube was mainly related to the ratio of operating flow rate to design flow rate [6,7]. Jacol et al. obtained the distribution characteristics of the pressure pulsation in the draft tube of a Francis turbine at different flow rates through field measurements of many prototype units, and also conducted a preliminary study on the causes of pressure pulsation in the draft tube under different operating conditions and improvement measures [8]. Nishi experimentally studied the relationship between the vortex excitation frequency and the operating head and flow rate, and proposed that the maximum pressure pulsation in the draft tube is related to the location of the vortex nucleus in the draft tube [9]. Iliescu measured the flow field characteristics inside the draft tube by PIV (Particle Image Velocimetry) and conducted a detailed study of the volume and center of the cavitation vortex zone with the variation of the cavitation number [10]. Wu Gang et al. investigated the relationship between pressure pulsation in the draft tube and the inlet flow field of the draft tube and the opening of the guide vane. On the basis of model tests, the results of water-pressure pulsation tests, flow field tests, and incipient cavitation observations were analyzed to study the effect of changes in the opening of the turbine's movable guide vane on water-pressure pulsation in the draft tube [11]. The use of experimental methods is subject to the uncontrollable influence of experimental methods and relatively large costs, and are not easily carried out on a large scale. In recent years, due to the development of computer technology and numerical simulation technology, there are now many versions of simulation software to study the phenomenon of pressure pulsation in the vortex zone of the turbine draft tube [12–15]. Most of these studies use CFD (Computational Fluid Dynamics) flow field analysis software to simulate normal operating conditions of reversible hydraulic turbines. Li Qifei et al. found that the instability of the draft tube was one of the major causes of vibration in reversible hydraulic turbines. When the reversible turbine deviates from the optimal working conditions of the turbine, the impact of its vortex belt on the wall will cause violent pressure pulsations in the draft tube area, which will propagate upstream and cause vibration of the whole unit, which is not conducive to the stable operation of the reversible turbine [16]. Significant draft tube back-flow has a direct contact with the runner surface. Induced low-frequency fluctuations of the hydraulic thrust and torque explain severe vibrations of the runner [17]. Further upstream propagation of the vortex rope instabilities increases pulsations inside runner channels by superimposing pressure pulsations into inter-blade vortices [18]. Every disturbance downstream can be found throughout the hydraulic circuit and can be a major cause of system instability in certain load ranges [19,20]. Guo Tao et al. analyzed the vortex band evolution in the turbine draft tube region under different incoming flows by means of a slip-grid technique [21]. Numerical simulations are too idealistic, so nowadays, a combination of experimental and numerical simulations is used to analyze the problem. Experts have also never stopped exploring to improve the stability of the draft tube. Theoretical studies on the structural safety of the turbine draft tube found that the area of the low-velocity zone of the runner exit flow field directly affects the size of the vortex band and the pressure pulsation [22], and the outlet flow rate and direction also interfere with the shape of the vortex band [23,24]. Not only that, the influence of the eddy current on the unit will also play a decisive role in the stability of the entire water transmission and power generation system [25]. A considerable number of scholars have also considered the method of adding additional structures to the draft tube to improve its internal flow field, all with varying degrees of progress [26–28]. The effect of changing the structure affecting the incoming flow on the internal flow of the turbine has also been continuously studied by scholars. Liangbeng Rong et al. used the NACA0018 parent model to obtain the numerically optimized airfoil type and solved it with the MRF (model Multiple Reference Frame Model, also called the composite coordinate system model, a method provided by Fluent software to solve the problem of coexistence of static and dynamic regions), giving the calculation results and comparing them with the parent model to examine the degree of improvement in the energy utilization of the turbine [29]. Qifei Li et al. found that the change of the radius of the head circle of the guide vane would also have different degrees of influence on the excitation force of the turbine runner [30]. This paper investigates the effect of transient flow field characteristics in the draft tube of a reversible turbine unit under normal operating conditions by changing the structure of the movable guide vane wing, focusing on the effect of the guide vane opening on the evolution of the vortex zone shape, pressure pulsation, and energy loss in the draft tube.

2. Design Process and Model Building

2.1. Activity Guide Leaf Design Process

As the water guide mechanism of the reversible turbine is cylindrical, the existing movable guide vane can be retrofitted by changing the wing profile of the movable guide vane. This paper refers to the excellent hydraulic design model of hydraulic turbine, the hydraulic turbine design manual guide vane section size table, and the original airfoil design data to modify the movable guide vane airfoil and design a new movable guide vane while keeping the original guide vane chord length unchanged. The "S" characteristic of the reversible turbine unit is improved by changing the direction and speed of water entry by changing the movable guide vane airfoil shape. The airfoil of the guide vane before and after the modification is shown in Figure 1.

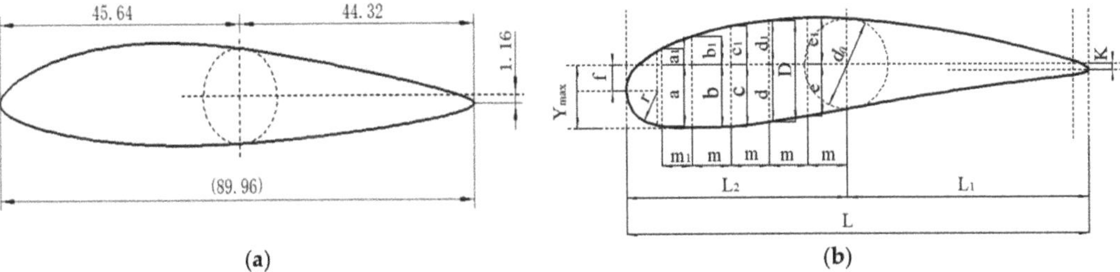

Figure 1. Schematic diagram of the modified anterior and posterior guide vane airfoils. (**a**) Prototype movable guide vane wing type (Unit is mm). (**b**) Modified movable guide vane wing type.

By changing the head circle radius r of the original airfoil and increasing the maximum thickness D of the original guide vane airfoil, the influence of the increase of the head circle and maximum thickness of the guide vane on the internal flow characteristics of the draft tube is investigated to provide a reliable direction for the optimization of the design of the next guide vane airfoil. The parameters of the guide vane after changing the airfoil shape are shown in Table 1.

Table 1. Airfoil design parameters.

Parameter Symbols	Numerical Values (mm)	Parameter Symbols	Numerical Values (mm)
D_0	638.639	e	9.271275397
a	11.06607099	e_1	8.236239902
a_1	2.851853121	d_0	16.09810526
b	11.29730233	m	7.465468788
b_1	5.021023256	m_1	5.890893513
c	11.16517013	k	0.902903304
c_1	6.815818849	f	4.71271481
d	10.2842886	r	6.012014688
d_1	7.718722154	L	89.96
L_1	47.34736842	L_2	42.61263158

In terms of the theoretical analysis, as shown in Figure 2 below, we analyze the velocity triangle between the new guide blade airfoil shape and the original airfoil shape, by Formulas (1) and (2). A refers to the cross flow area of the inlet side of the blade, the magnitude of the blade inlet angle is also measured by ANSYS CFD-POST (data processing analysis and visualization processing software), and the velocity triangle is derived from the blade inlet angle β, the absolute flow velocity V, and the axial surface flow velocity V_{1m}. Under the assumption that the direction and magnitude of the circumferential velocity do not change, if the angle of the inlet water flow of the rotor (the angle between the relative velocity of the water flow and the circumferential velocity) and the blade inlet placement angle β are equal, there is shockless inlet condition, and its hydraulic loss is minimal. Therefore, the greater the impulse angle α, the greater the hydraulic loss. The angle of rotation of the movable guide vane remains unchanged, the diameter of the distribution circle of the movable guide vane increases, the absolute velocity direction is changed and reduced by the influence of the implicated velocity, the incoming liquid and the impulse angle α of the runner blade decreases, the effective overflow flow increases, and the hydraulic loss of the runner decreases accordingly. From the triangular comparison of the exit velocity of the rotor. After changing the airfoil, influenced by the upstream disturbance, the flow and runner inlet impulse angle changes, the absolute velocity changes, the circumferential velocity component decreases, the water flow loop volume of the draft tube decreases, the centrifugal force becomes smaller, and the draft tube vortex belt theoretical analysis shows that the strength will be weakened by a certain degree.

$$U = \frac{n\pi D}{60} \tag{1}$$

$$V_{1m} = \frac{Q}{A} \tag{2}$$

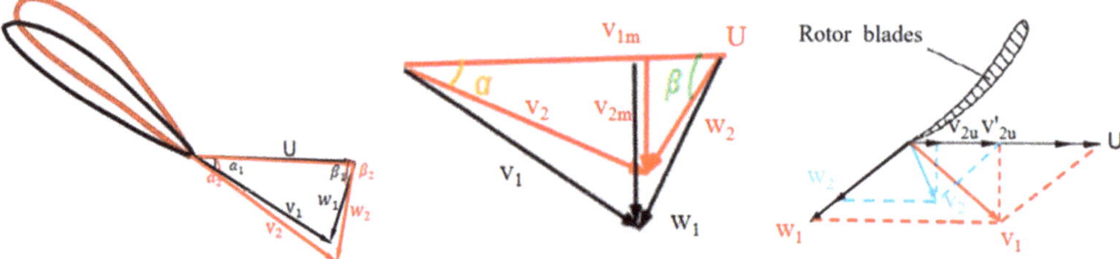

Figure 2. Comparison of the front and rear guide vane airfoil velocity triangles of the modification.

2.2. Mesh Classification and Model Building

The object of this study is a reversible hydraulic turbine model. The overflow components consist of spiral casing, fixed guide vane, movable guide vane, runner, and draft tube. The schematic diagram is shown in Figure 3 and the specific parameters are shown in Table 2.

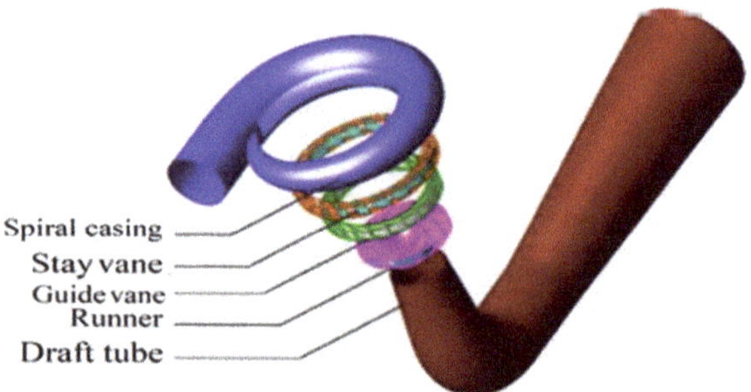

Figure 3. Model pump turbine calculation area.

Table 2. Geometric parameters of model pump turbine.

Parameter Name	Numerical Value
Number of blades/pc	9
Active guide leaf/pc	20
Rotor high-pressure side diameter/mm	473.6
Spiral casing inlet diameter/mm	315
Height of guide lobe b_0/mm	66.72
Number of fixed guide vane/pc	20
Height of guide leaf/mm	66.72
Rotor low-pressure side diameter/mm	300
Draft tube outlet diameter/mm	660

In order to ensure that the numerical calculation results are feasible and reliable, this meshing was carried out using the sub-function ICEM (professional pre-processing software that provides efficient and reliable analytical models) of the commercial software ANSYS for full-flow channel hexahedral meshing. The results of the meshing are shown in Figure 4. The wall function method was used to add boundary layer mesh at the wall location. As the flow rate in the runner area is larger than that in other locations of the basin of the reversible turbine, it was chosen to look at the y+ distribution in the runner area. Figure 5 shows distribution of blade Y+ wall surface.

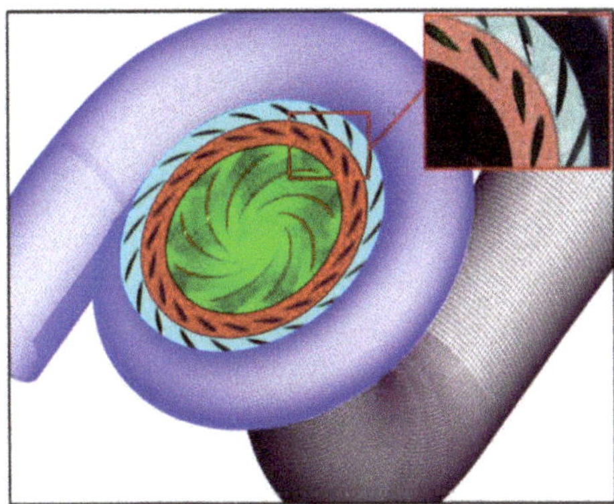

Figure 4. Local grid diagram.

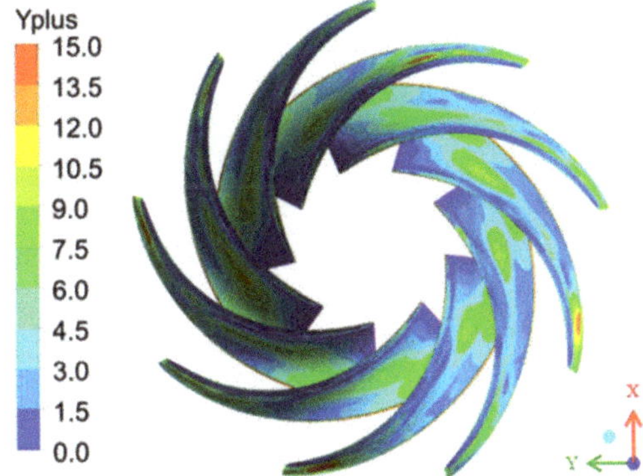

Figure 5. Distribution of blade Y+ wall surface.

After grid-independent verification, the number of grids reaches 5.5 million, the calculation results are within the error tolerance, and, as the number of grids increases, the reference value $H_m \cdot H_c^{-1}$ tends to level off, where H_m is the test head and H_c is the calculated head. When the ratio of the test head to the calculated head tends to be closer to 1, it means the more accurate the calculation results are. Figure 6 shows grid independence verification.

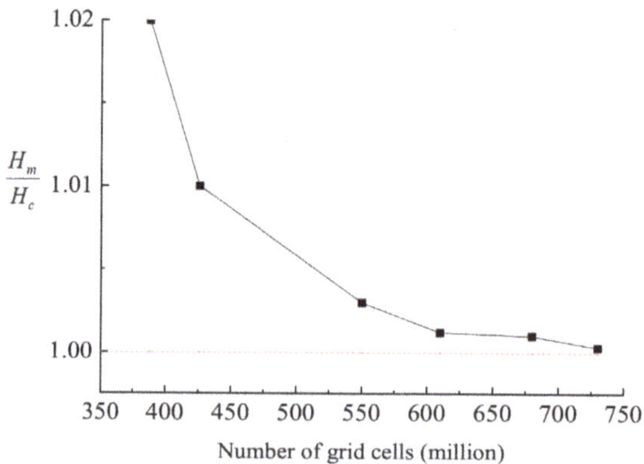

Figure 6. Mesh independence verification.

2.3. Introduction to the Test Rig and Numerical Calculation Methods

Introduction to the Test Setup and Conditions

In order to verify the reliability of the turbulence model and the feasibility of the modelling and numerical calculations, the numerical results are compared with the experimental results. The test stand is shown in Figure 7. The model was tested with a constant head of H = 30 m and the rotor speed and flow rate were measured using a torque meter and a flow meter and compared with the calculated results.

Figure 7. Pump turbine experimental device.

2.4. Turbulence Model and Boundary Conditions

Since the SST k-ω turbulence model (the method of studying the system of control body commonly used in fluid mechanics) can effectively capture the flow near the wall, especially for the complex geometric model of a reversible turbine with multiple guide vanes and blades, the SST k-ω turbulence model was chosen for this study to carry out numerical simulations [31]. The fluid medium is set to normal temperature water and the wall surface is set to a non-slip wall boundary condition. Inlet and outlet are set to mass flow inlet and free outflow, respectively. Data transfer between the stationary and rotational domains relies on INTERFACE boundary conditions. Using SIMPLE-C velocity-pressure coupling algorithm, the residual value is set to 10^{-6}, and the time step is set to $3.867 \cdot 10^{-4}$ s; 120 steps are needed to rotate one week, and each time step is rotated by $3°$ [32]. A total of 10 rotations of the runner rotation cycle are monitored, that is, T = 0.46404 s.

2.5. Calculation Results and Analysis

2.5.1. Reliability Verification

The model reversible hydraulic turbine movable guide vane a_0 = 33 mm (moving guide vane opening under rated conditions) was selected for this study to verify the reliability of the numerical calculation. Seven operating points were selected for the constant numerical calculation. The result of the numerical calculation is converted to unit speed and unit flow rate with the following equation [33]. Table 3 is the flow-rate test data.

$$n_{11} = \frac{nD_2}{\sqrt{H}} \tag{3}$$

$$Q_{11} = \frac{Q}{D_2^2 \sqrt{H}} \tag{4}$$

Table 3. Flow-speed test data.

Flow rate Q_{11} (m$^3 \cdot$s^{-1})	0.84	0.83	0.76	0.66	0.54	0.3	0.02
Speed n_{11} (r·min^{-1})	51.2001	56.8747	65.6563	70.8762	72.3342	70.4606	65.4471

In the formula: Q is the calculated flow, m$^3 \cdot$s^{-1}; n is the rotational speed, r·min^{-1}; D_2 is the diameter of the low-pressure side of the runner, m; and H is the test head, m.

Q_{11} and n_{11} were obtained by conversion, and then the n_{11}-Q_{11} characteristic curve was plotted. The n_{11}-Q_{11} characteristic curve obtained by conversion is compared with the test curve, and the results are shown in Figure 8. Through comparison, the two have a high degree of agreement and the error value is kept near 4%, which meets the requirements of engineering research. Therefore, the model selected for this numerical calculation has a high reliability.

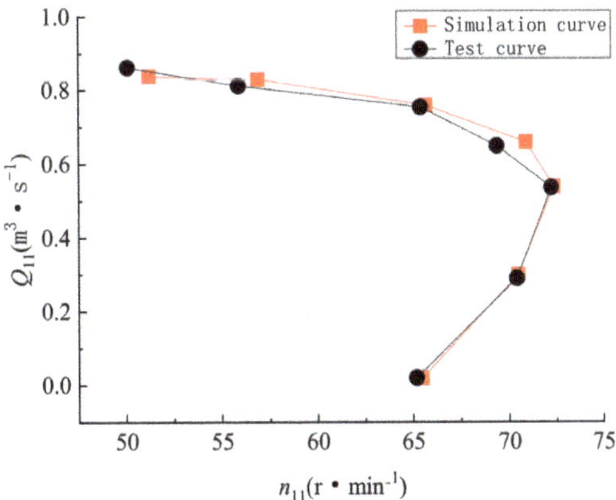

Figure 8. Comparison of experimental and simulation results.

2.5.2. Reliability Analysis of Pressure Pulsation

The pressure pulsations are measured over all operating ranges of turbine operating conditions and at the cavitation factor (also known as Toma factor—the ratio of the necessary cavitation margin to the head release of a centrifugal pump) of the power plant unit. It is necessary to measure the amplitude and frequency of pressure pulsations between the spiral casing, runner, and guide vane as well as between the top cover and the upper crown of the runner. The sensor arrangement should be located where the maximum pressure pulsation amplitude can be measured (such as near the blade inlet). The pressure pulsations are recorded and analyzed. Spectrum analysis should be performed on the collected data to determine the principal frequency and amplitude of the pressure pulsation. ΔH to represent the degree of pressure pulsation in the turbine/pump, while H is the turbine head/pump head and is the characteristic amplitude. The eigenvalues are statistically calculated and the values outside the probability range of the given probability range will be ignored, so the peak pressure pulsation uses the confidence method. The turbine pressure pulsation test is conducted in the full range of operating conditions, corresponding to the pressure pulsation amplitude for the confidence level of 97% peak. The experimental results of pressure pulsation in the draft tube at 393.63 Pa, a_0 = 33 mm are shown in Table 4. Each operating point and each measured signal of the turbine as a whole was analyzed and the experimental results are presented in Table 5.

Table 4. Experimental value of pressure pulsation of draft tube (a_0 =33 mm).

Monitor the Location	f (Hz)	f_n (Hz)	$\Delta H \cdot H^{-1}$ (%)	$f \cdot f_n^{-1}$
Upstream of the spinal canal	3.43	19.53	3.44	0.24
Downstream of the spinal canal	3.43	19.53	2.42	0.24
Inside the elbow tube	5.57	19.52	3.95	0.26
The outside of the elbow tube	37.6	19.53	1.9	1.98

Table 5. Hydraulic turbine working conditions pressure pulsation test.

Location of Measurement Points	Operating Conditions	Test Results ($\Delta H \cdot H^{-1}$)	Guaranteed Value ($\Delta H \cdot H^{-1}$)
Spiral case import import	At rated operating conditions	2.02	<3%
	Partial working conditions operation	2.14	<3%
Movable guide vane—between runners	At rated operating conditions	2.64	<7%
	Partial working conditions operation	5.76	<7%
Between top cover and runner	At rated operating conditions	3.45	<7%
	Partial-load or no-load operation	5.22	<7%
Draft tube	Optimum operating conditions	1.10	<2%
	Partial-load or no-load operation	6.39	<7%

After the test analysis, it is known that the pressure pulsation amplitude of each measurement point under the normal operating range of turbine conditions fully meets the design guarantee value requirements. Figure 9 shows the experimental observation map of the draft tube area.

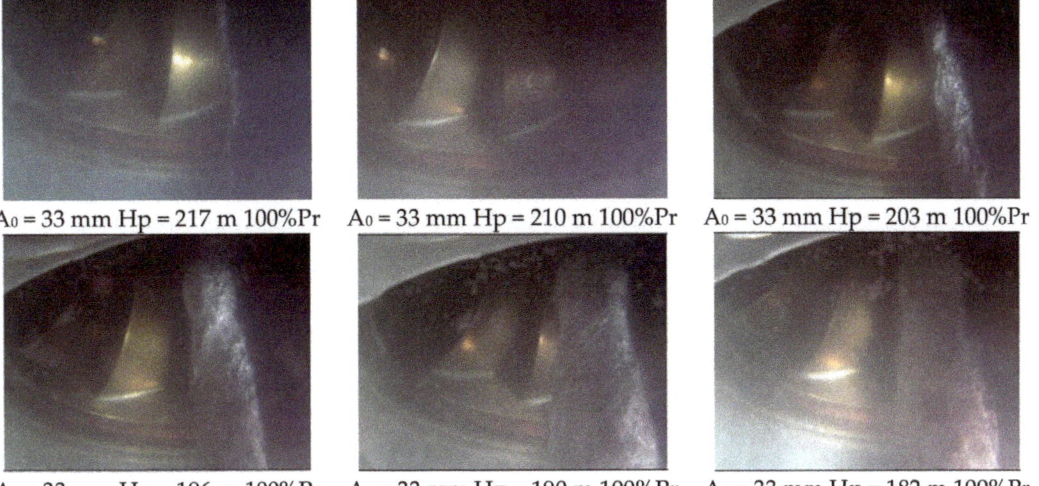

Figure 9. Experimental value of pressure pulsation of draft tube (a_0 = 33 mm).

3. Results

3.1. Draft Tube Local Flow Line Analysis

The streamline diagram can represent the running state of the water flow at a certain moment, and can roughly reflect the real movement of the water flow. In order to accurately analyze the internal flow characteristics of the draft tube area before and after the modification at low-flow conditions, the velocity flow diagram of the draft tube at the moment t = T was selected. It was found that the vortex appeared in the lower part of the elbow section and in front of the diffuser end of the draft tube. Therefore, two sections are selected at the inlet and diffusion section of the draft tube to observe the flow characteristics, as shown in Figure 10. From the dt_1 cross-sectional velocity vector diagram, it can be seen that there is a difference in the flow direction at the inlet of the draft tube, and it is the change in direction that improves the phenomenon of partial upward flow in the middle of the pipe. From the top view of the same section, the phenomenon that the recirculation area is concentrated on one side has also changed, and there is a trend of uniform distribution of velocity streamlines, which also reduces the impact of the water flow on the pipe wall, increases the radial flow rate, and makes the water flow more uniform. The flow is passed down more smoothly, reducing the impact of the water flow on the draft tube wall. The cross-sectional view of the diffusion section of the draft tube dt_2 also shows the phenomenon of local backflow, and the backflow intensity increases from the wall to the middle, and the upward flow is enhanced by the vortex. The cross-sectional velocity vector diagram as well as the overall flow line diagram can also show the vortex upward phenomenon caused by the change of direction of the vortex belt spinning into the draft tube. After the modification, the interference of the backflow to the normal water flow in the draft tube is weakened, the outflow of the turbine is increased, and the instability of the unit is reduced.

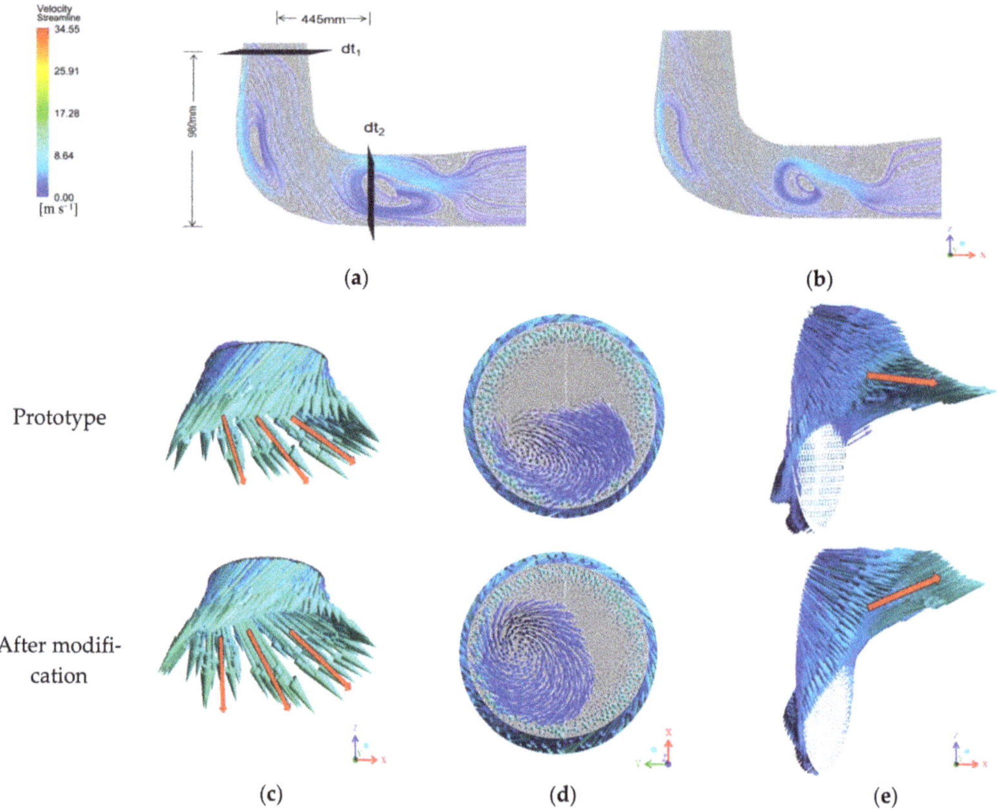

Figure 10. Overall velocity flow line of the draft tube before and after the modification and velocity vector diagram of the characteristic section at t = T (0.46404 s). (**a**) Prototype flow chart. (**b**) Modified flow chart. (**c**) dt_1 Cross-sectional streamline vector drawing. (**d**) dt_1 Cross-sectional flow vector top view. (**e**) dt_2 Cross-sectional streamline vector drawing.

3.2. Variation of Vortex Band Morphology in the Draft Tube under Different Guide Vane Airfoil Types

Draft tube vortex band is a symptom of unstable flow in mixed-flow turbines, which can seriously lead to fatigue damage of the unit [21]. Figure 11 shows the distribution of vortex cores in the wake tube region under non-fixed-length numerical calculations. It can be found that the distribution of vortex nuclei in the draft tube area mainly contains two types of vortex nuclei. The vortex belt in the centre of the draft tube appears to make eccentric circular motion around the rotating axis of the runner, and the lamellar vortex nuclei near the wall also change speed and shape continuously with the rotation of the runner. As time progressed, the vortex band morphology shifted. The lamellar vortex zone in the straight conical section gradually merges with the curved elbow region and continues to merge with the eccentric vortex zone in the central area of the draft tube up to the front of the diffuser section, with low-velocity lamellar vortex zones also appearing in the inner part of the curved elbow region. In terms of velocity, the velocity tends to increase and decrease from the straight cone section of the draft tube to the vicinity of the diffusion section. This is because the fluid (water) moving into the draft tube inherited the speed and vortex of the rotor outlet, due to the impact of the rotor area outflow, and the effect of its vortex hysteresis will affect the eccentric vortex zone in the center of the draft tube area causing the middle water velocity to be lower than the wall water velocity. Also, as the water passes through the elbow section, the velocity direction changes sharply due to the pipe restriction, causing the vortex belt to change its direction of rotation and causing an increase in velocity near the inside of the elbow section. The contrast in the draft tube of the $t_1 = 0.5T$ time modification becomes progressively clearer, with a break in the area of the vortex band in the middle of the runner and a significant reduction in the number of vortex cores attached to the lower end of the straight cone. The modification increases the speed of vortex belt separation and the collapse of the vortex nucleus in the diffusion section of the draft tube changes the process of top–down vortex belt spinning in and affects the winding and entrainment of the nearby water flow, reducing the pulsating effect of the water flow on the wall, and improving the safe operation of the reversible hydraulic turbine.

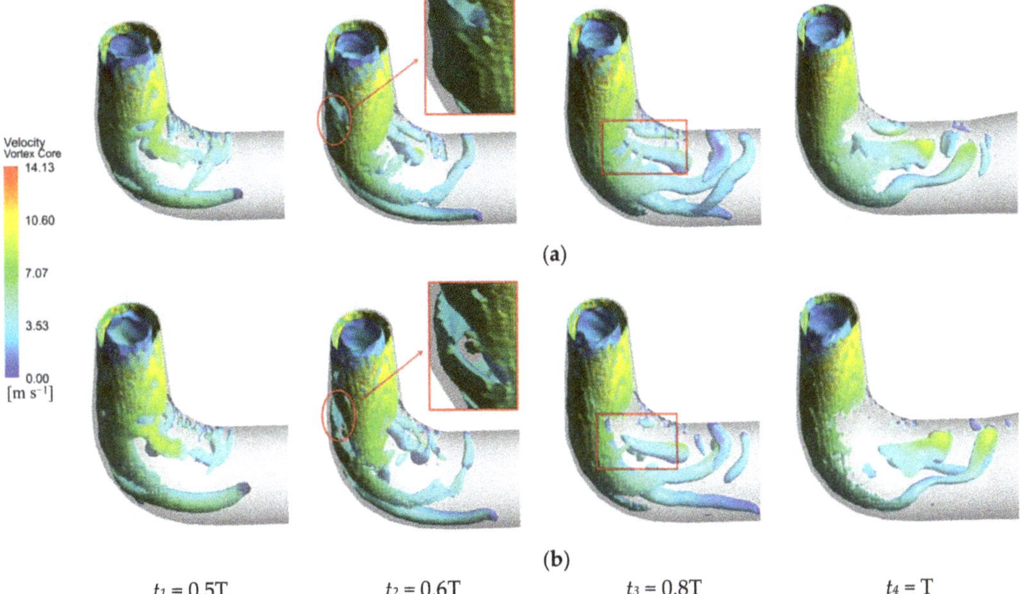

Figure 11. Comparison of the shape of the draft tube vortex before and after changing the airfoil. (**a**) Prototype. (**b**) After modification.

3.3. Analysis of Turbulent Energy of Draft Tube at Different Moments with Different Guide Vane Airfoil Types

Due to the irregular distribution of vortex bands, there will be different degrees of energy dissipation in the draft tube region. The Figure 12 shows a cloud of turbulent kinetic energy changes in the middle flow surface of the draft tube of a reversible turbine. The above analysis of the flow lines and vortex bands shows that there is significant vortex flow in the draft tube area, so the effect on the energy dissipation in the draft tube before and after the modification is analyzed. It can be seen that in this condition, along with the time change, the turbulent kinetic energy gradually develops from the straight cone section and the curved elbow section to the diffusion section, all concentrated in the front. The large energy dissipation at the near wall of the draft tube inlet is due to the high velocity of the water flow at this location and the directional distribution is more concentrated on the rotating side of the area. The vortex belt also rotates with a certain angular velocity in the exit area of the rotor, with the linear velocity increasing further away from the axis. The draft tube vortex belt interacts with the wall in its development and constantly impacts with the draft tube wall, thus causing noise and turbulent energy dissipation problems in the draft tube wall. The turbulent kinetic energy is mainly derived from the time-averaged flow, which provides energy to the turbulent flow through Reynolds shear stress work, i.e., the energy dissipation within the time-averaged is greater near the wall and in its vortex zone, which becomes the main part of energy loss in the draft tube. The turbulent energy dissipation problem is somewhat alleviated by the modification, with a more similar distribution pattern but a significantly smaller overall area of high dissipation (draft tube inlet wall, upper elbow area, and front of diffusion area).

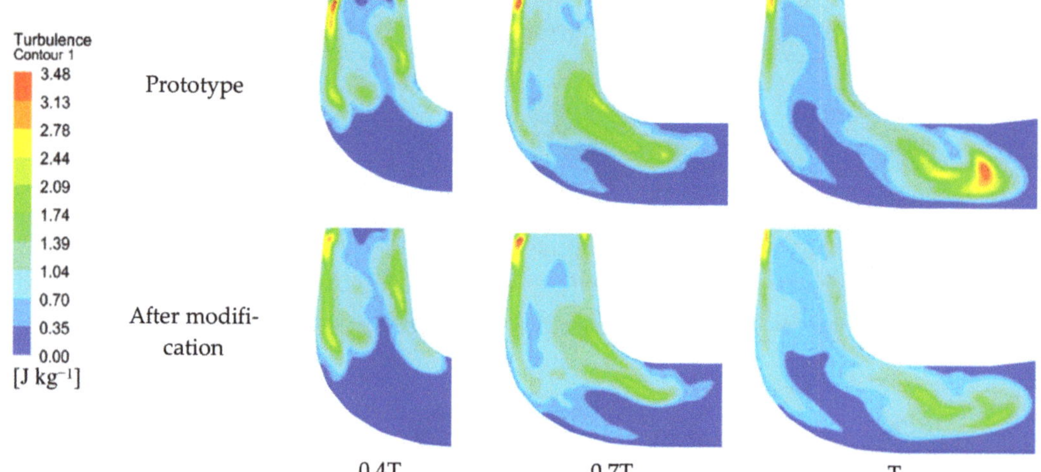

Figure 12. Comparison of turbulent energy evolution of draft tube before and after changing airfoil type.

In addition to the role of smoothly directing the liquid flow from the runner outlet downstream, the draft tube also converts the liquid flow energy from the runner outlet and the potential energy above the downstream tailwater level into additional vacuum, allowing excess energy to be recycled and, thereby, increasing the efficiency of the turbine. In order to further compare the draft tube energy losses, a draft tube energy recovery factor is introduced [34]. The calculation formula is shown in Equations (5)–(8).

$$\eta_w = \frac{\frac{v_2^2}{2g} - \frac{v_5^2}{2g} - \Delta h_w}{\frac{v_2^2}{2g}} \tag{5}$$

$$\Delta h_1 = 3.2\left(\tan\frac{\theta}{2}\right)^{1.25}\frac{v_2^2 - v_5^2}{2g} \tag{6}$$

$$\Delta h_2 = \frac{v_5^2}{2} \tag{7}$$

$$\Delta h_w = \Delta h_1 + \Delta h_2 \tag{8}$$

where v_2 is the average velocity of the draft tube inlet (m·s^{-1}); v_5 is the average velocity of the draft tube outlet (m·s^{-1}); Δh_w is the energy loss of the draft tube (m); g is the acceleration of gravity (m·s^{-2}); and θ is the diffusion angle.

As can be seen from Table 6, the modified movable guide vane can, indeed, achieve the effect of improving the flow condition in the draft tube and enhancing its energy recovery performance.

Table 6. Recovery coefficient of draft tube under different working conditions.

Before Modification		
Working Conditions Q_{11} (m$^3 \cdot$s^{-1})	Δh_w (−)	η_w (−)
0.83	0.510515	0.691242
0.66	0.387664	0.729541
0.3	0.306514	0.759853
After Modification		
Working Conditions Q_{11} (m$^3 \cdot$s^{-1})	Δh_w (−)	η_w (−)
0.83	0.551535	0.698520
0.66	0.397112	0.738467
0.3	0.329992	0.761053

3.4. Analysis of Pressure Pulsation of Draft Tube under Different Guide Vane Airfoil Types

Draft tube pressure pulsation is one of the most important reasons for the stability of reversible turbines. Eight monitoring points are evenly set at the inlet end of the draft tube and the elbow section. The positions of the monitoring points are shown in Figure 13. Unsteady calculations are performed before and after the modification, and the time-domain and frequency-domain diagrams of pressure pulsation are drawn.

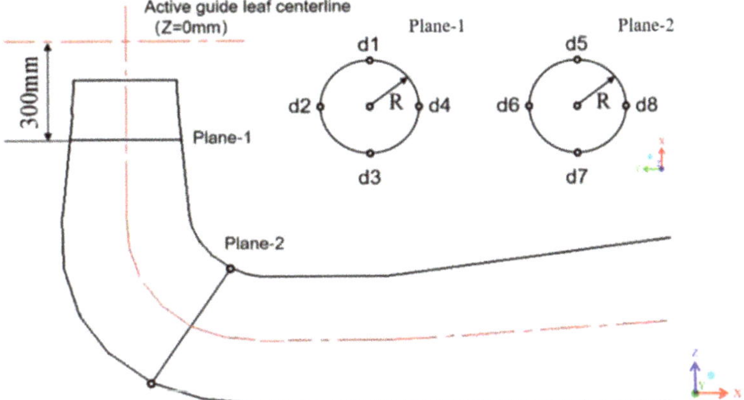

Figure 13. Schematic diagram of pressure pulsation monitoring points.

To facilitate better processing of monitoring point data, the dimensionless parameter $\Delta H \cdot H^{-1}$ is introduced as a parameter to quantify the intensity of pressure pulsation in the lobeless zone of the pump turbine, as shown in Equation (9). Then the fast Fourier transform (FFT transform) is performed on the time domain pulsation signal to make the pressure pulsation frequency domain diagram.

$$\frac{\Delta H}{H} = \frac{P_i - \overline{P}}{\overline{P}} \tag{9}$$

where: $\frac{\Delta H}{H}$ is the relative pulsation amplitude, %; P_i is the corresponding pressure monitoring value at point i, Pa; and $\overline{P}$ is the time-averaged pressure, Pa.

Figures 14 and 15 are the time-domain and frequency-domain graphs of the pressure pulsation at the monitoring points, respectively. The pressure pulsation pattern at each monitoring point in the draft tube is generally consistent with that before and after the modification. This is due to the fact that the change in wing shape directly affects the speed and direction of the inlet water in the rotor area, which, in turn, affects the action of the water on the draft tube. The pressure frequency domain plot highlights the effect of the draft tube vortex band on pressure pulsation before and after the modification as it is taken from the last two turns, reducing the interference from other frequencies. The dominant frequency of the frequency domain curve at the d1 monitoring point is 3.02 Hz with dimensionless $\Delta H \cdot H^{-1}$ of 0.02 and 0.19, respectively. Due to the downward development of the vortex belt the speed of the spin-in process decreases and the compound superposition of other disturbances, the main frequency of some monitoring points will change, the main frequency of d5 monitoring point is 6.04 Hz, which is about 0.25 times of the rotation frequency, which is basically consistent with the measured results also consistent with the vortex core rotation time period. The reversible turbine also has an increase in pressure from the water flow downwards at the same opening, and because of the complex morphology of the vortex belt in the draft tube it also causes radially asymmetric pressure pulsations at different locations in the draft tube wall. Overall, the water flow state inside the draft tube under low-flow conditions is complex, and the interaction of vortex bands and backflow in the straight cone section, bent elbow section, and diffusion section causes pulsation instability in the draft tube region, and the improved draft tube reduces the pulsation intensity in terms of pressure pulsation and increases the threshold of stable operation of the reversible hydraulic turbine.

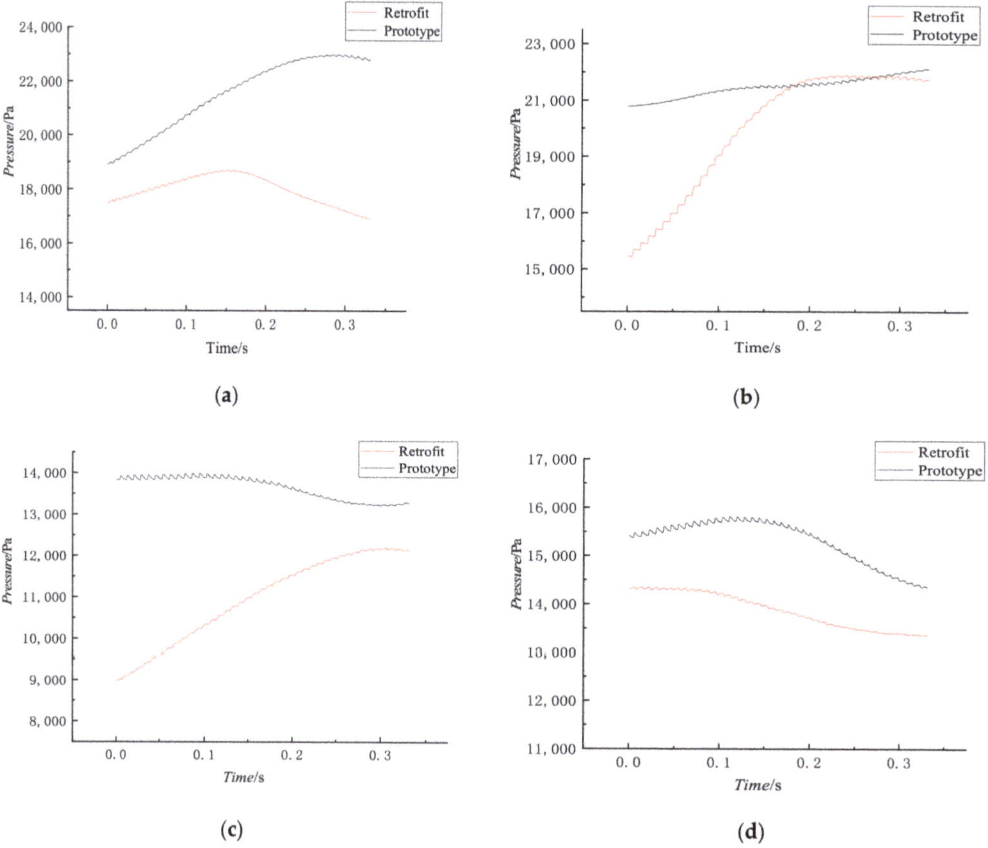

Figure 14. Time-domain diagram of monitoring point pressure pulsation. (**a**) d_1 Time domain curves. (**b**) d_3 Time domain curves. (**c**) d_5 Time domain curves. (**d**) d_7 Time domain curves.

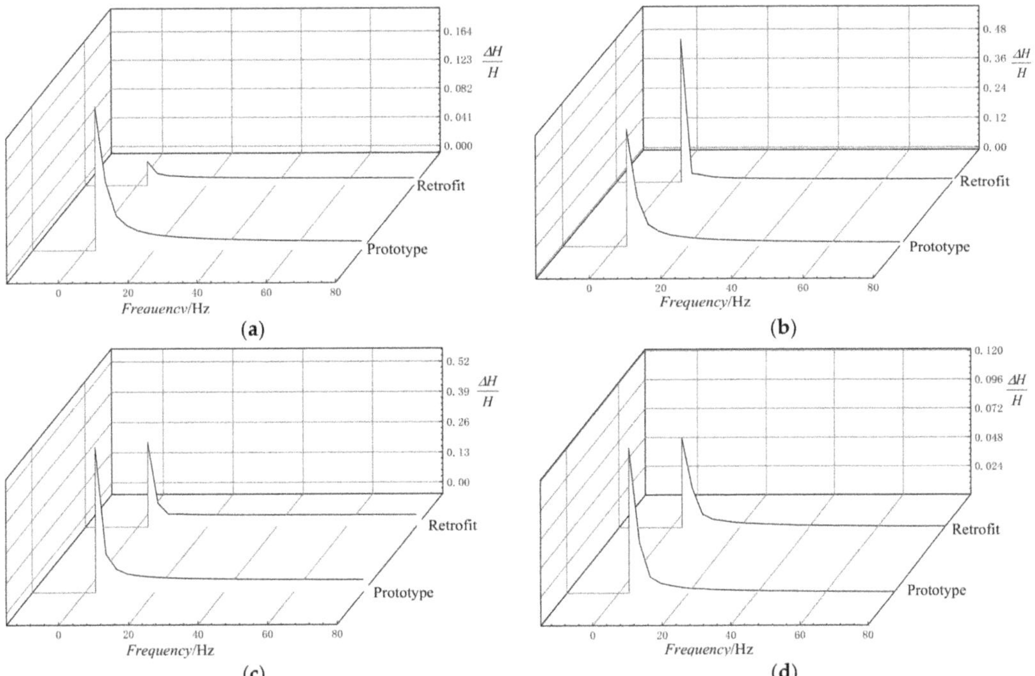

Figure 15. Monitoring point pressure pulsation frequency domain diagram. (a) d_1 Time domain curves. (b) d_3 Time domain curves. (c) d_5 Time domain curves. (d) d_7 Time domain curves.

4. Conclusions

In this paper, a pumped storage single-stage, vertical shaft reversible turbine is used as a model to explore the effect of changing the airfoil shape on the internal flow characteristics of the draft tube of a reversible turbine, while keeping other overflow components unchanged.

(1) Comparing the flow diagrams of the draft tube before and after the modification of the movable guide vane, it is found that the swirl area of the curved elbow section and the diffusion section of the draft tube is reduced after the modification. Further analysis of the vector diagrams of the dt_1 and dt_2 sections shows that the direction and velocity of the water entering the draft tube area (dt_1 section) have changed due to the change in the airfoil shape, and the axial outflow has also increased, reducing the concentrated impact of the water in the inlet area on the wall of a certain area. And combined with the top view of the two cross-sections found that the interference of the return flow to the normal flow of water in the draft tube is reduced, increasing the outflow of the turbine. The combined effect caused a reduction in the swirl region of the draft tube. The instability of the unit is reduced.

(2) The vortex belt evolution diagram of the draft tube before and after the modified movable guide vane can be found that the modified guide vane changes the direction of vortex belt rotation forward in a period of time, reduces the speed of rotation near the wall of the draft tube, and reduces the vortex density inside the vortex. t_1 to t_4 time to the wall of the bent elbow section and the middle of the diffusion section vortex belt accelerates the development of rupture, which makes the turbine operation more stable and reduces the risk of fatigue damage to the unit.

(3) The change of turbulent energy before and after the modification of the movable guide vane shows that the change of the airfoil shape causes the change of the direction of

the vortex belt in the wake pipe and accelerates the development of the vortex belt collapse, which reduces the turbulent energy near the vortex of the wake pipe. It is calculated that the change in airfoil shape improves the energy return coefficient of the draft tube by 0.69, 0.72 and 0.75 for Q_{11} = 0.83, 0.66 and 0.3 (m$^3 \cdot$s^{-1}), respectively, which improves the energy recovery level.

(4) By analyzing the pressure pulsation of the tailwater pipe area section before and after the modification, the main frequency of the straight cone section of both tailwater pipes corresponds to the vortex belt rotation period, and the pulsation of the tailwater pipe area is not stable. The modification reduces the draft tube vortex belt rotation and tailwater pipe wall collision reduces the pulsation intensity, which leads to a lower value of pressure pulsation fluctuation (6–9%) and improves the threshold of stable operation of the reversible turbine.

Author Contributions: Conceptualization, L.X. and G.X.; methodology, Q.L.; software, Q.W.; validation, L.X., G.X. and S.L.; formal analysis, Q.W.; investigation, L.X.; resources, Q.L.; data curation, L.X.; writing—original draft preparation, L.X.; writing—review and editing, L.X.; visualization, S.L.; supervision, Q.L.; project administration, Q.L.; funding acquisition, Q.L. All authors have read and agreed to the published version of the manuscript.

Funding: The National Natural Science Foundation of China (52066011); Study on the effect of reversible hydraulic turbine movable guide vane wing shape optimization on internal flow characteristics.

Institutional Review Board Statement: Not applicable for studies not involving humans or animals.

Informed Consent Statement: Not applicable for studies not involving humans.

Data Availability Statement: Not applicable.

Conflicts of Interest: The authors declare no conflict of interest.

References

1. Li, W.; Li, Z.; Qin, Z.; Yan, S.; Wang, Z.; Peng, S. Influence of the solution pH on the design of a hydro-mechanical magneto-hydraulic sealing device. *Eng. Fail. Anal.* **2022**, *135*, 106091. [CrossRef]
2. Li, L. Review and outlook of pumped storage power plant design. *Hydroelectricity* **2010**, *36*, 1–2.
3. Yu, J.; Cai, S.; Zheng, L.; Sun, J. Experimental analysis of pressure pulsation and velocity distribution of draft tube of Francis turbine. *J. Huazhong Inst. Technol.* **1981**, *6*, 63–70. [CrossRef]
4. Chang, J. Similar law of unsteady periodic water flow in draft tube of Francis turbine. *J. North China Inst. Water Resour. Hydropower* **1981**, *2*, 18–28. [CrossRef]
5. Cheng, L. On the hydraulic vibration of mixed-flow turbine. *J. Huazhong Inst. Technol.* **1981**, *6*, 57–62. [CrossRef]
6. Kubota, T.; Matsui, H. Cavitation characteristics of forced vortex core in the flow of a francisturbine. *Fuji Electr. Rev.* **1972**, *18*, 102–108.
7. Kubota, A.; Kato, H.; Yamaguchi, H. A new modeling of cavitating flow: A numerical study ofunsteady cavitation on a hydrofoil section. *J. Fluid Mech.* **1992**, *240*, 59–96. [CrossRef]
8. Nishi, M. Surging Characteristics of Conical and Elbow Type Draft Tubes. In Proceedings of the 12th IAHR Symposium on Hydraulic Machinery and Systems, Stirling, UK, 27–30 August 1984; pp. 272–283.
9. Jacob, T.; Prenat, J.E. Francis turbine surge: Discussion and data base. In Proceedings of the 18th IAHR Symposium, Valencia, Spain, January 1996.
10. Lliescu, M.S.; Ciocan, G.D.; Avellan, F. Analysis of the cavitating draft tube vortex in a Francis turbine suing Particle Image Velocimetry measurements in two-phase flow. *ASME J. Fluids Eng.* **2008**, *130*, 021105-1-10.
11. Wu, G.; Wei, C.; Tan, Y.; Zhang, K. The Effect of Gatage on the Pressure Fluctuation in Draft Tube of Francis Turbine. *J. Huazhong Univ. Sci. Technol.* **1999**, *9*, 27–29.
12. Konstantinov, A.; Staroselsky, I.; Orszag, S.A.; Yakhot, V. Renormalization Group-Based Transport Modeling of Premixed Turbulent Combustion: I. Incompressible Deflagration Model. *J. Sci. Comput.* **1998**, *13*, 229–252. [CrossRef]
13. Kurenkov, A.; Oberlack, M. Modelling Turbulent Premixed Combustion Using the Level Set Approach for Reynolds Averaged Models. *Flow Turbul. Combust.* **2005**, *74*, 387–407. [CrossRef]
14. Gao, Z.; Deng, J.; Ge, X. Numerical simulation of three-dimensional unsteady vortex rope turbulent flow occurred to the draft tube of a Francis turbine. *J. Hydraul. Eng.* **2009**, *40*, 1162–1167.
15. Zhang, C.; Diao, W.; You, J.; Xia, L. Numerical analysis of dynamic characteristics 0f the vortex rope in a francis turbine with splitter blades. *J. Huazhong Univ. Sci. Technol.* **2017**, *45*, 66–73.

16. Li, Q.; Zhao, C.; Long, S.; Quan, H. Study on the evolution of draft tube vortex band under flyaway condition of reversible hydraulic turbine. *Vib. Shock.* **2019**, *38*, 222–228. [CrossRef]
17. Fu, X.; Li, D.; Wang, H.; Zhang, G. Investigation on the fluctu ation of hydraulic exciting force on a pump-turbine runner during the load rejection process. In Proceedings of the IOP Conference Series: Earth and Environmental Science, Beijing, China, 16–19 November 2017; p. 012101.
18. Gao, Z.; Zhu, W.; Meng, L.; Zhang, J. Experimental Study of the Francis Turbine Pressure Fluctuations and the Pressure Fluctuations Superposition Phenomen on Inside the Runner. *J. Fluids Eng.* **2017**, *140*. [CrossRef]
19. Zhang, Y.; Zheng, X.; Li, J.; Du, X. Experimental study on the vibrational perform ance and its physical origins of a prototype reversible pump turbine in the pumped hydro energystorage power station. *Renew. Energy* **2019**, *130*, 667–676. [CrossRef]
20. Zhang, Y.; Chen, T.; Li, J.; Yu, J. Experimental Study of Load Variations on Pressure Fluctuations in a Prototype Reversible Pump Turbine in Generating Mode. *J. Fluids Eng.* **2017**, *139*, 074501–074504. [CrossRef]
21. Favrel, A.; Müller, A.; Landry, C.; Yamamoto, K.; Avellan, F. Study of vortex zone evolution and pressure pulsation in draft tube under upstream disturbance conditions. *Exp. Fluids* **2015**, *56*, 215. [CrossRef]
22. Chen, M. Study on the Correlation Mechanism Between Flow Field Structure of Mixed-Flow Rotor Outlet and Pressure Pulsation Characteristics of Draft Tube. Master's Thesis, Xi'an University of Technology, Xi'an, China, 2021. [CrossRef]
23. Li, C.; Pu, F.; Yang, J.; Wu, C.; Zhang, X.; Yang, H.; Yu, R. Analysis of the effect of different velocity angles of draft tube inlet on its internal flow field. *Yunnan Hydropower* **2022**, *38*, 58–62.
24. Wu, W.; Li, L.; Lv, R.; Yu, A. Research on pressure pulsation characteristics of pump turbine draft tube under different working conditions and improvement measures. *Hydropower Pumped Storage* **2021**, *7*, 11–19.
25. Yan, W.; Yang, J.; Zhang, Z.; Zeng, Y.; Yang, J. Effect of draft tube cavity vortex on the stability of pumped storage power plant system. *J. Hydropower Gener.* **2022**, *41*, 113–121.
26. Ma, Y.; Qian, B.; Feng, Z.; Liu, X.; Wang, B. Influence of cross-beam holes on the internal flow characteristics of the draft tube of axial-flow turbine. *People's Yellow River* **2021**, *43*, 122–124.
27. Yang, J.; Zhang, X.; Li, H.; Li, Y. Influence of draft tube deflector on the internal flow of centrifugal pump as a permeable level. *Hydraul. Pneum. Seals* **2021**, *41*, 60–60,63.
28. Yang, L.; Yan, C.; Zhang, J.; Deng, Z. The effect of draft tube with rectifier on its internal flow pattern. *J. Drain. Irrig. Mach. Eng.* **2018**, *36*, 124–128,135.
29. Rong, L. Hydrodynamic Analysis and Airfoil Optimization of Turbine Blades. Master's Thesis, Harbin Engineering University, Harbin, China, 2006.
30. Li, Q.; Ma, Q.; Xin, L. Effect of radius of guide vane head circle on hydraulic characteristics of water pump turbine. *J. Huazhong Univ. Sci. Technol. (Nat. Sci. Ed.)* **2022**, *50*, 94–100,115. [CrossRef]
31. Zeng, H.; Li, Z.; Li, D.; Chen, H.; Li, Z. Vortex Distribution and Energy Loss in S-Shaped Region of Pump Turbine. *Front. Energy Res.* **2022**. [CrossRef]
32. Li, Z.; Li, W.; Wang, Q.; Xiang, R.; Cheng, J.; Han, W.; Yan, Z. Effects of medium fluid cavitation on fluctuation characteristics of magnetic fluid seal interface in agricultural centrifugal pump. *Int. J. Agric. Biol. Eng.* **2021**, *14*, 85–92. [CrossRef]
33. Chen, X. Study on the Influence of Water Pump Turbine Guide Vane Wing Shape on "S" Characteristic. Master's Thesis, Lanzhou University of Technology, Lanzhou, China, 2020. [CrossRef]
34. Chen, A. Research on Vortex Band Characteristics of Water Pump Turbine Draft Tube and Improvement Method. Master's Thesis, Harbin Institute of Technology, Harbin, China, 2017.

Article

Research on Energy Loss Characteristics of Pump-Turbines during Abnormal Shutdown

Yuxuan Deng *, Jing Xu, Yanna Li, Yanli Zhang and Chunyan Kuang

BaiLie School of Petroleum Engineering, Lanzhou City University, Lanzhou 730071, China
* Correspondence: dengyuxuan@lzcu.edu.cn; Tel.: +86-159-0815-5884

Abstract: Pumped-storage hydropower (PSH) stations are an efficient emission-free technology to balance renewable energy generation instabilities. The pump-turbine is a core component of PSH stations requiring frequent start-up, shutdown, and working conditions for regulation tasks, making it prone to instabilities. Based on entropy production theory and vortex dynamics, we analyzed the energy loss characteristics for three working conditions of the pump, pump brake, and turbine when shutting down the pump-turbine. The results showed that the entropy production and vorticity of the spiral casing and draft tube remain almost constant, while the entropy production and vorticity of the runner region substantially change from the late pump braking to the late turbine condition. The entropy production and vorticity are derived from the guide vane transitioning to the runner flow channel through the vaneless space. The change law of energy loss through entropy production agrees with the change law of internal flow turbulence through vorticity. The entropy production analysis can quantify the energy loss and mark its location, while the vorticity analysis can quantify the degree of flow disturbance and show its location. The entropy production theory and vortex dynamics combination provide insights into the connection between undesirable flow phenomena and energy loss.

Keywords: pump-turbine; entropy production; vorticity; energy loss; numerical simulation

Citation: Deng, Y.; Xu, J.; Li, Y.; Zhang, Y.; Kuang, C. Research on Energy Loss Characteristics of Pump-Turbines during Abnormal Shutdown. *Processes* **2022**, *10*, 1628. https://doi.org/10.3390/pr10081628

Academic Editors: Santiago Laín and Omar Dario Lopez Mejia

Received: 28 July 2022
Accepted: 8 August 2022
Published: 17 August 2022

Publisher's Note: MDPI stays neutral with regard to jurisdictional claims in published maps and institutional affiliations.

1. Introduction

With the increasing proportion of renewable energy generation, deploying the peaking power of the corresponding scale is necessary to improve the safety and flexibility of the system operation [1]. Pumped storage hydropower (PSH) stations, which mainly undertake the tasks of peak and valley regulation and frequency and phase regulation, have been vigorously developed with their unique static and dynamic benefits. Therefore, they are practical and indispensable regulation tools for power systems with good development prospects [2,3].

As the "heart" of a PSH station, the pump-turbine must be started and shut down frequently according to different regulation tasks. Moreover, the operating conditions are changed frequently; thus, the unit passes through abnormal operating conditions, rapidly causing the vibration and efficiency reduction in the unit and endangering the safe and stable operation of the power station [4,5]. Liu et al. [6] introduced a reversible pump-turbine structure with S-shaped characteristics, which is essential for the transition process (start-up and load shedding). Yao et al. [7] combined the load shedding conditions of the Guangzhou PSH station under different guide vane closure laws and calculated the effect of guide vane closure laws. Walseth et al. [8] combined experiments and one-dimensional numerical calculations to study the pump-turbine flyaway process. Zhang et al. [9] used the VOF two-phase flow model to simulate the entire flow system numerically. The calculation results showed that the power station went through the pump, pump braking, and turbine conditions. Liu et al. [10,11] numerically simulated the transition process of the pump-turbine load dump and the pump-condition power failure shutdown. The authors found that the pressure fluctuation in the region near the runner is large during the load dump, and the pressure inside the runner drops significantly when operating in the pump braking mode. Moreover, they found that the reverse flow in the spiral case and the stall phenomenon in the flow channel substantially affect the change in the water head. Mao et al. [12] improved the stability of the fluid inside the pump-turbine by reducing the pressure change rate and improving the flow pattern. In particular, a coordinated valve and guide vane adjustment was adopted during load shedding. The results showed that this synergistic regulation method was practical for the operational stability of the system, especially for the final load shedding stage. In-depth research was conducted by Xia et al. [13] on the phenomenon of violent pressure fluctuation and unstable runner load during the runaway process of the pump-turbine. The authors found that the amplitude of the low-frequency component of the axial force was proportional to the amplitude of the flow rate of change. Li et al. [14] used a dynamic grid to simulate the guide vane closure process, verified the changing trend performance based on steady-state experiments, and analyzed the change law of performance characteristics such as head, flow and torque, pressure, and speed during the guide vane closure process. Zhang et al. [15] found that the inter-blade vortex structure at the impeller inlet shifted to the forward vortex structure and the return vortex structure when the pump-turbine operates in the S-shaped region. Moreover, they found that the forward vortex structure mainly dropped frequency drop and introduced low-frequency pulsation. In contrast, the return vortex structure caused unstable fluctuations. Yang et al. [16] described the evolution of the flow pattern during the rapid change in the pump-turbine operating conditions and analyzed the mechanism of pressure pulsation change on the flow pattern transition. Yang et al. [17] selected four different pump-turbines to simulate the runaway process. The authors analyzed the similarities and differences between flow patterns and pressure fluctuations, indicating that the transient processes controlling the mechanism of flow pattern transition and pressure change is a crucial element in flow path design. Zhao et al. [18] studied the stability performance of PSH generation systems during the transition from condensing to generating mode, developed a mathematical model to describe this dynamic transition process, and effectively predicted the dynamic response of the system parameters. Finally, Yang et al. [19] studied the relationship between the pressure pulsation value in the center of the pump-turbine draft tube and the wall pressure pulsation value using the CFD method. The study provided the critical value of the wall pressure pulsation when cavitation occurred, providing a reference for determining the cavitation phenomenon using the wall pressure pulsation value in practical engineering.

Furthermore, hydraulic losses due to friction and unsteady flow in a pump-turbine degrade its efficiency. Gong et al. [20] applied the entropy production theory in the hydraulic analysis of a hydraulic turbine. The authors also performed three-dimensional steady-state flow simulations and entropy production calculations for a hydraulic turbine. The results showed that entropy production theory is applicable to evaluate the performance of a hydraulic turbine and has the advantage of determining the amount of energy dissipation and the location where dissipation occurs. Li et al. [21,22] proposed the wall equation and used entropy production theory to obtain the detailed distribution of hydraulic losses of the pump-turbine in the pump mode, calculated the hydraulic losses in the wall region, and compared the hydraulic losses calculated by entropy production and pressure difference. In this way, the authors obtained improved results. In particular, the entropy production theory was used to directly reflect the energy dissipation in the hump region of the pump-turbine. Fu et al. [23] analyzed the energy conversion process, loss distribution, and flow mechanism inside the pump-turbine based on the numerical simulation, using the combination of entropy production analysis and flow analysis. The results showed that the pump-turbine load shedding process was an energy dissipation process, and the energy was converted between various energy forms. Li et al. [24] optimized the design of the runner blade and guide vane of the bulb tubular turbine by applying the boundary vortex dynamics theory. The authors also analyzed the effect of the vortex on unit performance and efficiency, and the results showed that the vortex distribution on the blade surface could effectively reflect the turbine performance. Li et al. [25] obtained the vortex dynamics parameters according to the vortex dynamics theory and analyzed the relationship between the vortex dynamics parameters and the hump characteristics. The results showed that the energy transfer between the flow channel and the fluid was performed through the vortex dynamics parameters, which is superior in evaluating the dynamic performance of the pump-turbine.

In summary, the traditional method of hydraulic loss analysis focuses on evaluating the pressure drop, which has some limitations and cannot determine precisely where the high hydraulic losses occur. However, entropy production theory and vortex dynamics analysis have strong practicality and accuracy in the energy loss in the pump-turbine transition process. Thus, to further verify the advantages of these two methods in the energy loss characteristics of pump-turbines, this study analyzes the energy loss characteristics of the pump-turbine shutdown process using entropy production and vorticity methods based on numerical simulation. In particular, the results of the two methods are compared to study their connection.

2. Numerical Model

2.1. Computational Method

In this study, the governing equations include the continuity equation of an incompressible fluid and the Reynolds-averaged Navier–Stokes (RANS) equation. The ANSYS CFX 17.0 commercial computational fluid dynamics software was used to solve the three-dimensional unsteady flow in a pump-turbine. In this study, the SST k-ω model is chosen as the turbulence model, considering the viscosity of the inner wall of the model compared with the standard k-ε model and the RNG k-ε model. This model has an improved turbulent shear stress transport, a more stable algorithm, and an improved simulation performance for the flow in narrow spaces [26].

For turbulent flow, the local entropy production is caused by the time-averaged motion and the pulsation velocity, which can be expressed as Equations (1) and (2), respectively.

$$\dot{S}_{\overline{D}}''' = 2\frac{\mu}{T}\left[\left(\frac{\partial \overline{\mu}_1}{\partial x_1}\right)^2 + \left(\frac{\partial \overline{\mu}_2}{\partial x_2}\right)^2 + \left(\frac{\partial \overline{\mu}_3}{\partial x_3}\right)^2\right] + \frac{\mu}{T}\left[\begin{array}{c}\left(\frac{\partial \overline{\mu}_2}{\partial x_1} + \frac{\partial \overline{\mu}_1}{\partial x_2}\right)^2 + \left(\frac{\partial \overline{\mu}_3}{\partial x_1} + \frac{\partial \overline{\mu}_1}{\partial x_3}\right)^2 \\ + \left(\frac{\partial \overline{\mu}_2}{\partial x_3} + \frac{\partial \overline{\mu}_3}{\partial x_2}\right)^2\end{array}\right], \tag{1}$$

$$\dot{S}_{D'}''' = \frac{\mu_{eff}}{T}\left\{2\left[\begin{array}{c}\left(\frac{\partial \mu_1'}{\partial x_1}\right)^2 + \left(\frac{\partial \mu_2'}{\partial x_2}\right)^2 + \\ \left(\frac{\partial \mu_3'}{\partial x_3}\right)^2\end{array}\right] + \begin{array}{c}\left(\frac{\partial \mu_2'}{\partial x_1} + \frac{\partial \mu_1'}{\partial x_2}\right)^2 + \left(\frac{\partial \mu_3'}{\partial x_1} + \frac{\partial \mu_1'}{\partial x_3}\right)^2 \\ + \left(\frac{\partial \mu_2'}{\partial x_3} + \frac{\partial \mu_3'}{\partial x_2}\right)^2\end{array}\right\}, \tag{2}$$

where T is the temperature (°C), and μ_{eff} is the effective dynamic viscosity of the fluid (Pa·s).

In the Reynolds time-averaged algorithm, the entropy production caused by the pulsation velocity is unavailable. Thus, ω in the turbulence model is correlated with the entropy production generated by the pulsation velocity [27], which can be expressed as follows:

$$\dot{S}_{D'}''' = \beta\frac{\rho\omega k}{T}, \tag{3}$$

where $\beta = 0.09$, ω is the turbulent vortex viscous frequency (s^{-1}), and k is the turbulent kinetic energy (m^2/s^2).

Then, the total local entropy production can be expressed as follows:

$$\dot{S}_D''' = \dot{S}_{\overline{D}}''' + \dot{S}_{D'}'''. \tag{4}$$

2.2. Computational Model

The main design parameters of the study object are as follows: nominal runner diameter $D = 349$ mm, the number of runner blades $Z_0 = 6$, the number of guide vane

$Z_1 = 28$, the number of stay vane $Z_2 = 14$, Table 1 summarizes the main technical parameters of a pump-turbine.

Table 1. Main technical parameters.

Parameter	Symbol	Unit	Value
Nominal diameter	D	mm	349
Number of runner blades	Z_0	(-)	6
Number of guide vanes	Z_1	(-)	28
Number of stay vanes	Z_2	(-)	14
Rated flow rate	Q_d	m^3/s	0.203
Optimal guide vane opening	α	$^\circ$	10
Optimal efficiency	E	%	92.61
Rated head	H	m	12

This study aims to investigate the energy loss characteristics of the pump-turbine when guide vane rejection occurs after a sudden power failure during pump operation. Thus, all the over-flow components of the entire pumped storage unit are considered the calculation domain, mainly including the spiral case, guide vane, stay vane, runner, and draft tube. UG NX software was used to model each different over-flow component in 3D, and the assembled full-flow channel calculation model is shown in Figure 1.

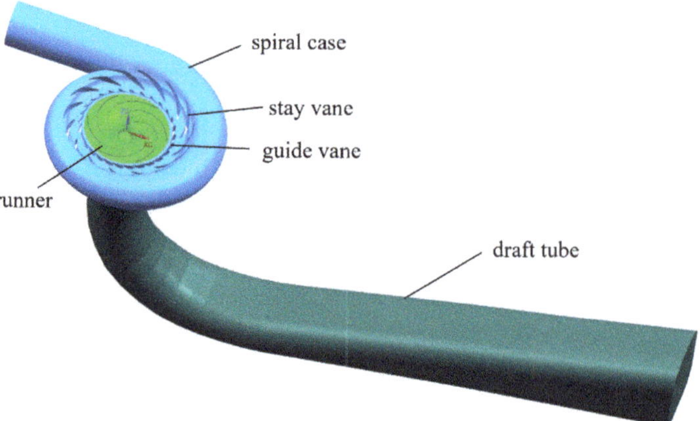

Figure 1. Computational model.

Considering the configuration of the computer (Intel Xeon E5-2650 @ 2.3 GHz with 64 GB of RAM) and the full-flow channel structure, this study was performed using the ICEM CFD commercial software because it required a high-precision structured grid for spatial discretization of the computational domain. For the components with narrow geometry, such as guide vanes and impellers, an "O-shaped" grid was used for the local encryption to ensure the uniform transition between the grids of each component and to capture the flow details accurately. The full flow channel and local grid are shown in Figure 2.

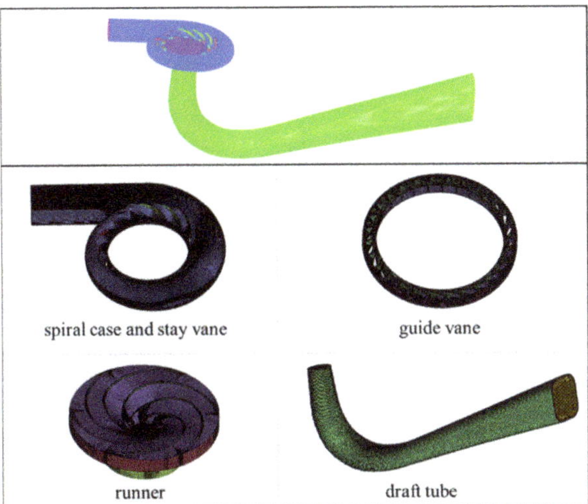

Figure 2. Computational mesh domain.

The grid density selected for the calculation should be as low as possible under the condition of satisfying the computational accuracy. In particular, the best efficiency point under the hydraulic turbine operating conditions should be selected for grid-independent analysis. The spiral case is the mass inlet boundary, which is set to 203 kg·m^3·s^{-1}, and the draft tube is the pressure outlet boundary, which is set to 111.13 kPa. The SST k-ω turbulence model was used, and the numerical solution of the control equations was obtained using the SIMPLEC algorithm for constant calculation. The residual convergence criterion was 10^{-5}. The quality of the grid was ensured using five sets of schemes with different grid numbers to validate and analyze with the head. The specific information of the grid is listed in Table 2, and the variation of the head with the number of grids is shown in Figure 3.

Table 2. Details of the grids.

Part	Nodes (Million)					Quality
	1	2	3	4	5	
Spiral case and Stay vane	1.58	2.05	3.78	4.82	8.36	0.30
Guide vane	0.76	0.94	1.09	1.83	2.48	0.42
Runner	2.11	2.11	2.11	2.11	2.11	0.33
Draft tube	0.75	1.00	1.41	1.89	2.70	0.38
Sum	5.20	6.10	8.40	10.65	15.65	—

As seen in Figure 3, the head variation is smooth when the number of grid nodes is greater than 8.4 million. Considering factors such as limited computational resources and computational time, the number of 8.4 million grid nodes was selected for numerical calculation.

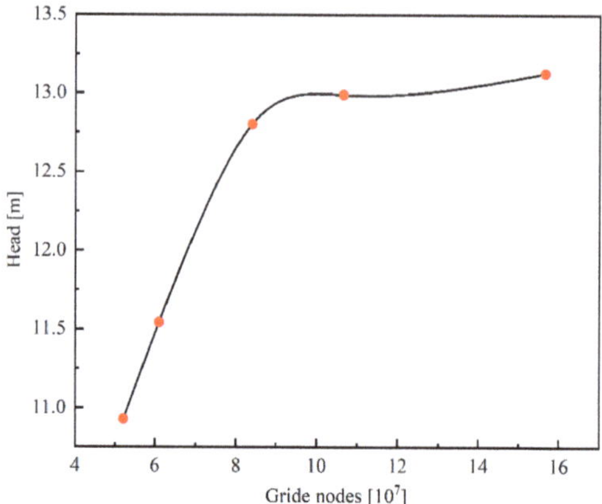

Figure 3. Grid independence verification.

2.3. Boundary Conditions and Setting

Figure 4 shows the N_{ed}-Q_{ed} characteristic curve of the pump-turbine in the transition process of guide vane rejection after the power failure in the pump condition. After the pump-turbine suddenly loses power in the pump condition, the flow rate decreases rapidly to zero along the pump condition direction. After entering the braking condition, the flow rate starts to increase in the reverse direction of the turbine condition, and the rotor speed starts to decrease rapidly to zero because the rotor blades lose power and are subject to fluid impact resistance. After entering the hydraulic turbine working condition, the runner starts to rotate to the hydraulic turbine working condition, and the impact of fluid on the runner blade provides power for the runner blade. The guide vane refuses to close the action, resulting in a continuous increase in rotational speed, and the unit finally enters the flyaway working condition area.

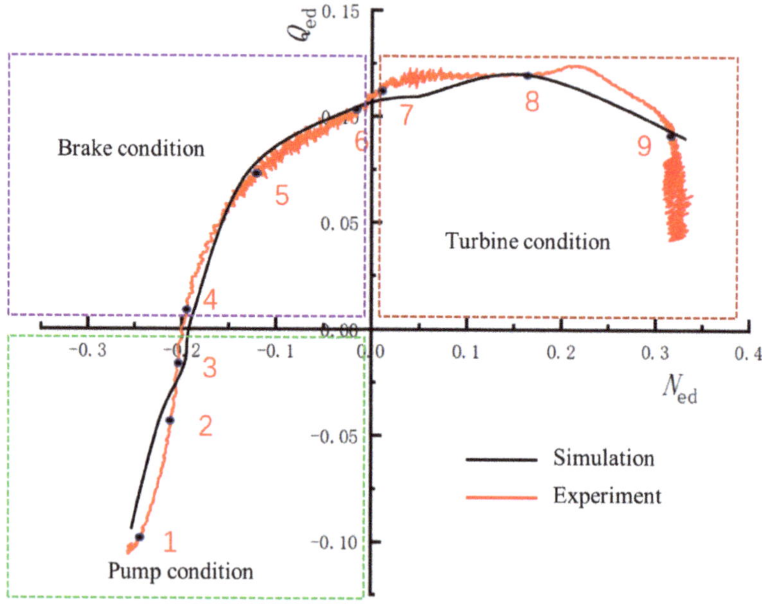

Figure 4. N_{ed}-Q_{ed} characteristic curve of pump-turbine.

In this study, nine working points in three working areas were selected for numerical simulation, and the numerical simulation, and experimental results were compared (where red is the curve obtained from the experiment and black is the curve obtained from the numerical simulation). As seen in Figure 4, the two characteristic curves basically match, indicating that the numerical simulation can reflect the flow characteristics of the pump-turbine during the power-off guide vane rejection transition.

The spiral case was selected as the mass flow inlet boundary condition, the flow direction was considered perpendicular to the inlet surface, and the turbulence intensity was 5%. The draft tube was used as the pressure outlet boundary condition, while the wall surface was used as the no-slip boundary condition. The runner speed was provided according to the experiment, and the connection between the spiral case, guide vane, runner, and draft tube was conducted using the interface. The numerical solution of the control equation was performed using the SIMPLEC algorithm. The transient calculation was performed using the constant calculation result as the non-constant initial field. The time step corresponded to the time of each $4°$ rotation of the runner, and the maximum iteration step was 80. The time step was the number of steps for ten rotations of the runner, which was set to 900, The calculated boundary conditions of each working condition are shown in Table 3.

Table 3. Boundary conditions.

Operating Point		Inlet(kPa)	Outlet (kPa)	Rotate Speed (rpm)	Turbulence Intensity	Time Step (s)
1	Pump condition	218.1	123.7	−461.7		−0.00014
2		193.7	125.5	−355.3		−0.00019
3		182.0	125.3	−319.0		−0.00021
4	Brake condition	181.5	125.3	−315.9		−0.00021
5		184.0	124.6	−211.9	5%	−0.00031
6		194.8	124.9	−18.8		−0.00355
7	Turbine condition	190.8	124.9	16.6		0.00402
8		218.4	125.4	339.0		0.00020
9		229.3	126.6	600.2		0.00011

3. Results and Discussion

3.1. Energy Loss Analysis

When the water flows through the flow channel, the kinetic energy and pressure energy of the fluid are converted into internal energy due to the viscous force, increasing the entropy production. The energy losses can be obtained indirectly by analyzing the change in entropy production of each component during the pump-turbine shutdown. In particular, vorticity is one of the main causes of fluid energy loss in viscous fluid mechanics. Moreover, the flow disturbance phenomenon due to various irregular flows inside the pump-turbine can be quantified by vortex analysis. Therefore, the entropy production and vorticity change law of each component during the pump-turbine shutdown are compared and analyzed to study their connection.

Figure 5 shows the change curves of entropy production and vorticity of each component of the pump-turbine during the abnormal shutdown. As seen in the figure, the entropy production and vorticity of the spiral case and draft tube remain almost the same under different working conditions. In contrast, the entropy production and vorticity of the runner change substantially from the late stage of the pump braking condition to the late stage of the turbine working condition. The main reason is that under the pump braking condition, the internal fluid flows along the direction of the turbine, but the runner still rotates in the direction of the pump condition. The fluid flowing downstream through the guide vane is impacted by the blade, and a vortex is generated at the inlet of the runner and the pressure side of the runner blade head, resulting in energy loss. At the beginning of the turbine working condition, the runner rotates along the direction of the turbine working condition due to the impact of the fluid. At this time, the flow rate is large, the speed is small, and the fluid impact loss is large. The following section focuses on the entropy production and vorticity of the runner part.

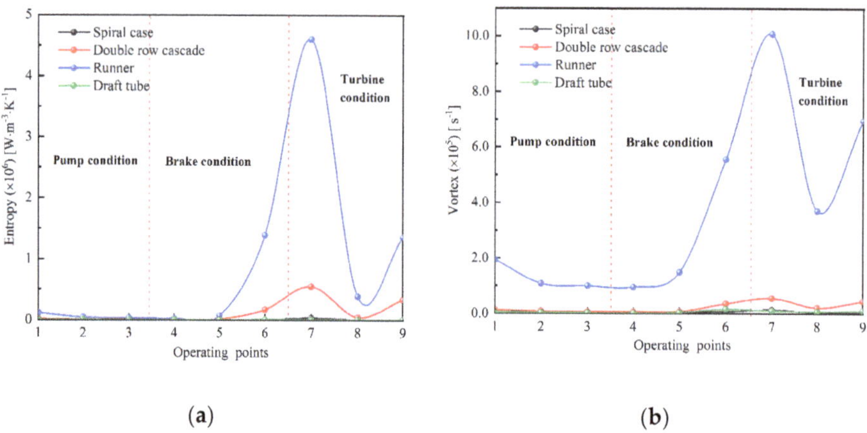

Figure 5. Energy loss during shutdown: (**a**) entropy production; (**b**) vorticity.

3.1.1. Entropy Production Analysis

Figure 6 shows the diagram of the entropy production distribution in the runner area under different working conditions. The diagram shows that the entropy production at the top of the runner blade is higher when the pump-turbine operates in the pump working condition. With the change in working condition point, the flow rate decreases, the fluid and blade impact weakens, and the entropy production reduces. The main reason is that after the sudden power failure of the unit, the guide vane refuses to act, and the fluid flows out of the runner and collides with the guide vane, resulting in part of the fluid flowing to the downstream runner area and blade impact. With the continuous reduction in flow, the fluid inside the unit reverses flow and starts to flow in the direction of the turbine. Moreover, the unit enters the braking condition zone, at which point the fluid flowing downstream through the guide vane is impacted by the blade, and a vortex is generated at the inlet of the runner and the pressure side of the runner blade head. When the working condition point 5 is reached, the entropy production of the front part of the runner is not uniformly distributed, the entropy production of the middle part is higher, and the entropy production of the middle part of the blade pressure side is also higher. When the working condition point 6 is reached, the entropy production of the top area of the blade is higher, and the high entropy production area of the runner is extended from the suction side of the front of the blade to the suction side of the tail, and the entropy production of the local area at the end of the runner is larger. After the rotational speed decreases to zero along the pump direction, the runner starts to rotate in the reverse direction and gradually increases along the turbine direction to enter the turbine working condition zone. When the working condition point 7 is reached, the front end of the blade is surrounded by the high entropy production area, and it extends along the front section of the runner to the end section. When the working condition point 8 is reached, the entropy production of the top area of the blade is higher, and the entropy production of the suction side of the blade is slightly higher than that of the surrounding area. Finally, when the working condition point 9 is reached, the entropy production of the entire vaneless space is higher due to the phenomenon of circulating flow in the vaneless space, and the entropy production of the front of the pressure side and the tail of the suction side of the blade is also significantly higher than that in the surrounding area.

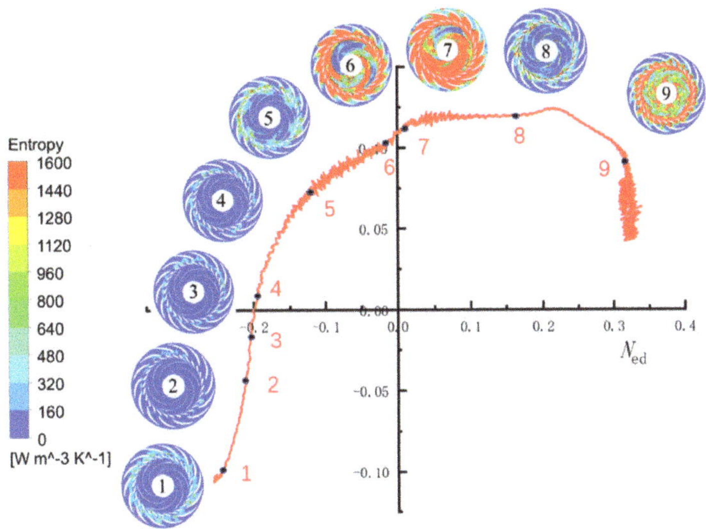

Figure 6. Diagram of the entropy production change in the runner area.

3.1.2. Vorticity Analysis

Figure 7 shows the vorticity distribution diagram in the runner area under different working conditions. The figure shows that the guide vane refuses to act after the sudden power failure, and the unit runs in the pump working condition area, the fluid flows out from the runner and collides with the guide vane, and part of the fluid flows to the downstream runner area and impacts with the blade. In this working condition zone, the vorticity at the top of the blade extends to the double-row impeller grille, and when it reaches the rear section of the blade, the pressure side vorticity becomes higher than that in the surrounding area. In the braking condition zone, the fluid flows in the direction of the turbine condition, and the runner rotates in the direction of the pump condition. When the working condition point 5 is reached, the vorticity on the pressure side of the blade head is substantially larger than that in the surrounding area. In particular, the vorticity distribution in the runner area is extremely uneven and gradually decreases along the runner. As the speed decreases, the flow rate increases, and the role of the blade on the fluid decreases and reaches the working condition point 6. In this condition, the high vorticity area only exists at the top of the blade, and the vorticity is much larger than the surrounding area. Moreover, the vorticity in the vaneless space is obviously reduced, the vorticity in the runner area along the center of the runner to both ends of the runner blade is decreased, and the vorticity at the end of the runner is larger. In the hydraulic turbine working condition area, the region with higher vorticity at the blade tip presents a U-shaped distribution. The area is more concentrated than the previous one, and the vorticity in the runner is small in the middle and large on both sides. When the working condition point 8 is reached, the vorticity on the suction side of the blade tip is larger, and the vorticity on the suction side of the runner near the blade is larger than that in other regions. Finally, when the working condition point 9 is reached, the high vorticity area at the top of the blade develops to the pressure side. Furthermore, the vorticity at the inlet of the runner is irregularly distributed, the vorticity value at the end of the runner is larger, and a ring-shaped area with high vorticity is present in the vaneless space.

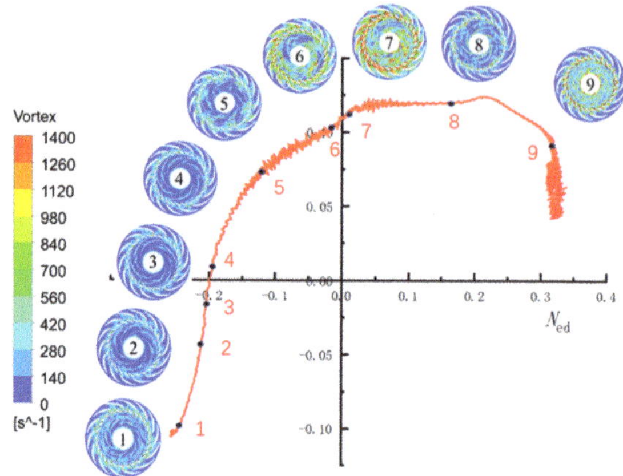

Figure 7. Diagram of the vorticity change in the runner area.

Figure 8 shows the curves of the entropy production and vorticity in the blade and double-row blade grille area during the abnormal shutdown of the pump-turbine. As seen in the figure, the entropy production and vorticity change trends are the same throughout the shutdown process. From the early stage of the pump condition to the middle of the braking condition, the entropy production and vorticity change slightly. Moreover, both curves rise rapidly from the middle of the braking condition, reaching the peak in the early stage of the turbine condition and decreasing rapidly. Subsequently, they rise rapidly after the middle of the turbine condition. The changing trend is consistent with the analysis results presented in Sections 3.1.1 and 3.1.2. Before the working condition point 5, the entropy production and vorticity slowly change. Then, they gradually increase from the working condition point 5 and reach the maximum value at the working condition point 7. Finally, both variables continue to rise to the working condition point 9 after dropping to the working condition point 8.

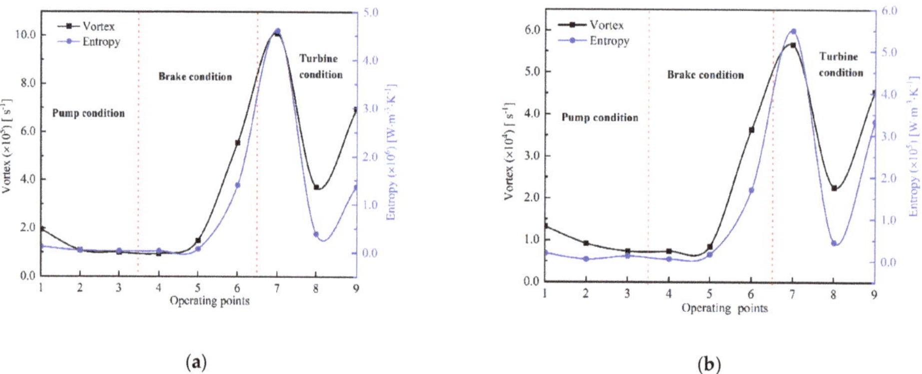

Figure 8. Comparison of the vorticity and entropy production: (**a**) blade; (**b**) double-row blade.

3.2. Distribution Law Analysis

The analysis presented above indicates that a certain correlation exists between the entropy production and vorticity of the pump-turbine in the process of the abnormal shutdown. In the following, the entropy production and vorticity of the runner area under the late braking condition, the pre-water turbine condition, and the late condition are analyzed to explore the internal flow mechanism of the distribution law of both.

Figure 9 shows the entropy production expansion diagram of the runner area under the three working conditions. As seen in the figure, most of the entropy production near the guide vane is high in the late braking condition. Only the entropy production at the head of the two guide vane in the A area is low, shown in the dashed box. Moreover, the distribution of entropy production in the bladeless space is chaotic; however, the overall intensity is high, and the entropy production starts from the guide vane and transitions to the impeller runners through the vaneless space. Nevertheless, in this condition, only the entropy production in the two runners is high and unevenly distributed. The remaining four runners have higher entropy production at the inlet or front end of the blade pressure side and have not started derived downstream. In the early stage of the turbine working condition, the entropy production in the guide vane area is uniformly distributed, and the intensity reaches the highest. The phenomenon of annular high entropy production in the vaneless region is enhanced, and only the entropy production in the B area shown in the dashed box is low. Moreover, the entropy production in the runner develops downstream, and the area of high entropy production in the entire runner increases significantly. The entropy production in two runners is high in intensity and uniformly distributed, while the entropy production in the rest is still in the transition process and disordered distributing. In the late stage of the turbine operation, the entropy production of the head of the movable guide vane (C area shown in the dashed box) exhibits a high and low intermittent distribution. The entropy production of the entire vaneless space to the front of the runner reaches the highest intensity and uniform distribution. In contrast, the entropy production of the middle part of the runner is low, and the entropy production of the pressure side of the rear part of the runner is high, with a regular overall distribution.

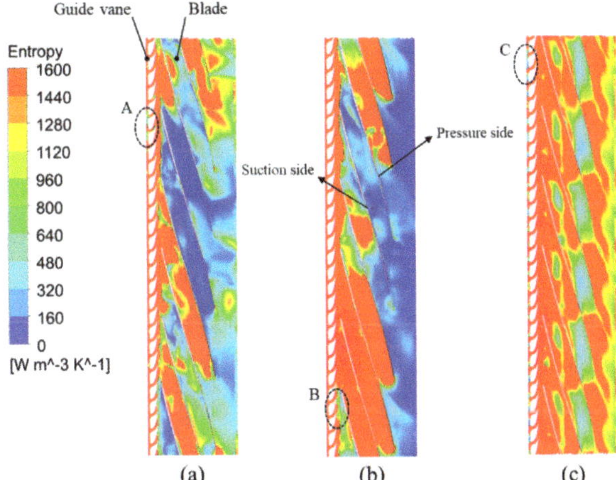

Figure 9. Distribution of the entropy production in the runner area: (**a**) late braking condition; (**b**) pre-water turbine condition; (**c**) late turbine operating condition.

Figure 10 shows the vortex diagram of the runner area under the three working conditions. Note that in the late braking condition, the middle and rear sections of the guide vane are all high vorticity areas, and some of them extend through the vaneless space to the head of the runner blade and occupy the pressure side flow channel entrance. Most of the blades have higher vorticity in the area near the pressure side front and middle section walls. Moreover, a small part of the blades has higher vorticity in the area near the suction side middle and rear section walls. Some of the blades have small-scale circular high vorticity areas at the end. In the early stage of the turbine working condition, the high vorticity area extends to the head of the guide vane, and the vorticity of the entire guide vane area is higher. Furthermore, the annular high vortex phenomenon in the vaneless space is enhanced, and the area of the high vorticity area of each part of the blade increases.

In the later stage of the turbine working condition, the vorticity near the head of the guide vane decreases, the high vorticity area only covers the middle and rear part of the guide vane, and the high vorticity area covers the vaneless space. The vorticity in the front and middle channels of the runner is higher on both sides of the blades and lower in the middle area. The vorticity in the middle area is low. Only a very small circular area in the rear part of the runner has high vorticity, and the overall distribution is more regular.

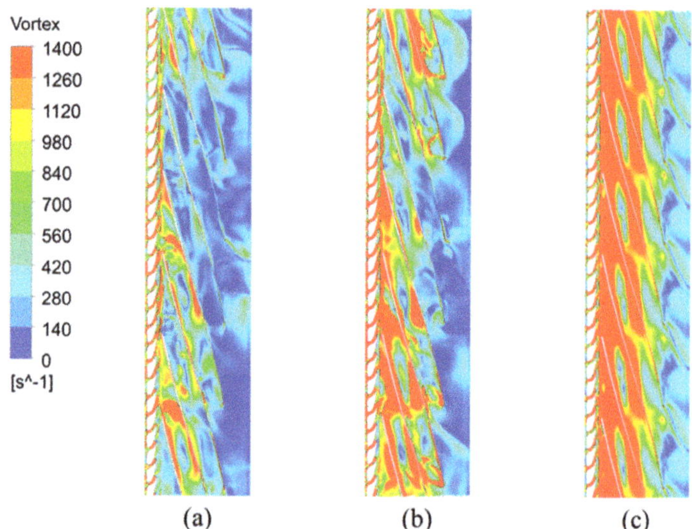

Figure 10. Vortex distribution in the runner area: (**a**) late braking condition; (**b**) pre-water turbine condition; (**c**) late turbine operating conditions.

Figure 11 shows the flow diagram of the pump-turbine in the runner area under different working conditions. Note that in the late braking condition, most of the guide vane head has swirling flow. In addition, the fluid flow velocity is larger near the pressure sidewall, and the high-speed circulation phenomenon appears at the tail wall of the guide vane pressure side. Almost all the runner channels are occupied by backflow and channel vortex, presenting a severe blockage. In the early stage of the hydraulic turbine working condition, the flow velocity increases, and the number of guide vortex decreases. Moreover, the high-speed circulation phenomenon in the vaneless space is weakened, and the number of vortex in the runner flow channel decreases. The fluid in part of the flow channel flows downstream at high speed along the suction sidewall of the blade. At the later stage of the turbine working condition, the number of swirling flows at the head of the guide vane increases, the scale increases, and the circulation phenomenon in the vaneless space becomes evident. Furthermore, the scale of the vortex in the runner flow channel becomes smaller, the number increases, and most of the swirling flows locate near the wall of the suction side of the runner blade. As the shutdown process proceeds, the swirling flow around the guide vane is produced easily due to the development of backflow. Moreover, the high-speed circulation phenomenon produces easily due to the vaneless space so that the fluid cannot smoothly flow downstream. In addition, in the runner backflow, off-flow and other phenomena developed from the vortex. Through the comparative analysis of the vorticity, entropy production, and flow line, the vorticity and entropy production distribution law is similar. In particular, the vorticity and the entropy production are larger in the parts with larger vortex flow or velocity gradient.

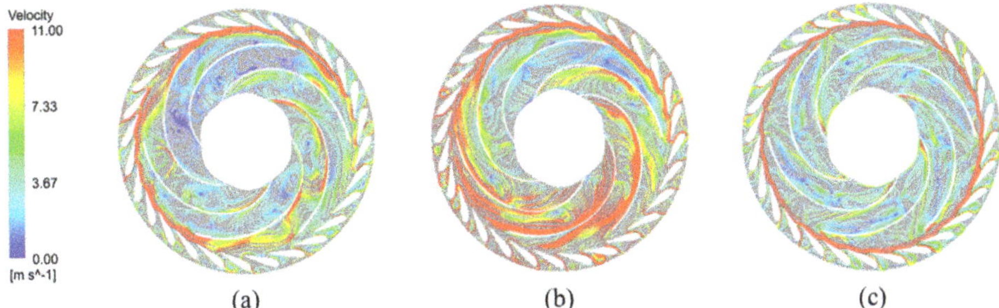

Figure 11. Flow diagram of the runner area: (**a**) late braking condition; (**b**) pre-water turbine condition; (**c**) late turbine operating conditions.

4. Conclusions

This study implemented the entropy production theory and vortex dynamics to study and analyze the energy loss characteristics in the process of pump-turbine shutdown. The following conclusions were obtained:

(1) During the abnormal shutdown of the pump-turbine, the entropy production, vorticity of the spiral case, and draft tube slightly changed. In contrast, the entropy production and vorticity of the runner area changed more drastically from the late braking condition to the late turbine condition, starting from the middle of the braking condition and rising rapidly to the peak in the early turbine condition and then decreasing rapidly. Subsequently, the entropy production and vorticity rise rapidly after the middle of the turbine condition.

(2) In the runner region, the entropy production and vorticity of the two distribution trends are similar. The head and tail of the movable guide vane rapidly produced the backflow phenomenon. Moreover, the vaneless area created a high-speed circulation phenomenon, the rotor blade wall near generated the leaf channel vortex. These irregular flow phenomena generated parts of the entropy production, with a high vorticity and energy loss.

(3) The analysis method using the combination of entropy production theory and vortex dynamics is practical. In particular, the entropy production analysis can quantify the energy loss and indicate the location of energy loss. Moreover, the vortex analysis can quantify the degree of flow disturbance and locate the flow disturbance, which can explore the connection between undesirable flow phenomena and energy loss.

Author Contributions: Conceptualization, Y.D.; methodology, Y.D.; writing—original draft preparation, J.X.; writing—review and editing, Y.L.; visualization, Y.Z.; supervision, Y.D.; software, C.K. All authors have read and agreed to the published version of the manuscript.

Funding: This research was funded by the Open Research Subject of Key Laboratory of Fluid and Power Machinery, Ministry of Education (Grant No. LTDL2020-005), Doctoral Research Foundation of Lanzhou City University (Grant No. LZCU-BS2019-07).

Institutional Review Board Statement: Not applicable.

Informed Consent Statement: Not applicable.

Data Availability Statement: All the data are already in the article.

Acknowledgments: The authors would like to thank all staff of the treatment plants for making.

Conflicts of Interest: The authors declare no conflict of interest.

References

1. Zhang, W.; Feng, D.; Qi, W.; Han, Z.; Guan, Y.; Sun, W.; Wang, W. Research on Peak Shaving Power Source Planning for Receiving-End Grid Considering High Proportion of New Energy and Large-Scale Outer Power. *IOP Conf. Ser. Earth Environ. Sci.* **2019**, *267*, 062009. [CrossRef]
2. Kong, Y.; Kong, Z.; Liu, Z.; Wei, C.; Zhang, J.; An, G. Pumped Storage Power Stations in China: The past, the Present, and the Future. *Renew. Sustain. Energy Rev.* **2017**, *71*, 720–731. [CrossRef]
3. He, Y.; Liu, Y.; Li, M.; Zhang, Y. Benefit Evaluation and Mechanism Design of Pumped Storage Plants under the Background of Power Market Reform—A Case Study of China. *Renew. Energ.* **2022**, *191*, 796–806. [CrossRef]
4. Kusakana, K. Optimal Scheduling for Distributed Hybrid System with Pumped Hydro Storage. *Energy Convers. Manag.* **2016**, *111*, 253–260. [CrossRef]
5. Yu, A.; Li, L.; Ji, J.; Tang, Q. Numerical Study on the Energy Evaluation Characteristics in a Pump Turbine Based on the Thermodynamic Entropy Theory. *Renew. Energ.* **2022**, *195*, 766–779. [CrossRef]
6. Liu, D.M.; Zheng, J.S.; Wen, G.Z.; Zhao, Y.Z.; Shi, Q.H. Numerical Simulation on the "S" Characteristics and Pressure Fluctuation of Reduced Pump-Turbine at Start-Up Condition. *IOP Conf. Ser. Earth Environ. Sci.* **2012**, *15*, 062034. [CrossRef]
7. Yao, Z.; Bi, H.L.; Huang, Q.S.; Li, Z.J.; Wang, Z.W. Analysis on Influence of Guide Vanes Closure Laws of Pump-Turbine on Load Rejection Transient Process. *IOP Conf. Ser. Mater. Sci. Eng.* **2013**, *52*, 072004. [CrossRef]
8. Walseth, E.; Nielsen, T.; Svingen, B. Measuring the Dynamic Characteristics of a Low Specific Speed Pump-Turbine Model. *Energies* **2016**, *9*, 199. [CrossRef]
9. Zhang, L.G.; Zhou, D.Q. CFD Research on Runaway Transient of Pumped Storage Power Station Caused by Pumping Power Failure. *IOP Conf. Ser. Mater. Sci. Eng.* **2013**, *52*, 052027. [CrossRef]
10. Jintao, L.; Shuhong, L.; Yuekun, S.; Yulin, W.; Leqin, W. Three Dimensional Flow Simulation of Load Rejection of a Prototype Pump-Turbine. *Eng. Comput.* **2013**, *29*, 417–426. [CrossRef]
11. Liu, J.; Liu, S.; Sun, Y.; Jiao, L.; Wu, Y.; Wang, L. Three-Dimensional Flow Simulation of Transient Power Interruption Process of a Prototype Pump-Turbine at Pump Mode. *J. Mech. Sci. Technol.* **2013**, *27*, 1305–1312. [CrossRef]
12. Mao, X.; Chen, X.; Lu, J.; Liu, P.; Zhang, Z. Improving Internal Fluid Stability of Pump Turbine in Load Rejection Process by Co-adjusting Inlet Valve and Guide Vane. *J. Energy Storage.* **2022**, *50*, 104623. [CrossRef]
13. Xia, L.; Cheng, Y.; Yang, Z.; You, J.; Yang, J.; Qian, Z. Evolutions of Pressure Fluctuations and Runner Loads during Runaway Processes of a Pump-Turbine. *J. Fluids Eng.* **2017**, *139*, 091101. [CrossRef]
14. Li, D.; Wang, H.; Li, Z.; Nielsen, T.K.; Goyal, R.; Wei, X.; Qin, D. Transient Characteristics during the Closure of Guide Vanes in a Pump-Turbine in Pump Mode. *Renew. Energy* **2018**, *118*, 973–983. [CrossRef]
15. Zhang, X.; Zeng, W.; Cheng, Y.; Yang, Z.; Chen, Q.; Yang, J. Mechanism of Fast Transition of Pressure Pulsations in the Vaneless Space of a Model Pump-Turbine during Runaway. *J. Fluids Eng.* **2019**, *141*, 121104. [CrossRef]
16. Yang, Z.; Cheng, Y.; Xia, L.; Meng, W.; Liu, K.; Zhang, X. Evolutions of Flow Patterns and Pressure Fluctuations in a Prototype Pump-Turbine during the Runaway Transient Process after Pump-Trip. *Renew. Energy* **2020**, *152*, 1149–1159. [CrossRef]
17. Yang, Z.; Liu, Z.; Cheng, Y.; Zhang, X.; Liu, K.; Xia, L. Differences of Flow Patterns and Pressure Pulsations in Four Prototype Pump-Turbines during Runaway Transient Processes. *Energies* **2020**, *13*, 5269. [CrossRef]
18. Zhao, Z.; Chen, D.; Li, H.; Wei, H. Performance Analysis of Pumped-Storage Plant from Condenser Mode to Generating Process. *J. Energy Storage* **2020**, *29*, 101286. [CrossRef]
19. Yang, J.; Lv, Y.; Liu, D.; Wang, Z. Pressure Analysis in the Draft Tube of a Pump-Turbine under Steady and Transient Conditions. *Energies* **2021**, *14*, 4732. [CrossRef]
20. Gong, R.; Wang, H.; Chen, L.; Li, D.; Zhang, H.; Wei, X. Application of Entropy Production Theory to Hydro-turbine Hydraulic Analysis. *Sci. China Technol. Sci.* **2013**, *56*, 1636–1643. [CrossRef]
21. Li, D.; Gong, R.; Wang, H.; Xiang, G.; Wei, X.; Qin, D. Entropy Production Analysis for Hump Characteristics of a Pump Turbine Model. *Chin. J. Mech. Eng.* **2016**, *29*, 803–812. [CrossRef]
22. Li, D.; Wang, H.; Qin, Y.; Han, L.; Wei, X.; Qin, D. Entropy Production Analysis of Hysteresis Characteristic of a Pump-Turbine Model. *Energy Convers. Manag.* **2017**, *149*, 175–191. [CrossRef]
23. Fu, X.; Li, D.; Wang, H.; Zhang, G.; Li, Z.; Wei, X.; Qin, D. Energy Analysis in a Pump-Turbine during the Load Rejection Process. *J. Fluids Eng.* **2018**, *140*, 101107. [CrossRef]
24. Li, F.; Fan, H.; Wang, Z.; Chen, N. Coupled Design and Optimization for Runner Blades of a Tubular Turbine Based on the Boundary Vorticity Dynamics Theory. In Proceedings of the Fluids Engineering Division Summer Meeting, Toronto, ON, Canada, 3–5 August 2022; pp. 603–609.
25. Li, D.; Gong, R.; Wang, H.; Han, L.; Wei, X.; Qin, D. Analysis of Vorticity Dynamics for Hump Characteristics of a Pump Turbine Model. *J. Mech. Sci. Technol.* **2016**, *30*, 3641–3650. [CrossRef]
26. Li, X.; Li, Z.; Zhu, B.; Wang, W. Effect of Tip Clearance Size on Tubular Turbine Leakage Characteristics. *Processes* **2021**, *9*, 1418. [CrossRef]
27. Kock, F.; Herwig, H. Local Entropy Production in Turbulent Shear Flows: A High-Reynolds Number Model with Wall Functions. *Int. J. Heat Mass Transf.* **2004**, *47*, 2205–2215. [CrossRef]

MDPI

St. Alban-Anlage 66

4052 Basel

Switzerland

Tel. +41 61 683 77 34

Fax +41 61 302 89 18

www.mdpi.com

Processes Editorial Office

E-mail: processes@mdpi.com

www.mdpi.com/journal/processes